UNCOMMON
PROPERTY

UNCOMMON PROPERTY

THE FISHING AND FISH-PROCESSING INDUSTRIES IN BRITISH COLUMBIA

Edited by
PATRICIA MARCHAK
NEIL GUPPY
JOHN McMULLAN

METHUEN

Toronto   New York   London   Sydney   Auckland

**This book has been published with the help of a grant
from the Social Science Federation of Canada, using funds
provided by the Social Sciences and Humanities Research
Council of Canada.**

Canadian Cataloguing in Publication Data

Main entry under title:

Uncommon property

Includes index.
ISBN 0-458-80990-X

1. Fisheries — Economic aspects — British Columbia.
2. Fisheries — Social aspects — British Columbia.
3. Fishery processing industries — British Columbia.
I. Marchak, M. Patricia, 1936-   . II. Guppy,
L. Neal, 1949-   . III. McMullan, John L., 1948-   .

HD9464.C33B75 1987   338.3′727′09711   C87-093900-9

Printed and bound in Canada
1 2 3 4   87   91 90 89 88

Contributors

Stephen Garrod is a graduate in sociology and education and has had experience working in processing plants. He is currently teaching in the Vancouver school system.

Neil Guppy is Associate Professor of Sociology at the University of British Columbia and has published studies on education and social inequality, as well as articles on fisheries.

Patricia Marchak is Professor of Sociology at the University of British Columbia and the author of *Ideological Perspectives in Canada*, *In Whose Interests*, and *Green Gold: The Forest Industry in British Columbia*.

John McMullan is Associate Professor at St. Mary's University, Halifax. He is the author of *The Canting Crew: London's Criminal Underworld, 1550–1700*, the co-editor of *Criminal Justice Politics in Canada*, and has written numerous papers on criminology, debt, and the state.

Alicja Muszynski is Assistant Professor, Sociology, at the University of Regina. She completed her Ph.D. thesis on "The Creation and Organization of Cheap Wage Labour in the British Columbia Fishing Industry" and has published papers on the union, women workers, and the early history of the fishing industry.

Evelyn Pinkerton is Research Associate, School of Community and Regional Planning at the University of British Columbia, and is involved in applications of fisheries co-management.

Keith Warriner is Assistant Professor, Sociology, at the University of Waterloo and has published work on environmental and energy issues, in addition to work on the fisheries.

Contents

List of Tables, ix
List of Figures, xi
Acknowledgments, xii
Preface, xiii

INTRODUCTION, 1
1. Uncommon Property, *Patricia Marchak*, 3

PART 1 CAPITAL AND THE STATE, 33
2. The Organization of the Fisheries: An Introduction, *John McMullan*, 35
3. Major Processors to 1940 and Early Labour Force: Historical Notes, *Alicja Muszynski*, 46
4. Competition Among B.C. Fish-Processing Firms, *Evelyn Pinkerton*, 66
5. The Production and Distribution of B.C. Salmon in the World Context, *Stephen Garrod*, 92
6. State, Capital, and the B.C. Salmon-Fishing Industry, *John McMullan*, 107
7. "Because Fish Swim" and Other Causes of International Conflict, *Patricia Marchak*, 153

PART 2 LABOUR AND ORGANIZATION, 171
8. Labouring at Sea: Harvesting Uncommon Property, *Neil Guppy*, 173
9. Labouring on Shore: Transforming Uncommon Property into Marketable Products, *Neil Guppy*, 199
10. Organization of Divided Fishers, *Patricia Marchak*, 223
11. Indians in the Fishing Industry, *Evelyn Pinkerton*, 249
12. Shoreworkers and UFAWU Organization: Struggles between Fishers and Plant Workers within the Union, *Alicja Muszynski*, 270

PART 3 COMMUNITY AND REGION, 291
13. The Fishing-Dependent Community, *Evelyn Pinkerton*, 293
14. Regionalism, Dependence, and the B.C. Fisheries: Historical Development and Recent Trends, *Keith Warriner*, 326

CONCLUSION, 351
15. Uncommon History, *Patricia Marchak*, 353

APPENDICES, 361
Appendix A: Notes on Sample Survey Methodology, 363
Appendix B: Papers and Publications, 365

Bibliography, 368
Index, 391

List of Tables

Table   4.1 : Catches and Landed Value — Salmon and Herring, 67
Table   4.2 : Salmon Canning and Freezing, Number of Firms: Selected Years, 1951–1984, 70
Table   4.3 : Total Salmon Production of Top Four Canning Firms as a Percentage of the Value of Salmon Production: 1972–1985, 71
Table   4.4 : Number of Firms Processing Roe Herring and Value of Herring Roe Production by Top Four Salmon Canners as a Percentage of the Total: 1974–1984, 85
Table   5.1 : Commercial North Pacific Salmon Harvest (All Species), 94
Table   5.2 : Commercial North Pacific Sockeye Salmon Harvest (All Species), 95
Table   6.1 : Distribution of Numbers and Gross Sales of Fish-Processing Establishments Classified by Value of Production, with Percentages, Atlantic and Pacific Regions of Canada, 1954, 121
Table   6.2 : Number of Fishing Vessels by Net Tonnage and Length Classes and Number of Fishers in the B.C. Commercial Fishery, 1953–1965, 123
Table   6.3 : Government Financing Used by B.C. Fishers by Purpose, 128
Table   6.4 : Loans Made and Claims Paid under the Fisheries Improvement Loans Act from Inception to March 31, 1981, 129
Table   6.5 : Loans and Repayments to Lenders under the Fisheries Improvement Loans Act from Inception to March 31, 1981, 130
Table   6.6 : Loans to B.C. Region and by Lender, 1975–1981, 131
Table   6.7 : Profile of Bank Involvements with Regard to F.I.L. Program, 1975–1981, B.C. Region, 132
Table   6.8 : Loans for Various Improvement Purposes and for Building and Construction, 1975–1981, B.C. Region, 133
Table   6.9 : Loans for Fishing Equipment, 1975–1981, B.C. Region, 134
Table   6.10: State Financing Utilized by B.C. Processors by Purpose (Selected Years), 137
Table   6.11: Chronology of B.C. Packers Consolidation and Control, 1980–1983, 143
Table   8.1 : Education Levels of Selected B.C. Workers, 177
Table   8.2 : Contribution of Fisheries Earnings to Total Personal and Household Income by Level of Education, 179
Table   8.3 : Comparing UIC Benefits for Fishers and other Workers, 180
Table   8.4 : Fishing Incomes and Earnings: The Accountants' Reports, 181

Table   8.5 : Salmon Landings by Highliners: Percentage Shares to the Top 20%, 185

Table   8.6 : Perceptions of Department of Fisheries and Oceans Policies and Regulations, 186

Table   8.7 : The Trend in Bonus Payments in the Salmon Fishery, 193

Table   8.8 : Canadian Fatality Rates in Selected Industries, 195

Table   8.9 : Percentage of Canadian Fishing Fatalities in B.C., 195

Table   8.10: Future Commitments to the Fishing Industry by Age and Position, 196

Table   9.1 : Establishments, Labour Force, and Wages in B.C. Fish Processing, 201

Table   9.2 : Indicators of Variation in Fish-Processing Work, 204

Table   9.3 : Comparison of Pacific and Atlantic Average Hourly Wages in Three Industrial Sectors for Selected Months (1984), 205

Table   9.4 : Input Costs and Value of Production in B.C. Fish Processing (in Constant 1976 Dollars), 207

Table   9.5 : Fish-Plant Jobs and Worker Discretion: A Summary, 214

Table   9.6 : Background Characteristics of Female and Male Fish-Plant Workers, 215

Table   9.7 : Average Earnings by Gender and Ethnic Background, 220

Table   9.8 : Paid Jobs of Respondents' Mothers, Fathers, and Spouses, 221

Table 13.1 : Chum Salmon Catch and Escapement, 1935-1983: Clayoquot Sound (Tofino Area), 300

Table 13.2 : Demographic Profile for Tofino and Ucluelet, 1950–1980, 305

Table 13.3 : Gross Income Ranges for Local Salmon Fishers: Selected Years, 1969–1981 (Tofino Area), 313

Table 14.1 : Shorework Labour Force for Regions for Selected Years, 330

Table 14.2 : Mean Landings, Vessel Characteristics, and Debt by Community Type, 341

Table 14.3 : Gears by Community Type, 343

Table 14.4 : Trends, 1967–1981, by Community Type, 344

List of Figures

Figure 5.1: International Trade in Salmon, 96
Figure 5.2: World Canned Export Salmon Trade, 97
Figure 5.3: World Atlantic Salmon Production, 99
Figure 5.4: Wholesale Values, B.C. Fisheries, 103
Figure 6.1: SEP Production Capacity by Species:
              Phase 1 Targets and Performances to Date, 146
Figure 9.1: Seasonal Labour Force Changes in Fish Processing, 202
Figure 13.1: Salmon Licences by Community, 295
Figure 13.2: Salmon versus other Licences (Clayoquot Sound), 312
Figure 14.1: Coastal Canneries, 1881–1984, by Region, 332
Figure 14.2: Fraser District and Puget Sound Pack, 1876–1954, 334
Figure 14.3: British Columbia Salmon Pack by District, 1881–1970, 336
Figure 14.4: Commercial Fishing Licences by District, 1881–1981, 339

Map of North Coast and Queen Charlotte
     Islands, British Columbia, 6
Map of Vancouver Island and Southwest
     Coast, British Columbia, 7

Acknowledgments

We wish to express our thanks to the fishers and processing workers who granted us interviews and to the many others — union and association representatives, bank managers, processing company officials, Department of Fisheries and Oceans personnel, members of communities in which we conducted field research, faculty members in the Departments of Oceanography, Marine Biology, Animal Resource Ecology, and History — who gave us information, permitted us to interview them, and aided us in numerous ways in our efforts to accurately describe and analyse the west coast fisheries in the early 1980s. Where there is no reason to keep names confidential, we have quoted such sources, but the formal survey interviews were guaranteed to be confidential, and no names are provided for survey data.

As well, we are most grateful to the Social Science and Humanities Research Council of Canada for funding over a three-year period, and for the encouragement and support of its officers. In particular, we wish to thank Patrick Mates for his continuing interest in the project. Assessors for a Social Science Publishing Grant were constructive in their criticism, and we thank them for their reviews.

Finally we wish to acknowledge our good fortune in having the support of our university and the aid, in particular, of the Deans of Arts, Graduate Studies, and Research Administration in 1981, Drs. Robert Will, Peter Larkin, and Richard Spratley, respectively, and of the President of the University at that time, Dr. Douglas Kenny. Their help in getting us started is much appreciated. Our department, Anthropology and Sociology, has been supportive and encouraging throughout the project, and we thank our colleagues for putting up with a great many fish stories.

Preface

In the fishery crisis of 1982, we were told this was the end of an era, the dying days of a great wild fishery. The enormous abundance of salmon in rivers was already just the memory of some natives and pioneer Europeans. "You could walk on their backs," a Bella Coola man said of the local river salmon runs in his childhood, and others had similar memories. In one mass meeting after another, fishers contemplated the future of the fishery and sought to understand why the resource was no longer sufficient to sustain them all.

In the bonanza year 1985, we were told that nature had played a trick. For unaccountable, unpredictable reasons, the salmon runs exceeded any since the 1930s. Some fortunes were made and most fishers had a good season, puzzled again, but now with the abundance.

This book is about the Canadian west coast fisheries in the 1980s, and so the problem of variable fish stocks worries the authors, who are inclined to believe that wild salmon stocks have been overfished and the fish habitat has been dangerously degraded. We assume there are biological conditions still not understood that give rise to unpredictable cycles of growth and decline, and we note that enhancement programs and conservationist measures are becoming more effective. But we are not marine biologists and have no specialized knowledge of fish migratory patterns.

What we do have, and what we want to share through this book, is some understanding of the social and economic structure of the industry. We know that whatever nature does, the social structure affects the outcome. The overcapitalization of vessels and plants throughout the 1970s had much to do with the decline in stocks and the crisis of the early 1980s. Overcapitalization, in turn, was connected to unusual market conditions and the licensing and regulation roles assumed by the Department of Fisheries and Oceans. The massive debt incurred by vessel owners and the many conflicts between capital and labour, independent vessel owners and unionized labour, shoreworkers and fishers all are social effects of the industry's structure and the role of the state. All influence the pursuit and capture of wild fish. The introduction of pen-reared fish is also a social development, possibly altering the genetic nature as well as quantity of salmon in British Columbia. Thus our concern with the political economy, sociology, and anthropology of fishing is a different approach to the problems of the west coast fishery than is normally advanced by marine biologists, but it is equally concerned with the resource.

THE "FISH AND SHIPS" RESEARCH PROJECT

The book is the product of a three-year research project conducted by members of the Department of Anthropology and Sociology at the University of British Columbia. The original members of the group that sought and obtained funding from the Social Science and Humanities Research Council of Canada included Neil Guppy, Patricia Marchak, John McMullan, Martin Silverman, Evelyn Pinkerton, and Brian Hayward. Silverman was obliged to take a less active role in the research because he was persuaded by his colleagues to chair the department during a particularly difficult period in the university's history, but the members of the research group remained in his debt for many insights, good advice, and much aid during the editorial process.

As faculty members, Guppy, Marchak, and McMullan continued with the project throughout. Pinkerton and Hayward were employed full-time as research associates. Graduate and undergraduate students were brought into the group once the project was underway, including Tom Burnell, Steve Garrod, Louise Gorman, Peter Lando, Susan Lee, James Leslie, Sophia Lum, Brian Martin, Alicja Muszynski, Scott Peterson, John Sutcliffe, Joanne Teal, and Keith Warriner. Field interviewers included Karen Anderson, Kathy Bedard, Betty Lou Benham, Jennifer Boyce, Faye Bromley, Debbie Cartwright, John Camparalli, and Diana Hall.

The group's objective was to do research on the political economy of fishing and the anthropological and sociological aspects of the enterprise. The Pearse Commission began its hearings shortly after our group received funding; necessarily, then, we became interested observers of the commission and critical readers of the report presented in 1982. The publication covered some of the topics we were investigating, and we have tried in this book to avoid repetition of material which readers can find in the Commission Report, *Turning the Tide*.

We published a number of articles in various professional journals in the course of our research. A list of these is given in Appendix B, to the time of this book's publication (others are still in press). We have not recovered the ground in those publications because readers may refer to them.

Research took the forms of intensive studies of communities by researchers temporarily residing in them; a sample survey of fishers, shoreworkers, and members of selected communities; analysis of data provided by Statistics Canada and the Department of Fisheries and Oceans; examination of historical records, particularly Sessional Papers; and records of numerous meetings of fishers in the aftermath of the presentation of the Pearse Report. In these meetings, members of our group occasionally took active roles, giving speeches by invitation and participating in workshops.

The survey instrument was designed mainly by Neil Guppy and Keith Warriner. Guppy took charge of the coding and computer processing. Warriner and Guppy also carried responsibility for processing of data obtained from the Department of Fisheries and Oceans and from Statistics Canada.

Examination of historical documents was conducted primarily by Alicja Muszynski. Her extensive notes on these documents were used by all members of the group.

Pinkerton and Hayward conducted the lion's share of interviews, and both took up residence in fishing communities: Pinkerton in Tofino, covering much of the west coast of Vancouver Island; and Hayward in Prince Rupert. Muszynski conducted preliminary community study in Steveston. Most members of the group resided for some period of time in these or other communities, gathering data or conducting interviews.

Garrod gathered data on international markets, and he and Sutcliffe also studied backward linkages between fisheries and other industries. These data informed our work, though the material on linkages is not included in this particular volume. Martin introduced information about fish-farming and mariculture, and the group remains in his debt for his contribution to our collective fund of knowledge. McMullan concentrated on collecting information on fishers' indebtedness, visiting bankers and processors as well as fishers. Marchak headed the project, a task made easier by the general good will expressed by the fishing community toward our enterprise.

The group met regularly throughout most of the project's history. Most chapters were subjected to group discussion at some stage of their evolution. All have benefited from information provided by others, and some have been based on original research done by persons other than the author. However, the group reflected the differences in the research field as a whole, both in theoretical orientation and identification of issues specific to the fisheries. These differences are evident in the written work, so that although this book is more integrated than the usual edited volume of essays, it does not give only one perspective on the subject. As well, with reference particularly to the history of the union and the perspective of native Indians, we have deliberately provided more than one perspective. The reader is warned that when there appears to be redundancy in recording of historical detail, the intention has been to demonstrate how these details are variously understood by the participants.

Social science theories, mainly within the political economy tradition and particularly of the state and of the labour process, are drawn on by all authors. These are introduced in the first chapter, and definitions are provided where the terms are not in general use. However, we have kept the general reader in our minds and have tried to avoid unfamiliar terminology.

ORGANIZATION OF THE BOOK

There are three parts to the book, plus an introduction and conclusion. Part 1 is concerned with the history of the industry, the role of the federal and provincial governments, international markets, significant differences in raw fish markets and their importance for the processing sector, and the international context for British Columbia fisheries.

In Part 2, we consider the labour process. This includes chapters on shoreworkers and fishers, with descriptions of their characteristics and their working conditions. It also examines their history of organization, the special place of native Indians in the fishery, and the perspective of history offered by the United Fishermen and Allied Workers Union newspaper.

In Part 3, we consider fishing communities: their viability when they are dependent on a diminishing resource and their responses to resource depletion.

INTRODUCTION

1

Uncommon Property

PATRICIA MARCHAK

The central actors in the British Columbia fishing and fish-processing industry are the Canadian federal government and its Department of Fisheries and Oceans; a large number of vessel owners and crew members who fish for salmon (the major species), halibut, herring, and assorted groundfish in the Pacific Ocean; two major processing firms (B.C. Packers and the Prince Rupert Fishermen's Cooperative Association) and several smaller firms together with their labour force; a union of shoreworkers and fishers (the United Fishermen and Allied Workers Union, or UFAWU); the Native Brotherhood of British Columbia, or NBBC; and a plethora of smaller fishers' associations linked to differing gear-types and geographical locations. This book is concerned with these many actors and their relationships both to one another and to fish.

The property status of fish and fishers is a central issue in our studies. Who has the right to fish, the right to manage the fishery, the right to exclude others, the right to profit from the sale of fish? Fish have been called common property, but there are issues buried rather than illuminated by that designation, and we argue here that these issues can be better understood if we treat the right to fish as state property. This chapter begins with that argument.

The role of the state, as played out by particular governments through specific periods of time, encompasses legislation pertaining to access rights, habitat conservation, licensing, capital, and markets. It also involves regulation of labour, unions, and the relationship between capital and labour in this as in other industries. Having examined the property issue, we move on to consideration of labour and the state's participation in its regulation.

This introductory chapter is concerned with the general and theoretical issues that inform our work throughout the book. Here, particular instances are cited by way of illustration rather than as detailed accounts of events. The detailed accounts follow, however, in subsequent chapters.

PROPERTY

Following the Rousseauian metaphor of history, private property was established when a man first staked out a plot of land, declared his sole right to control it, and others accepted his claim. Whatever the real origins of privatization, one may trace its history from control over land, through control over tools used to increase the fruits of the land and water, to control over the products of others' labour. In some regions, all of nature became the property of a few; much of European history, for example, consists of the progressive privatization of nature. Locke rested his case on the premises that men's "natural right" to property took precedence over civil law and that *the purpose of government was the preservation of that right.*

Privatization has not been a universal phenomenon. Prior to European settlement, B.C. Indians (like many other nonindustrial societies) allowed the environment to be owned and managed by kinship or territorially located groups. Ownership rights included allocation of access and harvest rights to fish and use of the caught fish, the exclusion of outsiders, and means required for proper management. Capture took place primarily in rivers and streams or in offshore locations close to villages. Since the fish at that time were plentiful, by all reliable accounts, such rights virtually guaranteed continuing supplies to owning groups. Not included in the rights, however, were sale or other alienation, and in this respect, particularly, common property differed from private property.

The term "common property" is central to a prolific literature on the fisheries; the argument is that, because fish are not privately owned until caught, they are therefore owned in common. The derived argument is that the depletion of a common property is inevitable because no one is charged with the responsibility for conservation, and the competing users ultimately destroy the common resource. However, this line of reasoning has flaws. To begin with, the concept turns out to have different definitions. In the case of property held in common by native groups, as described above, it implies the existence of a level of organization that, through usage and either tacit or formal agreements, provides the groups with exclusive right to given territories (including waters). If a group has the internal organization necessary to ensure that all members similarly enjoy the benefits of these rights in perpetuity and management is undertaken by the group as a whole, there is no logical reason why the property would become exhausted by group usage. This is the sense in which Ciriacy-Wantrup and Bishop, in an influential article on the subject, use the term: "[It] implies that potential resource users who are not members of a group of co-equal owners are excluded" (1975:715).

But this is not the same meaning as "the [general and unspecified] right not to be excluded" (Macpherson, 1978: 4–5, 201), or "everybody's right is nobody's right" (Gordon, 1954; Scott, 1955; Hardin, 1968; Ostrom, 1975). It would probably be helpful as a beginning to distinguish as communal property that variety noted above, implying a positive set of management obligations as well as *exclusive rights between co-owners*. For the sake of clarity, let us use the term communal and not common for this. Communal property has nothing in it that "inevitably" leads to resource depletion or "the tragedy of the commons"; or at least there is no evidence in history that groups so organized depleted their resources.

Exempting communal property, we are left with a negatively defined set of property rights, rights to nonexclusion, and with nonproperty, i.e., things to which no claims may be made or enforced. As Plourde (1975) and an ancient popular tune observe, the moon belongs to everyone, but such a right (until nations establish moon colonies) is both unenforceable and inconsequential.

Of these unenforceable property rights, Macpherson (1978) names, as examples, things created by states from public taxes to which not only the undifferentiated citizenry but as well noncitizens have access, such as parks, streets, and highways.[1] None of these is akin to fish in Canadian waters in the period since 1968 and in previous periods when access has been restricted in time, place, and quantity to those with licences granted by the state and subject to specified conditions. If the state licenses some but not all citizens, thereby granting a privilege to fish, then the right to fish becomes the private property of the fisher. This interpretation is strengthened if the fisher is able to sell that right to another, as has been the case in B.C. since 1968. Thus the term "common property" is at least technically inaccurate for the access to fish.

However, the fisher lacks management rights normally associated with property. There is a disjuncture between rights of access and of management that lies at the heart of the fisheries problem. If management rights include, as they do, decision making with respect to the resource, control of habitat and waterways, allocation of licences, and limitation on capture capacities, then the ownership rights associated with a licence are very limited. The problem, as identified by the Pearse Commission and many others, is that "there are too many boats chasing too few fish." But if this is the case, and licensing is enforced, then it follows that the manager has allocated too many licences. Since fishers cannot control licensing and since they are not permitted to manage a communal property, our attention should be focussed on the management of licensing access to fish, rather than on the licensed fishers. This means that instead of talking about "the tragedy of the commons," we should be concerned with the tragedy of mismanaged state property.

Fish Habitat as Property

To complicate the property problem further, we need to consider the total resource rather than only the wild fish. Fish live in seas, rivers, streams,

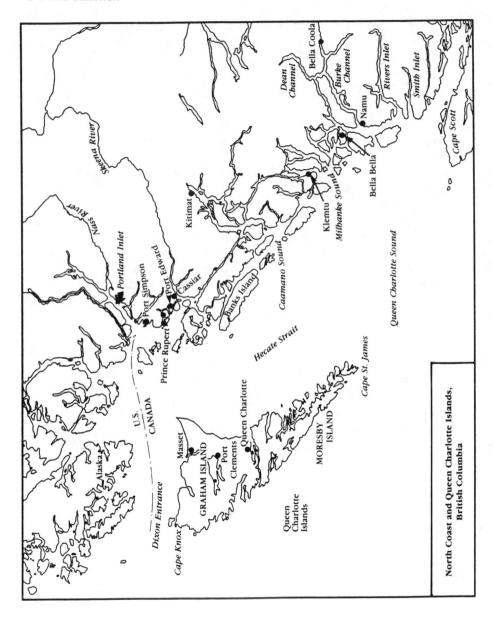

**North Coast and Queen Charlotte Islands,
British Columbia**

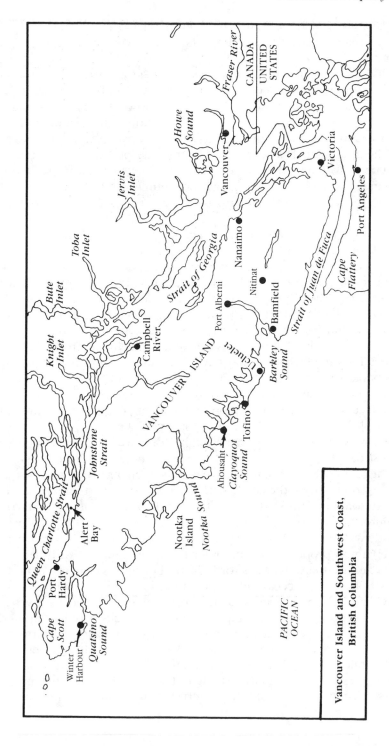

Vancouver Island and Southwest Coast, British Columbia

and lakes for which there are other commercial and noncommercial uses. As well, fish are subjected to impacts from whatever is in precipitation from the air.

Of the waterways, the sea appears to be the only location which might be designated as common property in Macpherson's sense, even though states do not create it. In general, no one person or group can exclude any others from the sea though nation states can delimit entry. Even so, an industrial plant located on or by a particular sea frontage can use the area in such a way that for practical purposes other users are excluded. Moreover, that plant may pollute the surrounding sea (and air) with impunity and considerable latitude, entirely on the basis of proximity. In B.C. this exercise of private property rights in a common property is significant as far as fisheries are concerned. The firms in question are mining and forest companies. Where they dump effluent into the sea directly or via land and stream or where they use the sea as a warehouse or transportation route (as with log booms), they have an enormous impact on the resources in the sea. The data on effects of pulpmill effluent are still inconclusive; the impact of mine tailings dumped in the sea are more clearly identified as lethal to fish; the known effects of acid rain are equally lethal, and much is yet to be learned of that form of pollution. The construction of hydro-electric dams affects all spawning streams linked to fluctuating water levels and may have downstream impacts on fish as well. This book is not about these many other causes of stock declines, but it must be noted that overcapacity of the fleet is not the sole cause, and the full effects of industrial pollution are not yet adequately researched to justify placing the blame exclusively on overfishing.

Governments grant contractual rights for specific industrial purposes to harvest or mine resources. In these cases (the majority of logging and mining properties), companies have exclusive rights to extract a resource, but they do not own the land itself. The right to harvest implies a right to declare some territory offbounds to outsiders and to use some territory for purposes of harvesting and transporting the harvest. The territory can include waterways and waterfrontage where fish spawn. This property is not common property: no one other than the company with harvesting rights can use the waterways for industrial purposes (unless a competing industry has equal access rights specified in its contracts with the state). Individuals may be permitted by the companies to use the waterways for limited purposes (picnics, camping), and companies may be required to permit that usage when logging in the vicinity is not actually in progress, but these individual access rights are both vague and rarely enforced. Thus, while forest and mining companies do not have fishing licences and have no vested interest in the fishing industry, their contractual property rights in fish habitats constitute powerful claims against the fish resource.

State Property Rights

In 1984, following the recommendations of the Pearse Royal Commission, the federal government proposed royalties as a means of defraying costs of excess vessel buy-backs and area and quota licences to limit excess capture. What is significant about these proposals for our present purposes is that the state assumed *the right to unilaterally alter the access and harvesting practices of citizens in what it still insisted was a common property resource.*

Up to this point, the state had limited its claims in the fisheries, even while fully practising management at public expense. Though Canadian governments have, in the three periods between 1889 and 1892, 1910 and 1920, and since 1968, limited access to fish by licensing fishers, they have not, thereby, guaranteed that any specified amount of fish may be caught by any licensee, nor have they charged rent or specified areas and quotas to fishers.[2] Since 1968 the state has further delimited not only access but capture by restricting the times and locations in which licensed fishers are permitted to ply their trade. In 1976, the federal government announced its intention to increase its interventions in the fishery and to control "the use of fishery resources from the water to the table" (Department of Environment, 1976:5).

The matter is of more than academic interest. Throughout the Pearse Commission report, fishing rights are called "privileges." Privileges are not contiguous with common property rights. A legal case was mounted by the Gulf Trollers Association in June 1984 based on the argument that the resource is common property in Macpherson's sense of rights not to be excluded. The association argues that neither the Fisheries Act nor the Constitution Act empowers the government to arbitrarily restrict fishing access on the basis of gear-type or area. The Department of Fisheries and Oceans argues, to the contrary, that the government is required to act in the interests of society at large, that any measures to conserve the resource, including arbitrary foreshortening of commercial fisheries, are permissible to that end, and that in fact:

> It is within the powers of the government of Canada to restrict the troll fleet, even if there is no conservation consideration. There is no right for trollers to any season at all. (Gunnar Eggertson, lawyer for Department of Fisheries and Oceans, *Vancouver Sun*, July 19, 1984:A8)

The federal court ruled that the fisheries department did not have the authority to allocate catch quotas by gear-type, and that it could allocate for conservation reasons only. Thus, if its restriction of trollers was imposed for purposes of aiding the net or sports fisheries to obtain a "fair share" of the catch, it acted improperly. Following this decision, a new, temporary

Fisheries Act amendment was introduced, providing the legislative means for the Department of Fisheries and Oceans to allocate by gear-type or for social and economic reasons. This reflected concerns with the Atlantic lobster fishery, as well as the Pacific salmon conflict, and does not solve a political dilemma that affects both fisheries. For the Pacific region, a formula was established for dividing the catch of chinook, sockeye, and pink salmon in the Gulf of Georgia, whereby trollers were permitted only 18 per cent of the chinook allocation and specified quantities of other species, the rest going to sports fishers. The court decision meanwhile was appealed and at the time of writing remains unsettled. The point here is that the issues arise only because there is a disjuncture between access and management rights, and overfishing in the gulf or any other region is related to how management is carried out, not to the assumed evils of common property.

A similar issue is involved in settlement of Indian land claims. Among outstanding claims are aboriginal rights to fish. British Columbia has a substantial native Indian population, comprising 5 per cent of the total population by official statistical measures, and a recent history of well-organized efforts by Indian communities to settle outstanding claims. Aboriginal rights were not abrogated by treaty for most of the native population. In the spring of 1986, the federal government passed by-laws under the Indian Act to permit the Gitksan-Wet'suwet'en Tribal Council to manage its own fisheries and to sell as well as catch fish for food. Immediately, other groups in the fishery objected, and the government suspended these by-laws. At the time of writing, this case also remains unresolved. The question again is, can the government unilaterally alter the rules of fishing so that some groups are given advantages? If that is possible, then clearly fish are not common property.

Significantly, in countries with monarchies, including Canada, state property is actually called crown property, indicating that property rights are not common even if the crown (including its property) is paid for by commoners. The crown may no longer refer literally to the monarchy, but it does imply a differentiation between ordinary citizens and the government they elect, together with the many vested rights and powers inherited by that government and institutionalized in its operation. If the state permits universal access or applies only such entrance requirements as may be universally obtained (e.g., automobile licences for road use), and where rules are universally applied (as in automobile traffic), a case may be made for the common property status. Common properties may be subject to commonly applied rules for purposes of protection (orderly traffic, public safety, or conservation). But, where the state restricts and allocates access rights, while simultaneously retaining extensive management rights affecting users in *diverse* ways, then the property is being treated as crown, not common, property.

STATE PRIORITIES AND PRIVATE
ACCUMULATION OF CAPITAL

If fishing is a privilege granted by the ultimate owner, the crown, then we follow a false lead by treating the depletion of the stocks as an example of "the tragedy of the commons." What we need, instead, is some understanding of how the state (crown) performs its ownership functions. Why has the state granted excess fishing privileges, and why has it permitted industrial and other users of the habitat to pollute it? Why did it first allocate special management rights to one tribal council and then bow to pressures from others to rescind these rights? How best can we interpret both the state's claim to the right to allocate differential rights to citizens and the contrary arguments of the courts and competing interest groups?

The Accumulation Process

The central difference between communal property owners, such as native Indian bands prior to European settlement, and contemporary commercial fisheries is that the users are not catching fish to eat; they are catching them to sell. Fish are now commodities. In addition to the fishers, there are processing firms and marketing firms that are able to operate as long as the cost of raw fish from the fishers is less than a retail market price. These sales, both from fishers and processing firms, continue year in, year out, so that both groups accumulate wealth over time; if they did not do so, they would seek alternative sources of income. As they accumulate wealth, they can choose to turn it into capital in the form of new and greater catching capacity (vessels, gear) or machinery for processing and freezing fish. Added capital increases the yield or the value of the commodity and thus the total accumulated wealth.

This accumulation process is the motive force of a capitalist system, differentiating it from the subsistence system within which groups can sustain communal property. Once accumulation, rather than subsistence, is the reason for catching fish (or cutting trees or any other activity), there is a need to define and defend property rights; without such definition, individuals and companies would be unable to ensure that they, rather than any others, should benefit from their investments and activities. One property holder could defend his or her property by force, as separate fiefdoms have done in earlier times. But a superior method of defining territory is a system of laws ultimately backed up by a single institution with a monopoly on force to which all members of the territorial unit are equally subject. This institution is the state.

The State and Accumulation

The state, in whatever form it takes in any one regional context, is endowed with the authority to enact legislation, govern, and regulate whatever within its territory may potentially affect the rights of property. To assume this is to acknowledge that the property rights preceded the state; the state is merely the instrument for ensuring their longevity. Its evolution is the result of the diversity of property interests that, because they engender conflicts, may lead to destruction of the social fabric. To forestall such conflicts, an agency has been created to ensure relatively peaceful negotiation of property claims. The same argument may be applied to international agencies: specifically, for our purposes, the United Nations in its Law of the Sea debates. There, too, existing property interests have gradually built up a supranational entity because their diverse claims bring them into dangerous conflict.

Property claims are diverse in nature and also in size. In the contemporary world, there is obviously a great difference in the economic power of global corporations and small fish-boat owners. The consequences of various actions taken by large units are obviously much greater: if they choose to locate plants, withdraw from regions, employ or fire large numbers of workers, the regional economies are dramatically changed. The power of large corporations necessarily affects and structurally constrains the choices governments make.

This, however, does not mean that governments always and in every action look after particular corporations nor that government is merely their instrument. If this approach to the role of government is valid, then it follows that governments must, insofar as they are informed and able to do so, administer the *system of property rights*, rather than the specific property holders at any one time, and the *system of accumulation*, rather than the accumulated wealth of any one group. Thus governments must concern themselves with many conditions pertaining to the longevity of capitalism, such as the health and welfare of workers, resolution of conflicts between competing smaller groups (e.g., the native Indians and other fishers), the viability of regions within their territorial jurisdiction, rising rates of unemployment, inflationary pressures, and the like. The maintenance of the system of property rights takes precedence over the particular interests of one or another property holder at any given time.

Property rights are not stationary. A capitalist system by its intrinsic nature is perpetually changing, because the cycles of accumulation and reinvestment persistently alter the nature of technology, the conditions of production, and the actors. To protect the system of property rights, then, means protecting the accumulation process even when the process adversely affects the particular property holders of a previous cycle.

This is succinctly argued by Offe, who examines the capitalist state in relation to the process by which profits are accumulated (1984). The state, he observes, has no authority to order or control production in sectors that accumulate profits, though it may initiate and sustain nonprofitable economic activity useful to the accumulating sectors. It has "not only the authority, but the mandate to create and sustain conditions of accumulation." This requires the state to prevent external threats or intracapitalist rivalries from disrupting the system. Its power depends on the continuing accumulation process; without that process, the state has no capacity to sustain itself. "Accumulation acts as the most powerful constraint criterion, but not necessarily as the determinant of content of the policy-making process." Finally, he argues, in common with numerous other theorists, the state must legitimize the accumulation process. It does this by conveying "the image of an organization of power that pursues common and general interests of society as a whole."

Some theorists regard the legitimizing function of the state as paramount. O'Connor (1973:6) for example, argues:

> The State must involve itself in the accumulation process, but it must either mystify its policies by calling them something that they are not, or it must try to conceal them.

This suggests that the state includes a component conscious of its alliance with property holders and deliberately engaged in obfuscating that relationship. But we need not make such an assumption in order to recognize the constraints on the state and will not make it in this study. Rather, we will assume that the liberal democratic state in a capitalist society is constrained by the economic context and by the demands of numerous groups in the society (not all of whom have property rights), and that it must describe its own actions to itself and to an electorate in the most favourable light.

Governments are fallible, in the sense that they rarely command total information, cannot accurately predict the outcomes of their actions, and consist of individuals who themselves have particular interests. As well, governments have short time spans, even if the state as an institution has a span identical to the property interests it supports and whose accumulating activities maintain it. Compared to some of those property interests, governments have limited economic power. And finally, governments in liberal democracies lean heavily on bureaucratically organized institutions wherein multitudes of individual workers have career interests. These state employee interests necessarily affect the execution of government directives, and sometimes the protection of employees becomes a significant concern of governments. This is particularly evident in the fisheries, as described later on.

Thus governments — the momentary embodiment of the state — operate within a constraint system that restricts their initiatives but does not dictate precisely how they should respond to specific situations. They must protect property rights. They must satisfy an electorate while ensuring the continued economic viability of the system. They must recognize popular values and deal with the needs and demands of various interest groups. And they must act, appear to act, and satisfy themselves that they are acting, to the best of their ability, in the public interest. These are considerable constraints on mere mortals: in arguing, as we do in this book, that fisheries policies have been inconsistent, sometimes irrational, and often contrary to the interests of either fish or fishers, we do not impute to various ministers of Fisheries and Oceans any evil intent. Rather we are trying to understand how the constraints have led to the responses, and how the responses — various legislative acts, regulations, judicial processes, and royal commissions — have affected the fishing industry.

Given this theoretical framework, our first question is, how important is the fishing industry in British Columbia, relative to other economic sectors? We ask this because the answer informs us about probable state priorities: if fishing were the leading economic sector, the defence of its property rights should be much more central to state management concerns than if fishing were a minor economic activity, especially if its major property holders' interests are in conflict with those of major industries in the same territory.

Economic Importance of Fishing

In British Columbia at the present time, unlike Newfoundland and parts of the Maritimes, the fisheries have relatively little importance in the overall economy. Forestry is the major industry, followed by mining. In 1980, aggregate landed values for all fish were estimated at $153 million; aggregate wholesale values at $264 million. By comparison, the value of shipments (current dollars) for all of the forest industries was $6,996 million, the estimated value of mineral production was $2,949 million. One seldom thinks of British Columbia as an agricultural region, nor, in comparative terms, would this be an accurate description of most of the province, but the overall returns from agriculture in B.C. are in fact higher than in the fisheries: the estimated receipts for fruit and field crops were $266 million in 1980, from livestock sales, $464 million (*B.C. Economic Activity, Review and Outlook*, 1980:17, 21, 26, 31).

These comparisons are constant over the postwar period. The year 1980 was a poor one, and the figures above are estimated to represent roughly a 53 per cent drop in revenues since 1979. But this is equally true in most other industries. If one takes another year, 1976 for example, when there were boom conditions in these industries, similar comparisons occur: in that year, the total value of shipments (current dollars) of forest products was $4 billion; mineral production value was estimated at $1,486 million. Total farm receipts were estimated at $400 million. And in fisheries, with reported "improved catches, higher values and strong market demand," the wholesale market value was estimated at $250 million (*B.C. Economic Activity, Review and Outlook*, 1976:31, 42, 45, 54).

Although the sports fishery takes only about 5 per cent of the total catch and its major interest is in chinook and coho, its claims for higher priority in such areas of the province as the Gulf of Georgia (near Vancouver) and the west coast of Vancouver Island, where wealthy tourists may charter vessels from fishing lodges and other companies, do not fall on deaf ears. The provincial government is particularly concerned with the development of the tourist industry; commercial fishing is an obstruction to its expansion.

The processing sector of the industry is "big capital" relative to any one small fishing vessel, the more so when that sector is ultimately owned by conglomerates outside the industry. This is evident even though the actual dollar investments in the fleet as a whole are much greater than investments in processing plants. The fact that they are made by many small investors delimits the economic power of vessel owners. But the term "big capital" is relative, and in comparison with other B.C. industries, no component of the fishery is an economic giant. The raw material, immediately usable and with very limited uses beyond food, does not lend itself to extensive processing; thus processing is not a high-value industry. It is not attractive to investors who can obtain higher marginal profits from other resource industries. As well, those investors who have participated have not developed the technological potential for greater production (Gordon, 1983). Prior to 1972, relatively little could be gained from extensive capitalization, and the brief period from then to 1979 was the only one during which the profits might have rewarded the investors for extensive technological development.

The commercial fishery is important for other reasons, however. It employs labour in coastal communities which have few or no alternative employers and provides a commercial livelihood for vessel owners and crew. It has some industrial linkages of benefit to B.C. shipbuilders and marine equipment suppliers, especially in the harvesting sector. While it has never been the major industry in B.C., it did have a more prominent position in the early history of the province and a paramount position in the precapitalist history of the coastal region. The economies and cultures of the coastal native peoples were built on the salmon runs, and one could not possibly describe the west coast of Canada without reference to fish and fishing. Yet in relative terms, the industry today has low priority on the provincial agenda; the fortunes of most of the population in the 1980s will not rise or fall with the ups and downs in fisheries.

Since fisheries are a federal responsibility while land and other resource industries are under provincial jurisdiction, the provincial government has an awkward, and easily avoided, role in the fisheries. Thus its legislation pertaining to forest tenures and mining rights affects habitat and fisheries, but a struggle over competing claims is avoided by referring conflicts to the federal state. In turn, the federal state is constrained by provincial priorities and legislation. It may make rules governing the capture of fish but it cannot control the context of fishing. Even were it able to do this, it would be unlikely that fisheries would have precedence, since for the federal and the provincial states the chief property owners and most powerful voices are in forestry and mining, not fisheries.

Both the federal and provincial governments are engaged in international negotiations respecting their more powerful industries along with fisheries. International negotiations are political events. What a nation loses in one arena may well be the quid pro quo for its gains in another. Fishing rights must be considered in a context that includes negotiations over seabed mining interests, lumber tariffs, offshore oil and gas exploration, ecological issues of air and water pollution, and general trade issues. In addition, since there are fisheries on both coasts, negotiations over an issue such as Georges Bank in the Atlantic must be weighed against those over Swiftsure Strait in the Pacific. Overriding these negotiations is the central fact of Canadian life: the United States has the greater economic power and can impose sanctions in unrelated trade areas if dissatisfied with agreements reached in the fisheries (Chapter 7).

This international and provincial context reduces the capacity of the federal state to introduce rational conservationist measures in the fishery or to act on behalf of any groups in the industry. If our general theoretical approach is valid, given these economic conditions, we would not expect the federal or provincial governments to design legislation in terms of any interests in the fisheries where such legislation would conflict with the interests of more powerful property rights. Further, we would expect that concern with the fisheries would have declined, and the rank of fisheries as a ministry within the federal government would have dropped over the course of this century, in line with the relative economic fortunes of the industry. And finally, we would expect a good deal of contradictory and inconsistent government action regarding the fisheries, because the competing interests of other sectors would necessarily push and pull this sector in directions quite contrary to those which would occur if fisheries were a primary concern.

Federal State Presence in B.C. Fisheries

Having established these constraints imposed by the relatively small role that the B.C. fisheries play in the total Canadian or B.C. contexts, we nonetheless note that the federal state is a major actor in the sector.

According to Pearse (1982:233-34), there are 1,231 Department of Fisheries and Oceans employees in the Pacific region. The total budget for the region was approximately $84 million in the fiscal year 1981-82. The total national budget for the Department of Fisheries and Oceans was $450 million, of which roughly 84 per cent is spent on fisheries. Of the Pacific region's $84 million, approximately $34 million was allocated for wages and salaries. Overall and including wages, about $50 million was allocated to Fisheries Management and Research. The remainder was allocated primarily to the Salmonid Enhancement Program, to which the provincial government added another $1.5 million.

When these figures are compared to the total value of the fisheries, one begins to recognize the significance of the state in the industry: the landed value of the catch is only twice as large as the cost of administration for the Pacific region alone. When one adds the costs of the main offices in Ottawa, some portion of which is reasonably attributed to the Pacific region, there is the distinct suggestion that the fisheries may cost as much in administration as they provide in landed value. The cost of administration now is equivalent to one-third of the wholesale value, and a realistic estimate that includes Ottawa costs attributable to the Pacific region might well increase that to one-half.

Certainly all industries receive a good deal of financial support, departmental aids, and infrastructure of various kinds from both levels of government. But other industries, with the possible exception of small-scale agriculture, provide higher returns from which such costs can be deducted. There is, then, a considerable imbalance here between state expenditures and economic returns. If it is the case, as argued above, that the fisheries are relatively unimportant to the provincial economy, why would the federal government invest so much in managing the resource?

The growth of employment in the Department of Fisheries and Oceans occurred throughout the postwar period, during which time almost all government departments similarly expanded. The DFO became a major employer for professional biologists, just as other government departments became significant employers of economists, psychologists, and geologists. In all these respects, governments everywhere were increasing their professional expertise in areas where they provided a service to private capital. These professionals did not manage the accumulation process. They did not, for example, intervene in managerial capacities within plants or on vessels, though they offered advice and research knowledge when it was requested or deemed useful. What they did in the fisheries was attempt to increase the yield to be captured by private vessel owners and processed by private processing enterprises.

Over and above this general condition, from the late 1960s to the late 1970s, the Pacific fisheries suddenly took on a new importance because of an unusual market. This occurred shortly after the introduction of a major conservation policy. The policy required constant policing of the resource capture process as well as research to increase resource supplies. This may be regarded as active intervention in the accumulation process, but in fact it was never intended to prevent the process from occurring; on the contrary, it was intended to ensure that, in the long run, the supplies would be maintained. In other industrial sectors, the state allocates resources through rents and exclusive property rights. In this industry, it used access licensing, an inadequate restriction on competing users. Thus it was obliged to introduce further allocative restrictions, assigning specified amounts of time and occasions for commercial fishing. This had no detrimental effect on the processing sector, which obtained the catch from one or another group of fishers regardless of state allocation rules.

One of the results was the creation of a substantial number of state employees with a vested interest in the continuation of the fisheries, though the unusual market demand has since disappeared. The state is less able than the private sector to dismiss employees, and a bureaucratic momentum develops that intensifies the state's involvement, independent of instrumental requirements of the accumulation process. There is a sym-biotic relationship between the fishers and the bureaucrats, but it is an essentially neurotic one. It requires that the bureaucrats steadily increase their watchdog role over the capture, and thereby frustrate the fishers and cause them a loss of income. But it also requires that the bureaucrats support the fishers in their attempts to salvage the industry.

This is but one in a series of contradictions — between the interests of the largest processor and those of commercial fishers, between native land claims and the claims of other fishers, between the demand for conservation and the demands of labour for jobs — institutionally struc-tured into the state. The several different ministers, each of whom has reorganized the top echelons of the Ottawa bureaucracy, have tried numer-ous experiments toward "solving" the contradictions. The minister's advi-sory council obliged the organized participants — the Fisheries Association, the union, the gear-type associations, the sportsfishers, and the Native Brotherhood — to devise their own solutions in the early 1980s when fishing fortunes were falling. But always the proposals were based on the assumption that the problem was too many fishers in a common property. The more fundamental problem of state management was not on the agenda.

By the mid-1980s, with high salmon runs and optimistic forecasts, the popularity of the Department of Fisheries and Oceans had increased. For the moment, the crisis was past and the contentious issues were shelved while everyone went fishing.

PROPERTY RELATIONS

If the accumulation process is to be maintained, the state must take an active role in containing conflict between participants. This conflict may be between enterprises, such as between the Japanese and indigenous processors in the 1970s. Or it may be between diverse interest groups defined in terms not specifically economic, such as native and non-native fishers (and similarly between numerous competing groups of fishers defined in terms of gear-types, regions, and particular fishery). In common with other industries, the state's intervention is required for containing conflict and providing a system of rules to regulate the relations between capital and labour; that is, between the propertied and the unpropertied classes.

Again, it is not the state's prerogative to become the employer in a profitable industry; but it is the state's mandate to maintain profitable conditions of operation for the industry, and, toward this end, the state usually establishes the ground rules for labour. In the fisheries, this has occurred primarily through legislation defining fishers themselves, so that they are independent businesspersons or coadventurers rather than labourers, and in actively enabling them to become independent owners both of vessels and of processing facilities. In Canada this has been especially difficult because the federal government has jurisdiction in the fisheries (and Indian Affairs), but the provincial government has primary jurisdiction respecting labour.

Definition of Labour

The typical production system within a capitalist enterprise involves owners, those who have invested capital in a plant, and wage workers who add value to a raw material through their labour. The added value, after the wage and plant costs are deducted, is referred to in social science literature as surplus value, meaning value not used for replenishing labour, but that becomes the property of the plant owner. Surplus value becomes profit when the commodities so produced are exchanged within a capitalist market. These profits may be reinvested in a new plant so that labour becomes more productive, and greater surplus value is created, which would occur only where greater productivity is expected to lead to greater profits in the market.

Workers in most industries over the past century have been progressively subordinated to capital. Where there were once households producing subsistence goods, some members became commodity producers for markets; where there were independent commodity producers, some became increasingly subject to controls by merchants in the marketplace; where there were dependent sellers of commodities, some became wage workers under direct supervision and control of employers. A fully subordinated labour force is one which has no independent power regarding the nature, pace, duration, content, end product, quality, or quantity of its work. Its production is geared entirely toward the creation of surplus value that will become the property of the owners of the plant and machinery in and with which labour produces the commodities.

In most industries today, we can identify "capital" and "labour" fairly easily, the word "capital" here referring not to the plant itself but rather the owners of plants. We have little difficulty recognizing B.C. Packers as a form of capital and shoreworkers as labour. As labour, B.C. shoreworkers are subject to a competitive labour market in which differences by gender, age, ethnicity, language, work histories, and job skills determine their chances of obtaining employment; their geographical location determines which kind of employment is possible.

Since the objective of capital is to make profits and since a large portion of profits must derive from surplus produced by labour, a major condition of profitability implies that labour receive as small a share of that surplus as possible and that capital receive as large a share as possible. Wages respond to conditions of labour supply: they tend to rise, because workers' bargaining capacity increases when there are fewer workers than jobs, and to fall when there are large supplies of competitive labour seeking scarce jobs. It is therefore in the interests of capital in general to have a surplus labour supply and to avoid a condition of full employment in any region or in any job category.

When the economy is expanding rapidly or the region has too few traditional sources of labour, capital must seek out new labour supplies, such as immigrants, temporary migrant workers, and previously untapped domestic sources, primarily women, rural farm populations, or minority ethnic groups not previously in the labour market. The state, in its immigration policies, labour legislation, and legislation regarding minority groups, inevitably becomes involved.

When the economy is not expanding, or when there are excess supplies of local labour, employers are able to make choices between different groups. Preexisting differences between groups, such as between men and women, can be tapped, and the lower wage groups may be preferred employees. The existence of a pool of reserve labour tends to push down wages for the employed, since the employer then has ready access to substitutes.

These strategies for reducing labour's cost and bargaining power are not simply theoretical possibilities. The history of capitalism is very much the history of the creation of a labour force followed by persistent cycles of capital relocation and employment practices that reduce any tendencies toward full employment and ensure the persistence of surplus labour competing for jobs. We see this history throughout the development of the labour force for shorework in B.C., and it is examined in detail by Guppy in Chapter 9 and Muszynski in Chapters 3 and 12.

Fishers: Labour or Capital?

Shoreworkers are clearly wage workers, but it is much less easy to identify Canadian west coast fishers. Neither vessel owners nor crew who are employed on a share basis instead of a wage are like shoreworkers. For vessel owners, the crux of the problem is that they have independence regarding the work process itself and they also have considerable control over the surplus value of their work once their commodity (fish) has been sold. What they do not control — or at least what most fishers cannot control, since there are differing degrees of latitude between gear-types and fisheries — is the market for their commodity. The union bases its argument for treating fishers as labour on the premise that legal control does not constitute real economic power when fishers are so hedged in by debts, preseason financing, contractual obligations, and the market controls of processors that they actually have few market options.

> Fishermen, with the exception of the larger vessel owners, are basically workmen. Their livelihood is earned by hard manual labour, long hours and the use of skill and judgment in contending with the elements and handling of boats, gear, and equipment. The fishermen's tools may be simply their boots, blankets, and oilskins, or they may include an individual's lifetime investment in a small boat and the necessary gear to enable him to ply his trade. Alone and unorganized, the fisherman cannot be on equal terms with the fish buyer. The unorganized fishermen are just as readily exploited by the wholesale fish dealers and fish-processing companies as are unorganized wage workers on shore. In fact, the perishability of the product of his labour may place him in an even worse position than his fellow workers in the plants. (UFAWU, quoted in Federal-Provincial Committee on Wage and Price Disputes in the Fishing Industry, 1964:80)

Many observers agree with this analysis. Clement (1984:7) for example argues that

> by maintaining a distorted form of petty production, capital shifts considerable capital risk and supervision of labour onto the producers themselves. Capital can exercise sufficient control through contractual obligations and market domination while minimizing capital investment and expenditure on supervision. Since small producers do not "own" the resource (fish) there is little pressure for capital to directly appropriate the means of production.

In other fisheries, there are instances of fishers being transformed into wage labour. In the Atlantic fisheries, the offshore company trawlers are akin to factories, and fishers on them are wage workers. Barrett (1979:131) argues that the transformation of Atlantic fishers occurred, in fact, much earlier. By the turn of the century, he says:

> What existed was . . . a form of oligopolistic control of primary production and offshore "schooner" fishermen who were dependent primary commodity producers. The price of fish, and all purchasing and selling transactions, did not reflect the unfettered operation of supply and demand, nor were profit calculations made for each vessel. Instead the logic of production reflected the profit requirements of an entire company's integrated operations.

Similarly, by the estimation of Gregory and Barnes, half of all fishers in the Alaska and Washington-Oregon coast fleets in 1939 were employed on company vessels (1939:17). All the trollers and most of the purse seines in Alaska and in the states were independents. But the traps, used extensively then in Alaska, and gillnets were largely under company ownership. In consequence, much of the analysis of labour in the industry offered by Gregory and Barnes is about subordinated wage labour.

The industrialization of the processing sector occurred within the past century, as well, in the Atlantic region of Canada and in many other places where fishing has existed over many generations. This has transformed the domestic unit that once produced salted fish into a passive agent, while its individual members became wage workers in external firms.

Peculiarity of the Pacific Fisheries

The Canadian Pacific fisheries have had a somewhat different history. Companies have not created industrial conditions through factory vessels. In the processing sector, they have certainly created a fully subordinated labour force and industrial conditions, but this has not been a transformation of an earlier domestic production system because such a system never took root in the relatively short period since the commercial fishery developed on the Pacific.

There was no period on the Canadian Pacific coast when the commercial fishery was wholly or even mainly based on the sale of raw fish, and no period when household production of dried or salted fish for sale was a general practice. As the fur trade penetrated Indian cultures, some portion of the caught fish was sold to traders (McDonald, 1985; Hawthorn et al., 1966) but the larger portion continued to be caught primarily for use by the local band or kin group. Local markets were served by independent fishers, and, during the final construction of the CPR, local markets constituted a significant demand for fresh fish. But for most of the early period, the coastal population was sparse. The great potential markets in Britain, the United States, and central Canada could not be served with fresh or mildly cured fish (Gregory and Barnes, 1939: 128–131); the distance dictated the need for a processing method that preserved fish for longer time periods.

This means that even had there been any substantial number of independent fishers on the coast in the 1880s, they could not have marketed any quantity of fish unless they also processed it. To process it they would have required investment funds sufficient to purchase knives, modest conveyance machinery, and tin for cans. The same individuals could not have fished and processed fish on board the simple vessels of the 1880s, nor could they keep fish fresh for many hours. They had to have either a cooperative method of dividing labour between the two major phases (similar to or encompassing the household division of labour on the Atlantic coast) or external canneries to which they would deliver a catch. The first of these was not possible for many fishers because they had no households, and for others (particularly natives) it was impossible because they had no capital. For these reasons, if for no others, a commercial fishery began not with fishers and raw fish but with processors and canned fish. At this point the perishability of the fish and the distance from European and North American markets had a greater impact on the nature of the fishery than the mobility and migratory habits of the resource. With the introduction of tin-plate manufacturing, it became possible to ship Pacific salmon to these markets (Ralston, 1968), and the objective of investors who established canneries was to discover means of processing the large quantities that were easily captured.

There is no evidence that any substantial number of independent fishers was actually in the wings in the 1870s. Canneries were established by brokerage firms and then by companies already embedded in the Atlantic fishery and in the United States, and these sought a fishing labour force through offers of daily wages and family employment: women and children at the cannery and men on the rivers or coastal vessels. These arrangements "captured" native Indians in northern areas and Japanese-Canadians further south. Chinese men as well were brought in on contract to work in the canneries (Muszynski, 1986a, 1986b, and Chapter 3). Thus whatever other reasons there may have been for a wage system of fishing in the early decades, the bottom line was that there was no other labour force available.

Once in place, canneries attracted fishers from elsewhere. Those with low ethnic status (particularly Japanese-Canadians) had little bargaining power despite their widely recognized fishing skills. Those labelled "white," if they had fishing experience and were familiar with the typical organization of independent fishers elsewhere, were unwilling to sell their labour for an average daily wage; others, without experience, were unwilling to work at Indian and Japanese-Canadian wages. The struggle for better conditions shortly led to demands for funding of independently owned vessels with access rights to fish.

When the first licences were introduced in 1889, the canneries obtained the majority. This continued to be the case in northern waters and in Alaska right through the 1920s. Gregory and Barnes (1939:17) note that in all United States Pacific waters, over half the catch was made by company vessels in 1937. Ralston (1965:33) records, however, that by 1900 there were some three thousand licensed fishers selling to forty-five canneries on the Fraser River, and that by that time daily paid fishers on company vessels were no longer in evidence. Independent vessel owners earlier dominated the Fraser fishery because the canners, who were more numerous and competitive with one another for the resource, had less easy access to a "captive" labour supply for company vessels but greater access to experienced European fishers.

One option canneries did not have was to capture fish in large, factorylike conditions. The large vessels now common in the offshore Atlantic fishery were not then technically feasible, but even a century later they would not be cost effective in B.C. inland waters where the salmon fishery exists. In fact, the early vessels were manpowered rowboats, and until suitable engines and then freezing techniques were developed, capture vessels were typically small (Hayward, 1981a). The only issue has been whether these vessels would be formally independent of the canneries.

The original reasons for allowing independent fishers entry to the industry were buttressed by other considerations which remained valid throughout the greater part of the next century and were, by 1930, as true of northern as of sourthern canners. With the single exception of the 1970s, when entirely peculiar conditions intervened and then disappeared again, the processors were aided by the supply of fish being caught mainly (though never exclusively) by independent fishers. These considerations were the tying up of capital in vessels and gear, producing profit only on a seasonal basis; the risk costs associated with obsolescence and loss of vessels; labour costs and the difficulty of supervising labour; variable harvests; and variable markets. Salmon have four-year cycles and the size of runs over the period is always unpredictable. Fishing is always dangerous, and weather conditions, uncertain. There are few economies of scale, even though there are advantages to increased capitalization in gear, so that investments in large factory vessels similar to those on the Atlantic would not be worthwhile. These conditions make the system of independents a sound strategy for capital in the processing sector, provided the commodity market can be controlled.

Although independents are the majority of fishers, companies retained ownership rights in some vessels as well as shares in vessels ostensibly owned by independents throughout the century. This practice ensured a guaranteed minimum catch and, perhaps more important, a way of providing processors with information about runs, cash buyers, and any other fishing conditions that might affect their supplies from independent but serviced fishers.

Fish are mobile, perishable, seasonal, and unpredictable. All these characteristics add to the risk of fishing and reduce the advantages of owning vessels if fish can be otherwise purchased in sufficient quantity by processors. But it would be a fallacy to blame fish for the unusual reversal of historical trends from wage work to independent commodity production. Contracting out phases of production to independents occurs in the oil, forestry, automobile, and electronics industries. What links these together is risks avoided by large capital, and risks may be associated with labour supplies and supervision, idle capital or costs of capital equipment, rapid obsolescence of designs and technology, and numerous other conditions. The situation, then, is not unique and is attributable to the resource only in the sense that the resource capture has similar risks to other commercial undertakings.

What is profitable at one stage may not be equally profitable at another. Monopoly control of the fur trade combined with independent hunters may have seemed, in its day, the only sensible organization of that industry. The resource was, like fish, mobile, migratory, and unpredictable. In its wild state this is still the case, but the fur industry is now organized largely around fur ranches that have altered the nature of the resource so that it no longer carries the same risk.

Similarly with fisheries, the particular organization of property rights and state management appears to coincide with the peculiarities of a mobile and perishable resource. But if the equivalent of fur ranches were created in the fisheries (fish farms, ocean ranching, or terminal fisheries would all be equivalents in some degree), then we might suppose that corporate capital would depend less on independent producers and more on corporate control of the harvest, other conditions remaining favourable to corporate activity.

Capital, State, and Union

The union's definition of fishers notwithstanding, neither capital nor a substantial number of fishers accept the argument. Indirect market controls are not, in their view, equivalent to direct subordination. This has led to numerous conflicts between fishers, as well as to persistent difficulties experienced by the union in its attempts to represent and bargain for fishers together with shoreworkers.

The UFAWU was established in 1945, following a period of high demand for fish during the war and thus a period in which labour had bargaining power. The 1945 organization (an amalgamation of previous ones, noted in later chapters) extended membership to vessel owners employing no more than two persons provided they were not simultaneously members of a vessel owners' association with which the union conducted collective bargaining and certain others in similar categories. At that period, all of the nearly 10,000 Canadian fishers under collective agreements resided in B.C., but 80 per cent of these, most fishers, were covered by entirely voluntary agreements, while shoreworkers were covered by union, shop, or preferential agreements. During the same period, fishers in the Atlantic region were attempting to organize. In 1947, a Nova Scotia Supreme Court ruling designated fishers as "joint-adventurers" rather than as "employees." Primarily on this definition, the court reversed a 1945 War Labour Board certification of Nova Scotia fishers (Steinberg, 1974:643), a decision that put into question the legality of a B.C. union that included fishers.

The legislation under which union affiliation for fishers was deemed problematic was the (Canada) Combines Investigation Act, which could be interpretated in such a way that "collective bargaining by fishermen and buyers is . . . held to be in restraint of trade." Antitrust law in the United States inhibited bargaining by fishers and companies on the same basis. If the courts interpreted that act to mean that persons selling commodities could not be classified as labour, and defined both vessel owners and their crews as "co-adventurers," collective bargaining by fishers was illegal. Companies were quick to challenge the union on these grounds, arguing that if they negotiated prices with the union they were engaging in illegal activity. A union representing only fishers would not likely have overcome this legal barrier, but a union which could mobilize shoreworkers to stop processing fish unless price negotiations had been undertaken was more persuasive.

In 1959 the companies refused, as they had on earlier occasions, to negotiate prices. The case went to court, as earlier cases had done. But faced with a strike threat by shoreworkers as well as fishers, the court, processors, union and Native Brotherhood achieved an accommodation in the form of a moratorium exempting agreements reached through bargaining during 1959 and 1960. This moratorium was extended annually for the next three years. Yet a long investigation of four main fisheries in connection with a challenge to the legality of bargaining resulted in a 1962 Supreme Court decision that either price fixing or restraint of competition, or both, were present in all of the instances of bargaining in the major B.C. fisheries. Further, the court argued that the disputes leading to these results could not be regarded as labour disputes, and the fishers' organizations were not unions, since the fishers were not employees (Steinberg, 1974:653).

The moratorium temporarily permitted negotiations to continue, but it did not prevent strikes. The increasing number and intensity of strikes induced the federal government to establish an inquiry into wage and price disputes stating that "in 1963 the fishing industry came to a virtual standstill when companies and the major union and Native Brotherhood failed to agree on prices, wages, and working conditions" (Federal-Provincial Committee on Wage and Price Disputes in the Fishing Industry, 1964:1).

The committee recommended in 1965 that price negotiations between salmon and herring fishers and fish-packing companies be legitimized, and, as a result, the moratorium of 1959 to 1965 was written into the Combines Investigation Act. At the same time, the committee recommended compulsory arbitration for the industry. Although there were further court cases regarding the legality of collective bargaining centring on the question of employer-employee status and on the question of jurisdiction for the federal government, the change in legislation removed a major obstacle to collective bargaining; simultaneously, it increased the barriers to strike action. In this way, then, the federal government intervened to reduce conflict and actions which could inhibit accumulation.

Other barriers had been erected meanwhile by the provincial government and, though these were not designed specifically for the fisheries, they had impact on the nature of bargaining, strikes, lockouts, and administrative machinery governing labour relations in that industry as in others.

The B.C. Labour Code, 1973, included the category "dependent contractor" in its definition of "employee," thereby providing the possibility that fishers could legally engage in collective bargaining. The Canada Labour Code, as amended in 1972, followed suit, but it failed to amend the definition of "employer" to include commercial processors in their relationship to crew members of vessels if the latter were regarded as dependent contractors. Several cases involving commercial companies and the Native Brotherhood or the UFAWU after 1972 failed to establish the rights of crew members to bargain under either code (e.g., see *Western Weekly Reports*, 1978,1:621–630).

However, the Combines Investigation Act uses the phrase "workmen or employees," a disjunctive noted in the case *Couture* v. *Hewison, Stevens, and Nichol* (UFAWU) (*Western Weekly Reports*, 1980, 2:136–148). Fishers are treated as workers under the Workers' Compensation Act, 1968 (B.C.) and the Unemployment Insurance Act, 1970, 1971, 1972 (Canada). The judgment regarding the collective bargaining rights of "workmen" who are "fishermen" was that they had the right to bargain; in the view of the presiding judge:

> I have been unable to come up with any logical reason for the placing of combinations or activities of fishermen for their own reasonable protection as fishermen in any less favourable position than combinations or activities of others in the labour force. (*Western Weekly Reports*, 1980, 2:143)

The problem of the legal relationship between these "workers" and companies that do not directly employ them was, however, not resolved, and these issues continue to present legal hurdles to negotiations between the union and Native Brotherhood (normally bargaining as a combined unit) and processors.

What we discover in this prolonged debate is periodic attempts by governments to clarify an issue so that strikes do not tie up the industry. Yet, at the same time, the legislation affecting labour and unions cannot be written primarily with reference to fishers; in B.C. the legislation is normally written with particular reference to woodworkers and miners. Thus it is frequently ambiguous for the fisheries.

While failing to resolve the problem of bargaining, the federal government has attempted to eliminate it altogether by enabling independent vessel owners to establish cooperative processing facilities, thus becoming owners, and to appeal to fishers as independent businesspersons rather than as labour. Since the Pearse Report and its highly contentious reception, the federal government has also invested substantially in the development of fish farms, a new method of fishing in which accumulation would occur. At the time of writing, this engages small property holders and few permanent employees and is not in the immediate interests of the major processing company, the union, or the small-vessel owners. Thus we see a classic, if small-scale, case of the government protecting and sustaining the process of accumulation, rather than particular property interests.

ACCUMULATION PROCESS AND COMMON PROPERTY

A major difference between a wage labour force and a fleet of independent commodity producers is that the independents can exert more control over the capital created from their own labour than can wage workers. When fish are few and markets are slow the costs of fishing may equal or exceed the value of sales, and small-business owners always run the risk of bankruptcy as the counterpoint to the risk wage labourers run of unemployment. But when fish are available and there is a market for them, both vessel owners and crew (whose payment is in shares of the price for sold fish) obtain returns beyond what is required to sustain them and the vessel. During the 1970s, the surplus returned to independent fishers was abnormally high. And a very great deal of it was reinvested in the vessels as fishers attempted to increase their catch through purchase of new and expensive technology. Norr and Norr (1978:169) observed that around the world, active fishers are more likely than nonfishing investors to reinvest surplus in vessels. This is not surprising if fishing is seen by fishers to be a way of life, whereas for others it is one of many possible investment opportunities. The problem is that the sum of these individual investments is so great and the increased capture capacities so enormous, that by the late 1970s the B.C. fishing fleet was vastly overcapitalized relative to the supplies of fish.

The point at issue is not whether overcapitalization occurred. It is why it occurred, for there is no doubt about its existence. Nor is the fishery alone in this predicament. Forest firms also overinvested in this period, and overcapacity afflicts them in the same way as it afflicts fishers. Yet forests are subject to "contractual property rights" and under the effective control of private capital. Overinvestment was equally common in the automobile industry in the 1970s, and automobile manufacturing is certainly not common property. In all these industries, overinvestment occurred in response to specific market and financing conditions and was aided by government legislation and loans. We are not addressing here "the tragedy of the commons." The problem is much less poetic. It has to do with the inherent difficulties of managing an accumulation system where numerous, competing property interests strive to maximize their profits in volatile markets.

Yet the ideology of common property is of immense importance here. It motivates fisheries biologists employed by the state to conserve and enhance the resource. And it motivates fishers to accept state management even when this frustrates their fishing efforts. In these respects it is a positive ideology from the perspective of both capital and state. But it has an obverse side: fishers believe it, and thus when the state abruptly assumes the rights of ownership, as in attempting to reallocate the resource, fishers mount opposition. Since the ideology has been publicly stated so frequently, it takes on a life of its own and begins to have legal implications. Law courts are obliged to take it into consideration; political parties assume its reality. Thus we arrive at the 1980s with a depleted resource, state managers striving to reduce the fishery, native fishers demanding aboriginal rights to greater allocations, sports fishers claiming special rights more in line with the increasing economic importance of their industry, and numerous disunified groups of commercial fishers mounting legal battles against these special claims on the specific grounds that the resource belongs to everyone. It was into this context that the Pearse Commission was appointed: its task was to develop recommendations for dealing with what everyone defined as "common property."

The Pearse Commission

The stated objectives of the Pearse Commission were to conserve the resource, maximize the benefits of resource use, improve incomes in the fisheries and develop economic opportunities for coastal communities and Indian people, enhance social and cultural development, provide better returns to the public, increase flexibility of policy, and make the administration of the fisheries more efficient. These objectives are not compatible, but the contradictions were not noted either by Pearse or by participants in the process. The full development of opportunities demanded by Indian fishers, for example, excludes either better returns to the public or flexibility of policy. Conserving the resource is not compatible with improving fishers' incomes. Improving returns to fishers would require either elimination of well over half of the fishing capacity or creation of an entirely new market demand. Even so, Pearse argued that fleet rationalization, habitat improvement, and a system of royalties and licence fees would achieve the stated objectives.

He provided a most valuable set of recommendations for improving the habitat, but his mandate did not permit him to conduct a full-scale investigation of the extent to which other industries' use of the resource impinged on the possibilities for a continuing fisheries industry. Nor was he in a position to investigate the role of the banks in financing the over-capitalized fleet. Thus the report, and the ensuing debate, took place without its total context, and inevitably both report and debate concentrated on the fisher/fish ratio, though Pearse repeatedly noted the deficiencies of past government policies (see Chapter 10 for major recommendations).

Pearse describes the stock decline as a result of an "inexorable tendency" created by competition. If such an inexorable tendency exists, then one naturally asks, why has the state selected to keep the property available to competitive users whose diverse private capacities lead to their collective ruin? The conventional answer is that the fish are mobile and therefore cannot be subject to capture guarantees or rights, which is also the conventional explanation for its status as "common property." But there were always alternatives to a state that had the rights originally granted it by propertied interests.

Specifically, the state could exclude everyone from capture by prohibiting access or could exclude all but preferred groups in the same way as it excludes most citizens from the rights to harvest timber. It could have done this at any time since its own creation with the legal definition of crown rights inherent to it. Capture cannot be enforced or guaranteed, but even so it can be limited through limiting access, imposing quotas, rents and royalties, or creating rights of access for only those vessels with equipment ill-suited to large catches. Fish may be caught in weirs or traps at terminal fisheries as they were in Alaska up until the 1930s, and then the trapping capacities could be reduced or managed through conservation methods or legislation. Overall, the fisheries could lean far more heavily on farming and ranching methods. None of these methods is being recommended here; the point is rather that the mobility of fish is important to the commercial fishery only because the crown has positively selected, at least up to 1984, to permit excessive access and capture capacities for wild fish. Its inconsistent attempts to conserve the resource have been frustrated by its prior commitment to a process of private accumulation rights and by its own dissemination of a "common property" ideology.

Fishers, contemplating potential loss of livelihood and recalling the contradictory statements of the past that induced them to overcapitalize, might well feel that insult is added to injury when their independent and competitive behaviour is judged to be the cause of a tragedy, while simultaneously they are informed that fishing is a privilege granted by the state.

NOTES

1. In fact Macpherson appears to give the state two quite different characters in these descriptions of different property rights. Where the state keeps property rights to itself, it is not the executive of society but the arm of

specific and propertied interests; yet where it creates common property, it is translating traditional social understandings into socially acceptable practice. There is no explanation as to why it would do this for some things and not for others, an omission especially distressing in view of the general argument regarding the changing definitions of property and justifications of rights in different social eras.

2. Annual licences were required at other periods, but these were not limited in number. Their main functions were raising funds for the crown and keeping track of numbers of fishers.

PART 1

CAPITAL AND THE STATE

2

The Organization of the Fisheries: An Introduction

JOHN MCMULLAN

This chapter is a descriptive historical account of the different fisheries and their gear-types. It documents the evolution of the B.C. fishing industry from its inception as a single fishery (salmon) utilizing one type of gear and its growth into multiple fisheries using several separate gears, vessels, and crews, but without strong connecting linkages. This period lasted until approximately World War II. The current stage is characterized by multi-purpose vessels and gear with the capacity to roam the coast, with labour power able to exploit many fisheries, and with firms capable of multiproduct processing.

GEAR-TYPES IN SALMON FISHERIES

Gillnet Fishery

Salmon gillnetting is the oldest commercial fishery, and the fishing technique is much the same as it was in the 1870s. A net, with mesh size large enough to allow passage of undersize fish, and small enough to capture larger salmon, is strung from behind a boat across rivers, inlets, passages, and channels. The salmon, on their spawning migration, encounter the net, get entangled, and drown. After "the drift," the gillnet is pulled on board, the salmon retrieved, and "another set" is made. This is a passive style of fishing, requiring patience, stamina to withstand long sleepless periods, and knowledge and skill to make a "good drift." Because the location of gillnet fisheries is often in river mouths, inlets, and channels, the weather conditions are more stable than in the open sea. As a consequence, the gillnetter uses a small vessel.

The technology involved with gillnet fishing is not complex; only three basic innovations having been adopted over the years — motors, power-driven net drums, and nylon nets. However, it is a fishing method requiring regular investments and support services. Packing services to move the fish from the grounds to the canneries, storage and transportation of nets along the coast, and financial assistance to cover seasonal start-up costs are persistent and normal requirements of the gillnet fleet.

The fish traditionally caught by gillnetters are sockeye, pinks, and chum salmon. They frequently are marked and bruised by the nets, and once boarded they are not usually dressed or packed in ice. Much of their production must be canned quickly for this reason. Each canner relies upon a supply from the gillnet fleets' production, and each gillnetter is reliant, for sales and markets, on the canning company for which he fishes. For this reason, gillnet fishers, even though independent by virtue of boat ownership and fishing licence, have remained linked in a dependent relationship with the cannery processors.

Until about 1900, most salmon fishing was done with two-person crews, fishing from small, open boats. Between 1900 and 1910 the industry underwent considerable change. By the end of that decade most of the southern fleet of gillnet boats were motorized, crews became single-handed, and mobility was increased away from the mouth of the Fraser River.[1] In the 1930s, the power drum on which the net was rolled was installed, and by the 1960s, the introduction of nylon nets increased catches in the clearer ocean waters, away from river estuaries (Miller, 1978).

In the last twenty years, the Department of Fisheries has dramatically curtailed the time and areas for gillnetting. Consequently, new hull designs and more powerful motors have been used to enable gillnet fishers to travel coastwide to fish openings in as many areas as possible. Also, there has been a large growth in the number of gillnetters who have added trolling gear, enabling them to troll when gillnet openings are denied. Statistical data on landings of salmon (1967 to 1981) show that in the years before 1975, landings by combination gillnet/troll vessels comprised around 9 per cent of the total salmon fleet. After 1975 there was a 40 per cent increase in the number of combination gillnet/troll vessels, raising their catch to 13 per cent of the total salmon fleet (Pearse, 1982).

Purse Seine Fisheries

Purse seiner sets a net around schools of fish, and then closes off the bottom of the net with a purse-line. This prevents the fish from swimming out of the mobile trap — somewhat like the closing of a pouch-style purse with a drawstring. The net's size, its heavy mesh, and the leadline used to sink the web make it a cumbersome apparatus. It is difficult to fish without manpower or powerful machinery to unload, set, and load the net and to scoop the fish out of the net's enclosure. Until the early 1950s, these tasks were done largely by hand. When fishing salmon, crews of seven were required, and up to four more crew members when fishing herring or pilchard because of their compact mass and weight.

In 1886, purse seining was introduced to the west coast (Philips, 1971:3C). The seines were set from scows and skiffs; the scows powered by tugs, and the skiffs hauled out the net by oars. In 1902 this cumbersome and inefficient technique was improved with the building of the first power seine boat (Philips, 1971:3c). By 1910, the addition of living quarters eliminated the need for fishers to return nightly to their fish camps or ports. The seiners' range of operation was expanded greatly, and by 1912, they were fishing at the western end of Juan de Fuca Strait and delivering to canneries in Puget Sound, the San Juan Islands, the Fraser River, and Vancouver.

While the early gillnet technique required two persons per vessel, the purse seine technique introduced the captain and multiple crew structure into the salmon fishery. A twelve-portion share system was introduced for dividing the value of the catch. The vessel received five shares, and the captain and crew each received one of the remaining equal shares. Extra payments for nets, captain's share, or even an engineer's share were taken from the vessel's share. In 1941, after a salmon seine crew strike, the share system was modified and based on eleven shares — four for the vessel and seven for the crew. This division remains in the salmon industry to the present day.

The large boats required to handle nets of this size and the necessary crew complements ranged from forty-five to sixty-five feet and were used to fish halibut, trawl for groundfish, and fish or pack salmon, herring, or pilchard. By the 1950s and early 1960s, technological innovations, such as the power-block to hoist the net and the power drum to roll the net, reduced the crew size to four or five. In the 1970s the running line and bow thruster gave the seiner increased speed and manoeuvrability in making sets.

The salmon seine fishers also have been dependent on the cannery processors. Until the 1970s, much of the seine boat fleet was company owned; the captain and crew of these vessels were company employed. The particular conditions of this fishing method — the large volume deliveries, the undressed product, and the absence of ice chilling systems — dictated a contractual agreement that demanded immediate delivery and processing. In other words, seine fishers have been in a similar situation to gillnet fishers. Independent boat owners, captains hired on company vessels, and crew have had to maintain close ties with the processors, and their income has been directly related to the prices paid by the processors. Seine boat crew, then and now, are the least independent sector of the fleet. Their situation most closely resembles that of employees on piece-rate.

Troll Fishery

The troll technique made its commercial appearance about 1910. It is distinctly different from the net method of capture. Fishing lures are towed behind the vessel at various depths. They are attached to lines that lead from poles extending out the sides of the fishboat. Fish are attracted to the lures and are captured when they strike at the hook. Upon surfacing, they are killed, bled, and, dressed. Vessels that stayed at sea for more than one day preserved their catch by packing it in flaked ice. Some were stored for up to ten days. Some modern trollers, however, have freezer systems, so the fish can be frozen and glazed at sea.

Spring and coho salmon have traditionally been the targets of the troll fleet and are better suited for the higher-valued fresh, frozen, smoked, and salt-cured consumer markets. This demand for a high quality, fresh salmon product paralleled the development of markets for Pacific halibut in the metropolitan centres of eastern Canada and the United States. With the emergence of these new but small markets, the troll fishery was established as the best method of capturing the favoured species of salmon and delivering them in quality condition. Moreover, the oil content of coho salmon encourages canning procedures, especially at the beginning of the season when the softness of their flesh makes salmon unsuitable for high quality markets.

The troll fishery began in the Gulf of Georgia and the northern Queen Charlottes, but the method was readily adopted on the west coast of Vancouver Island. It has remained the least dependent upon cannery services and financing. Since the early 1970s, a small number of troll vessels have installed freezers, allowing them to stay at sea for as long as their supplies of fuel and food last. Until recently they seldom employed deckhands, did not fish as part of a group, and roamed far out to sea. Troll fishers have been known as the "lone wolves" of the industry, but this image has changed recently. The technological advances of telecommunications have entered the troller's wheelhouse, enabling frequent and reliable contact with the entire fleet on a host of radios.

Since 1970, experimentation and adaptation of their gear have allowed troll fishers to compete with the net fleets to capture sockeye, chum, and pink salmon species. In turn, this has caused a considerable change in the economic opportunities in the troll fishery. Once known as the "poverty-sticks fleet" (referring to the fishing poles projecting out the sides of the boat), the trollers are now viewed with apprehension by other sectors of the fleet because of their ability to locate and catch the net fish out at sea before they arrive in waters suitable for net fishing. The troll fleet is in direct competition with the net fleets, and disputes are legion about the allocation of "fair" catch quotas.

The competition does not end with the capture. Net fleets now watch salmon, formerly caught in their nets, receive better prices because of their troll-caught quality. These traditionally canned fish are now entering the lucrative fresh, frozen, and smoked salmon markets. For trollers the value of their fish has always been subject to dockside grading. The relative strength and place of markets for troll salmon has never been negotiated by the union. Yet recent prices paid for troll fish have equalled and often exceeded net prices.

The outlets for troll-caught fish also have not been subject to oligopsonistic control. The troll sector has been able to ignore the many conflicts in the industry that have traditionally revolved around efforts by the UFAWU to establish minimum prices for net-caught fish. Yet, the welfare of the troll sector is threatened every time the cannery association negotiates minimum prices with the union, because these net prices may be subsidized by lower prices for the troll-caught fish. This is very pronounced with the larger canning companies, for whom a steady and uninterrupted supply of net fish is of the utmost necessity.

THE HALIBUT FISHERY

While the salmon industry was establishing its commercial viability, there was also a recognition of the commercial potential of halibut. Halibut are caught by the longlining method. A long, set line is lowered to the sea bottom. Attached to it are short lines and hooks, regularly spaced and baited. The line is anchored and buoyed at its origin and end. When it has been down long enough for the halibut to feed and be captured, the line is reeled up on a small narrow drum. Halibut, some weighing two or three hundred pounds, are hauled on board, gutted, cleaned, and iced down. As early as 1878, halibut packed in ice were transported to markets in San Francisco.

By the 1880s,

> . . . a small group of two-man sailing sloops that could carry about 3,000 pounds fished halibut to supply the local markets of Victoria, Nanaimo, Vancouver, Port Townsend, and other Puget Sound points. (Bell, 1981:21)

At the same time, Atlantic halibut stocks were failing on the east coast and fish merchants were seeking alternative sources of supply. In the late 1880s, a number of schooners from Maine and New England began fishing from Tacoma and Seattle. In September of 1888, the first railcar of iced halibut from the Pacific arrived in poor condition in Boston.

> The difficulties of securing ice, the relatively high transportation costs, and the long distance to the volume markets in the East slowed down the development of the fishery in Puget Sound. (Bell, 1981:77)

When the New England Fish Company (N.E.F. Co.) from Boston entered the halibut fishery in 1894 it chose Vancouver as its headquarters. The city was the terminus of the Canadian Pacific Railway, and it had, for two years, shipped chilled halibut and salmon eastward. Unlike the railroad shipping from Seattle and Tacoma, the CPR boxed iced halibut in their fastest transcontinental trains, and waived shipping charges if they did not arrive in quality condition. They constructed six refrigerated rail cars for this market. Bell notes:

No Pacific Northwest railhead port had the facilities the Union Steamship Co. possessed. Its coastal freighters . . . first chartered as fish packers and later permitted to become dory fishing vessels and its dock and warehouse that were used for 10 years by the New England Fish Co. for unloading and transshipping their halibut to railcars were vital elements in the early success of the halibut fishery out of Vancouver. (1981:81)

The New England Fish Co. chartered two steamers and initiated the dory/mothership technique of longlining halibut. In 1897 the company constructed its own fast steamer, designed to fish the Pacific coast halibut banks and capable of carrying twelve dories and a crew of thirty-seven. By 1904, they had constructed additional steamers, and they acquired two more with the purchase of the Canadian Fish Co. in 1909. By 1911 this company was producing 80 per cent of the halibut landed in Canada. In contrast, most of the halibut landed by the American fleet sailing from Seattle were not caught by a company-based fleet. There, few large steamers prevailed.

Over the next fifteen years, conditions evolved that greatly strengthened the independent halibut fleet in Canada and allowed Prince Rupert to flourish as a halibut port. By 1911, the largest cold storage plant in the world was constructed. In 1912, N.E.F. Co. acquired their own cold storage plant on the city's waterfront, and the completion of the Grand Trunk Railway in 1914 directly linked these plants to the eastern markets. In 1913 the Underwood Tariff treaty allowed halibut to enter American markets duty free, and by 1915 this privilege was extended to fish dealers. Recognition of the international nature of the markets was consistent with the international nature of the fishery.[2] Both American and Canadian crews were sailing on Canadian vessels, and these vessels fished both Canadian and American waters.

The development of storage and transportation infrastructure and the removal of territorial barriers to markets were accompanied in 1925 with the International Fisheries Commission seasonal closure plan. Each year they curtailed fishing for the three most inclement winter months. This reduced the need for large steamer vessels capable of withstanding the sea conditions of winter. These steamers employed salaried crews of eleven to thirteen, as well as the dory fishers. This meant high fixed wage costs at a time when the halibut stocks were declining and full trips were doubtful. The closures, which were introduced as a response to the reduced catch levels, undercut the company steamer fleets. The new type of halibut vessel, which appeared after World War I, was built on the seine design.[3] This added to an earlier trend in B.C. for smaller, multipurpose vessels. Some of these vessels were built by B.C. Packers and the Canadian Fishing Company. However,

> . . . former steamer fishermen oftentimes aided by capital from supply houses or ship chandlers invested their savings in vessels of varying size. (Bell, 1981:39)

The effect of this restructuring was the decentralization of company ownership and control. There developed a fleet of independently owned halibut vessels, and factors like the extension of tariff privileges to fish dealers meant that outlets other than Canadian Fish and B.C. Packers were present on the docks.

As a result of fleet and shore outlet changes, a system of halibut exchanges was established whereby shore buyers bid on deliveries as vessel catches were recorded on the exchange "board." This auction system enabled on-the-spot sellers and buyers to determine the price of a load of halibut. This was decided by such factors as markets, fish size, the length of time the catch had been stored, and the care taken in icing down the fish. Once a price was determined, the captain, in consultation with the crew, either rejected the bid, put the catch on the exchange again, or sailed to another port and tried to make another deal. The system allowed the many variables determining the value of the fish to be applied jointly and in open competition with all buyers. It usually saved the fishers from sailing wharf to wharf or port to port trying to find the "right" price while their catch lost quality.

In 1941 the Deep Sea Fishermen's Union, representing only halibut fishermen, and vessel owners joined the Prince Rupert Fishermen's Co-op Association. They contributed financial aid to build a liver and viscera reduction plant to take advantage of the wartime market for vitamin A oil. This amalgamation with the salmon trollers marketing cooperative doubled the membership in the organization and ensured the affiliation of nearly all the full-time halibut vessels to the co-op. By the end of the war, seventy-four halibut vessels were delivering to the co-op. It was the largest producer of vitamin A oil, and it was marketing chilled and frozen halibut products (Hill, 1967:92–96). The processing companies responded to the loss of these large halibut vessels by encouraging their own fleets to fish halibut to augment salmon earnings. They also introduced facilities to receive landings of halibut at their outlying salmon camps.

By 1940 the majority of halibut landed in Canadian ports was from Canadian vessels; by 1963, the landings from Canadian vessels was the majority of the total catch on the Pacific coast. This lasted until 1974 when the depletion of the rich stocks on B.C.'s coast brought the Canadian portion down below the Americans (Bell, 1981:100). In 1979 and 1980, the closure of American waters to Canadian vessels removed the opportunity for a separate and economically viable Canadian halibut fishery. Dependence on harvesting other fishery stocks was now necessary.

The halibut fishery was a major contributor to the evolution of an integrated structure in the B.C. fishing industry. It was the halibut fishery that brought the New England Fish Co. (later, the Canadian Fish Co.) to the west coast. Twenty-four years later, it initiated salmon production. Later, the halibut fishery was crucial to developing the economic base that enabled the Prince Rupert Fisherman's Co-operative Association to compete with the major processors and to create an integrated organization involving all fisheries.

PILCHARD, HERRING, AND GROUNDFISH

The pilchard, herring, and groundfish fisheries also have been important in the growth of the total industry. Initially each of these fisheries was distinct and peripheral to the salmon fishery. Unlike halibut, the final products of these fisheries, except for herring roe, do not have a mass market, nor have they been able to compete against similar products from Canada's east coast. Their economic viability has been inconstant and irregular.

By 1918, the purse seine fishery had extended to the pilchard stocks, which were numerous in waters off the west coast of Vancouver Island.[4] These fish were unsuitable for dry salting and the market for the canned product was limited, but the volume was huge, and purse seiners supplied many reduction plants with thousands of tons of pilchard, which were turned into fish-meal oil. Indeed, by 1926, nineteen reduction plants were operating on the west coast of Vancouver Island. Similar to the purse seine herring fishery, cannery processors owned a portion of the fleet and employed a crew on a share-payment basis. The stocks of pilchard disappeared after World War II, and herring emerged as the mainstay of the reduction industry.

Herring was fished well before the turn of the century. It had markets in a variety of salt-cured forms: as halibut bait and as reduction stock for its oil and nutrient content. By the turn of the century, purse seine herring fishing found a major dry salted market in the Orient. By 1907 it was a most important fishery, supplying bait for the burgeoning halibut fishery and serving foreign consumer demand (Canada Sessional Papers, 1907–08). By 1925, the Department of Fisheries relaxed its prohibitions on herring reduction (Annual Report of the Fisheries Branch, 1926–27:63) Not surprisingly, the seining of pilchard and herring for the reduction industry stimulated the growth of the seine fleet. Between 1923 and 1927, the number of vessels increased by 128 per cent to 332 vessels. After World War II, the pilchard stocks disappeared, and the herring seine fishery became the only source of reduction fish. By 1967 fears of resource depletion and overfishing led to the closure of herring fishing.

Stocks recovered by 1972 and limited exploitation of a completely new fishery was allowed: the herring roe fishery. Herring about to spawn were captured by the traditional purse seine method[5] or by gillnets strung from small, open aluminum skiffs.[6] The product is the mature herring roe. They must be harvested when the roe is at its most mature prespawning stage. Regulated fish openings are sudden, short, and intense. Some have been as brief as twelve minutes! One of the priorities in the recent construction of seiners was to build vessels capable of holding immense volumes of fish. This ensured that herring fishing would not be delayed by details like inadequate hold capacity and has resulted in an increasingly integrated purse seine fleet. During the 1970s, the inflationary spiral that occurred in the value of herring broke all records in the industry. The lucky seine captain and crew who looked down into a net of one hundred tons of herring understood the term "silver gold." The enormous returns from the herring roe were, in part, the reason for the 29.5 per cent increase in the number of seine boats between 1968 and 1977, and for the 2.9 ton increase in average vessel capacity (Hayward, 1981b:39–51). By 1982 the seine fleet was estimated to have doubled the catching capacity required to harvest the available stocks of salmon and herring (Pearse, 1982:76).

Up to World War I, the groundfishery was limited, as was local market demand. The main methods used were hand lines (similar to trolling with lures except that the vessel remains stationary), and the beam-trawl technique of dragging.[7] Beam trawling involves towing a weighted net along the seabed, which is attached to the vessel with warps (rope or wire lines). The mouth of the net is spread open by the extension of a beam sticking out from one side of the vessel and from which one of the warps is led. It is a cumbersome method of fishing and is limited by the ability of the beam to support the excessive weight of a full net or a very deep warp. The method was marginally successful in groundfishing, although it is still a method for shrimp trawling. During World War I the otter trawl technique was used to catch halibut. It employed

> . . . two large, heavily weighted doors so rigged that when towed through the water, they spread the mouth of the net laterally, the doors acting as underwater kites. The vertical spread is accomplished by buoying up the top or head rope and weighting down the foot rope so that the net will remain on or in close proximity to the bottom. (Bell, 1981:134)

This technology was better suited for groundfish, as it could support the heavier gear needed to reach the deep and rough terrain of this type of fishery. It was first used in 1917 when the Canadian Fish and Cold Storage Co. of Prince Rupert retooled two of their North Sea halibut longliners to trawl for groundfish on the inshore banks adjacent to Prince Rupert (Bell, 1981:134). Throughout the 1920s and 1930s until World War II, the markets for groundfish were local, based primarily on the lower B.C. coast.

Another type of groundfish fishing also developed during World War II. This was the market for dogfish livers, used to produce vitamin A oil and reduction meal. The predominant fishing method was weighted and sunken gillnets in which bottom-feeding dogfish were captured. Trawl vessels also adapted their gear to fish this species. However, the invention of artificial vitamin A after the war ended this short-lived industry.

By the mid-1960s, heavy investment in shore processing facilities for groundfish consumer products supported a considerable fleet of trawlers.

When B.C. Packers opened its modern groundfish plant in 1966, it did a
boom business with a large number of boats making groundfish deliveries.
Since that time, uncertain markets and escalating production costs have
contributed to the decline in the number of vessels and in the value of
production. Many types and sizes of vessels have finished off their salmon
season by groundfish trawling, or "dragging." In the main, the fishery has
been limited to the larger boats with the capacity and horsepower to tow
heavy gear and retrieve sizable hauls of fish. In the past, the groundfishery
had few restrictions regarding seasons or locations. The prevailing wisdom
was that the stocks could not be fished out because so much undersea
terrain was not accessible. The only restrictions on the dragger fleet, aside
from undersize stocks, were those imposed by the weather. Even this
weighed lightly on the mind of many a resolute captain. The economic
return on groundfish was and remains low relative to fixed costs. The
fishery is only viable when harvests are made in large volume. Deliveries
of 50,000 to 80,000 pounds were not uncommon. Of the approximately
eighty vessels in service today, at least one has capacity for landing 225,000
pounds of groundfish. The full-time trawl vessels of this size require the
employment of crew to operate the machinery and heavy gear and to
properly sort and store the fish.[8] Smaller vessels often carry one or two
crew to assist the captain while larger boats need up to six or seven crew
members.

The groundfish trawl fishery has been rather different in the value
and variety of fish caught, the length of season fished, the nature of tasks
performed, and the types of vessels employed. However, these differences
have not kept the crews and vessel owners out of the fishing organizations; many have been flexible or broad enough to accommodate their
political interests. Chapter 10 will discuss the incorporation of trawler
crews into representative organizations.

NOTES

1. A restriction prohibiting motorized gillnet vessels on the northern half
   of the coast was in effect from 1911 to 1923.
2. The international nature of the halibut fishery was formally recognized
   in 1924 with the establishment of the International Fisheries Commission
   (renamed the International Pacific Halibut Commission in 1953). In
   contrast to the role that the Canadian state has played in the management of the salmon resource, the operation of the I.F.C./I.P.H.C. has
   been characterized by a strict adherence to its mandate to protect the
   resource, irrespective of particular aims or interests coming from the
   political arena. It is entirely likely that because of the successful operation of the I.F.C./I.P.H.C., the international boundary between the U.S.
   and Canada has been largely irrelevant for halibut fishers until 1979.

This has been important for the Canadian fleet, since after 1960 the majority of the Canadian-caught fish has come from Alaskan waters (Bell, 1981:97). The closure of these waters to Canadian fishers after 1980 spelled the virtual collapse of the big boat halibut fleet.

3. Until the 1940s when the seasonal restrictions began to limit the available fishing time more severely, vessels and crews could be assured of a nine-month season. This meant little movement among the large halibut vessels into other fisheries, as halibut took up the whole of the season they wished to fish.

4. Pilchard are often known as "California sardines."

5. The purse seine roe herring fishery entails a crew of five in addition to the captain.

6. The gillnet fishery involves two persons per skiff, but often the pooling of labour and skiffs around one "support ship" on which the fishers live means that a dozen or so independent fishermen may join together to fish as a group for the duration of this very short and intense fishery.

7. By 1916 Japanese fishers supplied most of the fresh cod for local markets, using hook and line-hand fishing for which a licence was not required (Canada Sessional Papers, 1917).

8. This entails sorting the many varieties caught in the net, returning those that are protected species or undersize, and properly storing the catch in the hold in ice for the balance of the trip, or up to two weeks.

3

Major Processors to 1940 and Early Labour Force: Historical Notes

ALICJA MUSZYNSKI

The history of fish processing is marked by cycles of competition and concentration. Many small canneries were supplanted by fewer large canneries when labour shortages (especially skilled labour) and bottlenecks on the assembly lines led to mechanization of canning lines and pushed up the costs of operation after the first decade of the century. The subsequent development of refrigeration techniques enabled new processors to become established by opening small, cold-storage plants. Other technological developments have similarly affected the ease and cost of entry over the past century.

In addition to technological changes, there have been fluctuations in markets that affected the viability of smaller firms, sometimes easing their entry, other times forcing them into bankruptcy. And there have been varying conditions of labour supply in different regions, and at different times, which have influenced the kind of technology introduced into processing plants.

THE PROCESSORS

On some occasions it has been relatively easy for small firms to enter the industry. In the formative period they constructed small canneries on rich salmon-producing streams. Mechanization of canning lines, accelerating after the first decade of this century, made it increasingly difficult for these small operators to find the capital, and, by World War II, this avenue was closed. But another avenue had opened in the interval. Development of refrigeration techniques enabled entrepreneurs to open small cold storages and take advantage of periodic cycles of strong fresh/frozen markets.

In addition, most of the fish runs are cyclical and erratic, and markets are notoriously unstable. During financially unstable periods, the larger companies try to buy out the competition, closing plants in an effort to dominate the field.

A few medium-sized firms have been established and survived, though events of the 1980s are reducing their survival capacities. They operated a few multiline canneries, realizing enough profits in the good years to offset the bad.

Since 1902, one company, B.C. Packers, has made periodic bids for industry control. Today it has a larger market share than at any time in the past, but it has lost its autonomy in the process, becoming a subsidiary of Weston, a large, food-products conglomerate.

Precommercial Fishery

The native peoples exploited the rich fisheries resources long before a commercial fishery was established. They developed a number of techniques to preserve various species of fish, and thus maintained a winter food supply as well as trade with inland tribes. These trade patterns were altered to accommodate the entry of the fur-trading companies. When posts were established in the early nineteenth century, salmon became a staple for personnel stationed in the two colonies of what later became British Columbia and was also exported salted in barrels to company crews in the Sandwich Islands and Australia. Along with fur, native peoples supplied salmon, using traditional capture methods. Demand for salmon in this processed state was insufficient, however, to allow the Hudson's Bay Company to market it commercially on an extensive scale.

Early Commercial Fisheries, 1860–1902

The introduction of a fully capitalist enterprise was dependent on further developments in processing techniques. A large market for *canned* red salmon developed in Great Britain, a result of the industrial revolution and a working class that could not grow its own food. American entrepreneurs were the first to apply the canning technique to Pacific salmon and to find a large market for the product. Entrepreneurs in British Columbia followed the American model. They were forced to can a species similar in colour, texture, and taste to the red kings. After initial experimentation with red springs, B.C. canners exploited the sockeye. Eventually, other processing techniques were adopted. Other species of salmon, and other types of fish, notably halibut and herring, were marketed. But the capital originally employed in salmon canning enabled those entrepreneurs to gain a significant degree of control over the entire fishing industry.

The B.C. canning industry developed in the 1870s, independently of the earlier Sacramento and Columbia River plants. Local commission merchants with direct trade connections to Great Britain provided financing. The Fraser River, the largest sockeye-producing stream in the province, was the first to be exploited.

> Before the advent of the limited companies in the 1890s, the Fraser River's industrial organization was characterized by low levels of industrial concen-

tration, small firms run by individuals or partners, and by a high incidence of local proprietorship. . . . Long-run operating capital, which was especially important to the industry because the salmon market had an eighteen-month cycle from the time the tinplate was ordered until the season's pack was sold, was supplied by commission agents, who made advances in the form of overdrawn accounts on goods in transit. These agents provided canning and fishing supplies and a distribution system to the market as well as capital. (Stacey, 1982:6)

By the mid-1880s, both the Columbia and Fraser Rivers experienced overexpansion, and canners began searching for salmon-producing streams in Alaska and northern British Columbia. At that time, the Fraser River canning industry consisted of thirteen firms, each tied to brokerage houses (Reid, 1981:323). Victoria was the financial centre of the province, and in the years between 1871 and 1891, salmon canning, sawmilling, and the north Pacific seal hunt replaced the fur trade and gold mining "as the leading staple industries in the Victoria-centred B.C. economy" (McDonald, 1981:371). By 1881, salmon canneries and sawmills employed the largest labour forces and, ten years later, 85 per cent of provincial exports were products of the mines, fisheries or forests. Fraser River canned salmon was the fastest growing export industry. In the twenty-year period ending in 1896, "the value of canned salmon exports shipped to external markets by sea increased five times as fast as the value of forest product exports" (McDonald, 1981:371–72).

Between 1876 and 1896, there was a change in the nature of financial control of the industry. One of the early cannery men, T. Ellis Ladner, noted that business was highly profitable for the principal agent shareholders in the canneries, but the companies themselves had little direct control. The interests of the two differed. The agents were not concerned with competition between canneries. They wanted financial control of the product and a commission appropriate to their investment.

The advent of eastern Canadian banks to British Columbia changed the situation for those canneries not already too involved in the old order. The more independent of the fiscal agent the cannery man was, the more he was able to control his own business. He could purchase materials at the lowest price and he could finance the introduction of modern plant methods and increase his profits through improved operations. (Ladner, 1979:92)

The move of Canadian banks to British Columbia spelled a shift of fiscal control from Victoria to outside the province. The banks also tended to locate regional offices in Vancouver, close to the new transportation terminals. Subsequently, Victoria declined as the financial capital of the province. The Bank of Montreal and the Canadian Bank of Commerce became the principal backers in the salmon-canning industry. By 1901, this new financial capital source enabled companies to achieve independence from financial agents through incorporation.

The shift in financial control was paralleled by a change in transportation from the ocean to the railroads. By the late 1880s, transcontinental lines were completed across Canada and the United States. American freight rates were cheaper than Canadian ones. There was a much larger population in the eastern and southern states, and these markets became much more valuable to American salmon canners. British Columbia canners could not compete, and the maritime market to Great Britain remained central to the B.C. salmon-canning industry. However, the growth of Vancouver as a railway terminus did stimulate other exports. Canned salmon failed to hold its dominant place and was replaced by forestry products bound for prairie markets (McDonald, 1981:383).

The ABC Company

The Anglo–British Columbia Packing Company (ABC), incorporated in April 1891, in London, England, was the first attempt at control of the industry. Henry Bell-Irving, the company's agent and chairman of its local committee, acquired options on nine canneries which he promptly sold to ABC. The "English syndicate" began by subscribing large amounts of its capital in Canada. At this time, a boat-licensing program was in effect on the Fraser River. By acquiring additional canneries, the firm could pool boat licences and reduce competition from both canners and fishers. Bell-Irving put it succinctly:

> My company do [*sic*] not intend this year to work all its canneries because we cannot get enough boats to supply all the canneries with fish — it is proposed to run half the canneries on the Fraser River and use the fish from those boats of canneries not running to put in the other canneries and double up, thus reducing expenses, but I think it most essential that there should be a fixed number of licenses to the canners . . . so there should be no danger of being frozen out by any combination of fishermen, as canners have money invested and not the fishermen, and if it was not for the canners the fishermen would have a very small market indeed — the local market and which is a mere nothing to them. (Sessional Papers, 1893, XXVI, 7 [10c]:330)

ABC acquired two additional firms, and the company became "the largest producer of sockeye salmon in the world" (Ralston, 1965:25). This merger was one in a series. In 1889, the British Columbia Canning Company was incorporated in London. Its principals included the pioneers on the Fraser River: Findlay, Durham, and Brodie, who held four canneries (three of them in the north). Another pioneer, Alexander Ewen, also expanded, and by 1889 he had the largest cannery on the Fraser. In February 1891, Victoria Canning Company of British Columbia Limited Liability was incorporated (Ibid.: 26). J. H. Todd, a Victoria merchant and also an original entrant, remained outside these mergers. Thus, by the beginning of the 1891 season, five major groups were consolidated on the Fraser. All the American-owned concerns (representing 30 per cent of total fixed capital by 1881 were bought out by ABC [Ralston, 1981:300]). Local entrepreneurs were involved in the formation of both British-backed companies. ABC and Victoria Canning Company controlled over 60 per cent of the Fraser River's sockeye pack. All the companies, except ABC, were linked to Victoria. As Ralston notes:

> From the very start of the canning industry the commission merchants of Victoria (later also of San Francisco) provided the finances; the growth of the industry, in fact, depended on their ability to carry the producers until the pack was sold. At a later period, their advances were secured by chattel mortgages on the pack and cannery. (Ralston, 1965:19)

The acquisition of boat licences, thus securing fish supplies, was a major reason for buying and consolidating canneries and was the objective of the English syndicate as well as its three major rivals. The English syndicate's attempt at oligopsonistic control developed from conditions of overcrowding on the Fraser River. The number of fishers had risen dramatically, and, in the period 1872 to 1888, the number of canneries increased from three to twelve. The introduction of licence limitation on the Fraser increased the pressure to consolidate. In 1889, 1890, and 1891, the number of licences was limited to five hundred, and each cannery was allotted an average of twenty (Stacey, 1982:13). The program met with understandable resistance from both canners and fishers, and, four years later, it was abolished.

The end of licence limitation, however, gave rise to new entrants, both fishers and canners. Since technological development at this stage was minimal, operations were labour intensive and labour was cheap. Gregory and Barnes (1939:30) noted:

> The greater number of canneries prior to 1893 was owned by single proprietors or partnerships. . . . Except for the more elaborate ones they could readily be moved from one site to another to adjust to changing fishing conditions and competition. They were devoid of much machinery; their costs were low, and most of the early packing and handling prior to 1903 was done by hand. . . . Often the canneries were enlargements of salteries that preceded them.

Competition "was often of the severest type, having as its mainspring the desire for maximum packs and the elimination of rival concerns whose product was flooding the market." Of the estimated 132 canneries constructed in the province before 1920, 64 were sold, dismantled or destroyed (Ibid., 92). On both sides of the border, there was a trend towards a fish pack divided between a small number of large companies (who dominated the industry) and a large number of small firms. To return to the original discussion, while ABC purchased and then closed a number of plants, within ten years the situation had come full circle. "By 1901 the level of concentration had reverted to its pre-merger position" (Reid, 1981:320).

The Formation of British Columbia Packers

The availability of bank capital enabled canners to assume a greater degree of control over their operations. Three of the large firms, Victoria Canning Company, Alexander Ewen, and George Wilson, took advantage to make a second bid at control of the industry. They were joined by Aemilius Jarvis and Henry Doyle. When he established Aemilius Jarvis and Company, Investment Bankers, in 1892, Jarvis established important connections with financiers in central Canada. Doyle, on the other hand, had developed detailed knowledge of the salmon-canning industry through his association with the Doyle Fishing Supply Company of San Francisco. And because he had operated in the United States, he appeared to provincial canners as an impartial partner, one who did not have interests in specific plants. Before these principals could obtain the needed financing, they had to obtain control of over 60 per cent of the operating plants. Commitments were sought from canners leaving the industry (after selling to British Columbia Packers) specifying that they would not participate in the industry for at least seven years. Plants were bought, but this did not prevent their owners from later re-entering the business. J. H. Todd is a case in point. Henry Doyle sought and obtained his active support for the new merger. But, at the last minute, when most of the plants were purchased, Todd pulled out of the agreement. After Doyle had eliminated most of the competition, Todd took advantage and remained in business, thus gaining substantially from the new balance of power (Reid, 1981:315–319).

In 1899, T. B. McGovern had made an unsuccessful attempt at a similar amalgamation. He represented a New York commission brokerage firm, owner of the "Clover Leaf" brand name, later the B. C. Packer's trade mark. McGovern failed to persuade local canners to join in his scheme, thus marking the importance of Doyle's efforts and expertise in securing their cooperation. Doyle's participation in the industry in the United States led him to undertake a provincial amalgamation modelled on the two American combines. In 1893, the Alaska Packers Association was formed as a successor to the merger of 1892, when 90 per cent of the producers combined their operations. This was followed, in 1899, by the formation of another combine, the Columbia River Packers Association (Lyons, 1969:230–232). Since Doyle had no direct investment in provincial canneries, he appeared neutral in the B.C. venture.

The heaviest provincial pack carryover occurred in 1901, resulting in many canners becoming indebted to the banks. The Bank of Montreal held half of the salmon canners' accounts, while the Canadian Bank of Commerce held another 40 per cent. Molson's Bank (taken over in 1942 by the Bank of Montreal) held the remainder. In 1902, Doyle secured the approval of these three banks for amalgamation. Jarvis had already formed a syndicate and was acquiring subscriptions from central Canadian businessmen. He undertook the formal organization of the new company, chartered April 8, 1902, in New Jersey, and called it The British Columbia Packers' Association of New Jersey (Lyons, 1969:233).

In a letter addressed "To the Salmon Canners of British Columbia," Doyle explained the reasons for amalgamation. These included reorganization to ensure increased and sustained profits; closure of a large number of operating plants; enhancement of packing capacity for those remaining; and the realization of large economies of scale in the cost of supplies, pack handling, and world marketing (Reid, 1973:48; Doyle gave further details in a letter addressed to A. G. Kittson and Co.). An additional motive was lowering the cost of raw fish "by doing away with the excessive competition for the fishermen" (Ibid., 49–50).

The relaxation of American laws (the United States had originally legislated against such combinations) led to these mergers in Alaska, on the Columbia River, and in British Columbia. The modern corporation emerged. States competed in bidding for corporate mergers, with New Jersey in the forefront. Similar legislation enacted earlier in Great Britain had made possible the ABC enterprise (Reid, 1981:311). "The Victoria merchant community, traditional source of capital for the coast canning industry, was the principal casualty of the reorganization, with canneries previously controlled on Vancouver Island now owned by the larger corporation centred in Vancouver" (McDonald, 1981:389).

Inflated costs of raw inputs and poor market prices further facilitated this merger. In the period between 1881 and 1891 and 1896 and 1899, the cost of raw fish per case rose 240 per cent, while the price of canned salmon on world markets fell 25 per cent (Reid, 1981:313–314). "Over 80 per cent of Fraser River output was sold in European markets where it was generally conceded that the Alaska pack set the market price" (Ibid., 325). In the 1890s, new low-cost salmon-producing areas emerged in Alaska and on Puget Sound in the United States. Puget Sound production intercepted runs headed for the Fraser River. Reid estimates that in 1890, 97 per cent of Fraser fish were canned on the river, but by 1900 the proportion fell to below 40 per cent (Ibid., 326).

With lower Fraser River production, ownership of northern plants became very important. As Sinclair observes, "The northern canneries had shown historically much larger profits per case than those on the Fraser. Any company, therefore, wishing to control the Fraser River fishery would be financially stronger if it also possessed northern plants" (Reid, 1973:ii). B.C. Packers took possession of twenty-nine of forty-eight canneries on the Fraser, and an additional twelve in the north.

By 1902, with B.C. Packers' takeovers, one-third of the earlier canneries were closed, and remaining plant capacity was doubled by concentrating machinery and equipment from idle plants. The commencement of mechanized lines, which increased the amount of capital needed to enter canning, allowed combination of lines in one plant. The multiline cannery was superior to the single-line operation. Production did not have to stop when a change was made to a different-sized can, and surplus buildings in idled plants were ideal for storage. By 1905, according to Stacey (1982:19) fifteen plants were nearly equal in capacity to the twenty-nine purchased three years earlier on the Fraser River.

Because the company was classified an "alien corporation," because of its American registration, it could not own or operate any steamers or other vessels in Canada. Therefore, it immediately incorporated a subsidiary, The Packers Steamship Company Limited. In 1910, British Columbia Packers' Association of New Jersey was re-incorporated as a provincial company. The problem now was that it could not do business outside British Columbia, and most of its markets were in exports. The British Columbia Fishing and Packing Company Limited, a new corporation, was created in 1914, under dominion charter. Changes were made in provincial laws leading to a final provincial incorporation in 1921, when the company acquired its present name, B.C. Packers Ltd (Lyons, 1969:239).

J. H. Todd & Sons Limited

Reid (1981) argues that medium-sized firms in the salmon-canning industry could, for a variety of reasons, operate more efficiently than smaller or larger operations. Two examples of such firms are J. H. Todd and Nelson Brothers, both family-run operations that never aspired to domination in the industry and gained a solid reputation among their peers.

The founder of J. H. Todd & Sons was Jacob H. Todd, a native of Ontario who, in 1862, moved to British Columbia. He was a merchant in Barkerville until 1868, when his enterprise was destroyed by fire. Subsequently, he moved to Victoria and, in 1875, established the wholesale grocery firm of J. H. Todd & Sons Limited, supplying goods to the Okanagan district (Lyons, 1969:168). The firm became involved in the salmon-canning industry in 1881 and was the first Victoria firm to enter the business. It operated until 1954, when its operations were divided between B.C. Packers and Canadian Fish Co.

The firm operated canneries distributed strategically on major salmon-producing streams.[1] It began with Richmond Cannery on the Fraser River, and, toward the end of the 1880s, acquired Beaver Cannery located in Steveston on the Fraser. In 1902, it took control of Inverness on the Skeena River, the first cannery built in that area (in 1876). On the central coast, it operated Klemtu, located on Finlayson Channel. On Rivers Inlet, Provincial Cannery was operated by the Provincial Canning Co., closely connected to J. H. Todd. A unique operation was the Sooke traps on Vancouver Island, which supplied Empire cannery, built in 1905 at Esquimalt.

In the 1890s, the industry on Puget Sound began to pose formidable competition to Fraser River canners and fishers. Both industries intercepted the same sockeye runs, but the operation of traps was not prohibited on the American side of the border as it was in British Columbia. Puget Sound canners could therefore catch fish far more cheaply than could their Fraser River counterparts. A fishery commission investigated the question in 1902, and, as a result of that inquiry, in 1904 an order in council allowed the use of trap nets in the Strait of Juan de Fuca. The two companies that took advantage of the situation were J. H. Todd and Sooke Harbour Fishing & Packing Co., a subsidiary. In 1935, American traps in that area were finally prohibited, and fishers pressed the government to ban Canadian traps, since continuation of their use might encourage American interests to do likewise. A number of traps were being operated in Alaska, and Alaskan fishers were also opposed to their use. A commissioner was appointed to investigate the matter, but, in 1940, he recommended the continued operation of the Sooke traps because they provided employment to the community. After the war, J. H. Todd & Sons had a difficult time re-establishing their brand names on the market. Their canneries were in run-down condition, and the new Todd generation appears to have taken little interest in the fishing industry. The company was offered for sale and was taken over by the big two companies. The traps were abandoned around this time.[2]

Nelson Bros. Fisheries Limited

The company was not incorporated until 1929, but two of the brothers, Richard and Norman, had been part of the industry for many years. They began trolling on the west coast of Vancouver Island in 1919 and subsequently bought a packer to transport their catches. Soon they were transporting fish for others as well, out of Kyuquot. In 1933, they purchased their first cannery, St. Mungo on the Fraser River (built in 1899), and, in 1934, they took over Ceepeecee in Hecate Channel on Vancouver Island. That plant was built in 1926 for fish reduction, and the brothers added a cannery to the operation. In 1937 they bought the plants and assets of Massett Canners Limited and extended their northern interests in 1940 when they began operations in Prince Rupert. However, their plant was located on Ocean Dock and was taken over by the U.S. Army during the war. Thus, in 1943 they bought Port Edward from B.C. Packers. Like J. H. Todd, they operated at strategic locations throughout the province and acquired a solid reputation as a medium-sized operation. In 1955, they closed St. Mungo and built Paramount Plant in Steveston (Lyons, 1969:405, 459, 522). B.C. Packers acquired the assets of Nelson Bros. at the end of the 1960s, in a manner that received little publicity at the time. Paramount is today part of Imperial Plant, while Port Edward was the major northern operation of B.C. Packers until Oceanside was acquired from Canadian Fish, rebuilt, enlarged, and renamed Prince Rupert Plant, at which time Port Edward was closed.

Canadian Fish Company

In 1909, an American firm, New England Fish, acquired a halibut fishery company called Canadian Fish Company. Retaining the name of the subsidiary, New England Fish built up the Prince Rupert halibut plant but did not enter salmon canning until 1918 when it purchased the Home plant in Vancouver. Canadian Fish became the major rival of B.C. Packers. The two companies dominated salmon canning for most of this century.

World War I and the Depression

Significant changes occurred in the industry between the formation of B.C. Packers in 1902, and 1928, when the industry underwent a second round of merger activity. World War I created inflated prices and market demand, thus encouraging new entrants. B.C. Packers' dominant position was increasingly eroded.

Prior to the 1900s, only sockeye and coho were utilized; the British consumer accepted only red salmon. However, by 1911, all five species were canned. Americans found a ready market for the cheaper canned grades in the southern states, while British Columbia firms tried to develop markets in the Commonwealth, using their United Kingdom connections to negotiate special tariffs and duty concessions. World War I separated the fish-supplying nations of northern Europe from their western European markets, and Canada stepped in to fill the need. As meat became scarce in the war-torn countries, fish was substituted, further improving market demand.

With the export of canned fish to Europe, North American consumers turned to fresh and frozen salmon, halibut, and, to a lesser extent, other groundfish. Cold storage facilities were constructed, applying refrigeration technology. Prince Rupert emerged as the second major urban centre in the provincial fisheries. In 1914, a second transcontinental railroad was completed, terminating at the port of Prince Rupert. Given its proximity to rich, unexploited halibut banks, its ocean and rail links, and the infusion of private capital investment in cold storage facilities, the city became the centre of the halibut fishery.

World War I also created a heavy demand for canned herring and pilchard. Coupled with the market for the cheaper canned fall salmon, Vancouver Island, rich in all three fisheries, emerged as a third major fishing area. However, because sockeye production there was poor, the area remained relatively undeveloped. The wartime boom was short-lived, and, with the sudden signing of the armistice, all three markets collapsed.

Wartime prosperity eroded the dominant position of B.C. Packers through excessive competition and overexpansion. Federal fisheries officers stationed in the province noted the trend with great alarm.

> It would appear, however, that the investor and those who think they can earn a living by entering the fishing industry are turning their attention exclusively to canning operations as being a medium for getting rich quick, but it must be remembered that whilst canneries no doubt produce profit not equalled in many other lines of commerce, still they have their off seasons . . . unlimited canneries would mean unlimited fishing, with the result that the fisheries would be depleted, and the smaller investor would go to the wall while only the big companies would remain in operation.
>
> The prevailing price for canned salmon can hardly be called normal, and when commerce again assumes normal conditions, the prices to the fishermen and manufacturers will no doubt reach a level. (Sessional Papers, 1917, LII, 21[39], Appendix 9:244)

At the start of the war, B.C. Packers produced 25 per cent of the provincial pack. In the following decade its share declined. Between 1919 and 1925, it accounted for a mere one-sixth of the canned pack (Gregory and Barnes, 1939:95). The three combines of the Pacific northwest fisheries, Alaska Packers Association (Alaska and Puget Sound), Columbia River Packers Association (Alaska and Columbia River), and B.C. Packers (British Columbia) continued to exert considerable power, but throughout the war they found stiff competition from smaller firms. Expansion was severely curtailed by two recessions in the 1920s that were especially severe in this industry and by the Depression in the following decade.

The 1920s was the last decade in which firms could compete with small manual operations. For example, of approximately forty-four provincial canneries inspected for insurance purposes in 1923, almost half operated without iron butchers. Of these, only one ran a three-line cannery. Even the majority of those with iron butchers ran only one or two canning lines. Six plants still made their own cans (Insurer's Advisory Organization, Folder 9–7, University of British Columbia).

The collapse of markets and the recessions forced many of these small operations out of business. All three combines moved in to buy up the smaller concerns, but their expansion put them in jeopardy.

The combines experienced a different set of difficulties from the small firms. In economic hard times small operators could cut losses because their equipment was leased. The large concerns, however, held huge inventories in the form of pack carryovers, equipment, and cannery properties. The effect on B.C. Packers was described by Gregory and Barnes (1939:102):

> The British Columbia Packers Association . . . approximated the unfortunate experience of the American companies during these post-war years. An old firm, it purchased a large number of high-priced canneries in the late 1920s, paying for them in newly issued stock and also in cash. The consolidation proved unsuccessful and the company was forced into bankruptcy in the early 'thirties'. It was taken over by banks and canning manufacturing companies which had advanced considerable amounts of credit. Subsequently a number of its canneries were closed.

Recuperation: the 1940s

In 1930, quick freezing methods were introduced in fresh/frozen fish processing. It was not until the wartime demand of the next decade, however, that the technique was widely adopted. Fish boats and packers, equipped with refrigeration units, could now transport fish over large distances. Economies of scale became feasible since it was no longer necessary to establish processing facilities close to resource capture. Instead, operations could be combined and concentrated in urban areas, close to marketing outlets. This technology provided a new avenue of entry for small operators.

World War II, like its predecessor, created artificial markets. The United Kingdom relied heavily on the B.C. fisheries for canned salmon and herring, vitamin A supplements from fish reduction, and fish meal for fertilizer used in domestic food production for the war. Beginning in 1941, the British Ministry of Food negotiated with the Canadian government for guaranteed packs and prices. Two-thirds of Canadian production of canned salmon was procured. The following year, the British purchased the entire provincial pack of canned salmon and herring (Muszynski, 1984).

The diversion of canned salmon to wartime markets stimulated domestic and American demand for fresh/frozen fish. This became the new entry point for small processors. Capital investment and labour costs were much smaller than in canning. Where, in the pre-Depression period, small operators established largely manual operations on remote streams, now they built small cold-storage facilities to take advantage of demand for fresh/frozen fish. However, a revived B.C. Packers and Canadian Fish emerged as the strongest competitors, by virtue of new canning capacities. They were not dependent on the prices for fresh/frozen products as were small operators who generally lacked large cold-storage facilities in which to store the product. The large firms had consolidated their operations in multiline and multiproduct plants. The core of their business revolved around established brand names for canned salmon. B.C. was never a price setter in the industry, and both the large and small firms had to deal with volatile markets. But the large firms had flexibility. If fresh/frozen market demand was strong, they could divert the product from the can. If demand weakened, they could concentrate on canning. Unlike frozen fish, canned salmon would keep indefinitely, allowing processors to wait for price and market upswings and to supply the product year round.

Guaranteed packs and prices were effective until 1948, when the British could purchase only a fraction of the pack because of a dollar shortage. The Canadian government intervened and bought a portion of the canned herring pack for overseas relief aid. During these years, only tiny portions of canned salmon were released to bolster domestic markets. Demand continued through 1948, when domestic consumers bought two-thirds of the pack, an increase in excess of 70 per cent of prewar demand (B.C. Packers Annual Report, 1948). Canners were less successful in disposing of their canned salmon packs the following year and collectively engaged in a "no brand" advertising campaign to reintroduce the produce to Canadian consumers.

Prince Rupert Fishermen's Cooperative Association

Several fishers' cooperatives were established in the late 1920s and 1930s, which chartered and later purchased fish packers and sold fish directly to American buyers. The Prince Rupert Fishermen's Cooperative Association, established in 1931, was one of these.

The PRFCA expanded in 1939 by merging with the North Island Trollers Cooperative Association and by extending its operations into the longline halibut fishery and, later, into trawl and seine salmon fisheries. Up to this point, none of the cooperatives was engaged in processing. However, in 1940, the PRFCA, together with the Deep Sea Fishermen's Union, built a fish liver oil-reduction plant in Prince Rupert.

Summary

By the end of the 1940s, the major processing techniques included canning, reduction, filleting, and cold storage for fresh/frozen markets. B.C. Packers began to close its outlying canneries and to concentrate processing operations in three central facilities located in Steveston, Prince Rupert, and Namu (central district). The co-op was established and ready to expand its operations. Canadian Fish Company was also expanding. Nelson Brothers and J. H. Todd remained independent, medium-sized companies until their takeover by B.C. Packers and Canadian Fish Company.

LABOUR SUPPLIES AND THE LABOUR PROCESS: EARLY HISTORY

Discovery of the method of rolling out thin sheets of metal into tin plate formed the basis of the canning industry. Canning involves the organization of labour for assembly production of a standardized product. Early salmon canners recruited factory workers from two groups: native peoples (especially women and children) and Chinese men.

The Labour Market, 1870–1890

There was a shortage of cheap labour throughout the Pacific northwest during these years, partially solved by the immigration of single Chinese male peasants. B.C. canners employed them, contracting for their labour with agents working primarily out of Victoria or San Francisco. Since this was a pioneering labour force, these men became skilled in the various processes of manufacturing tinned salmon. A common practice was to hire a contractor who supplied the entire cannery work crew, both Chinese and native. The most common practice was to pay the contractor by the case of salmon packed. The contractor, in turn, paid his labour force. In addition, he provisioned the Chinese workmen.

> The Chinese employed by canneries in British Columbia were brought from China by labour contractors, or "big merchants." They were indentured to the contractors, who advanced trans-Pacific fare and the admission tax. (Ladner, 1979:114).

They found it difficult to pay off their indentures because most work was of a seasonal nature. Since they could speak no English, they were dependent on contractors to find them work and also to tide them over periods of unemployment, especially in winter. Those unable to work became further indebted to the contractors. Chinese crews were paid a variety of rates. The skilled were hired for the entire season. However, crews worked only when required.

> Common labourers receive $1.10 for a day of 10½ hours, with half an hour for the midday meal. Overtime was paid at the rate of time and a half. Some canneries, however, arranged with contractors for Chinese labour at a rate per case of salmon packed. Also, a few Chinese were paid by the month. (Ladner, 1979:115)

Racism was growing during this period, and in 1885 head taxes were imposed to limit entry. In 1900, and again in 1903, these were increased to further restrict immigration. Canners stated repeatedly at commission hearings held on the subject that they could not have engaged in the business without Chinese labourers. However, the skilled European working class felt threatened by the employment of this cheap labour force. Increased restrictions on immigration, coupled with new employment opportunities resulting from increased cannery construction, improved the bargaining position of Chinese workers. Canners responded to the reduced numbers of indentured workers available and the increased bargaining power of the remainder by mechanizing the canning lines. However, between 1870 and 1900, Chinese men became the core cannery labour force.

Around this core, flexible cheap labour had to be on hand, ready to process a fluctuating volume of fish. It was especially important to the canners to have enough workers available at the peaks of the runs. The ideal way to fill this need was to persuade whole native villages to relocate at the canneries each season, and cannery owners paid native "recruiters" to undertake this. Generally, the native men fished for the companies or made nets, and the native women worked in the plants or repaired nets. The elderly and older children looked after infants. During peak periods, these extra people were also called in to work. Often the canners paid one person (usually the female head of the household) for the work of the entire family at the end of the season. And sometimes the Chinese contractor paid the native labour force, depending on who hired it.

Recruitment of entire native communities was especially prevalent in the north, where the native economy centred on the fisheries. It was relatively easy for these people to incorporate cannery work into their seasonal migrations. Many villages became part of the cash economy, dependent on plant employment for economic survival in the emerging capitalist society. At the same time, because fishing formed the basis for their subsistence, these villages were able to retain many of the features of their traditional ways within the new capitalist economic structure (Muszynski, 1986b:Chapter 4).

Racial distinctions in job assignments were reinforced through company housing. In general, employees of European descent (plant managers, bookkeepers, foremen, and tradesmen) had the best accommodation. Native families were segregated in their own row housing, or, where enough flat land was available, put up in cheap housing in a small village. Chinese men generally had the worst housing. They were placed in bunkhouses, geographically isolated from the others, often with ten or more men to a room, with either appalling or nonexistent washing and toilet facilities.[3] Not only was housing segregated, but each group also had its own diet and separate cooking arrangements. Chinese contractors were responsible for feeding the Chinese crew. In a poor season, when earnings were low, contractors often tried to make up their losses by cutting back on provisions. A popular myth was that the Chinese did not need as much food as white men; that they could subsist on next to nothing.

Another industry myth is that the Chinese were docile labourers. Such does not appear to have been the case. In 1904, they formed the Chinese Cannery Employees Union, with the intention of dealing with contractors who, after payment by the canners, absconded to China without paying their crews. The union may also have been formed in response to the establishment of the Chinese Contractors Union, organized by contractors in an attempt to negotiate collectively with cannery operators (Gladstone, 1959:296–97). Both endeavours appear to have been short-lived. By this date, Chinese cannery workers were both a scarce and a skilled labour force. They were therefore in a stronger bargaining position than they had been twenty years earlier. When the costs associated with employing Chinese skilled labour began to rise, canners responded by mechanizing the more skilled labour processes.

The Work Process in Early Canneries[4]

The canning season on the Skeena River began around the middle of June. Windsor cannery will be used as an example. During the winter, the manager, based in Victoria, contracted with a Chinese firm to make the cans and process the salmon. When the river was clear of ice, around the beginning of April, the Chinese labourers were sent by steamer to Windsor.

Windsor Cannery had two main sections: a cleaning house with two double cleaning tables and a filling room containing three tables with benches at the ends, at which native women worked. At the first cleaning table, the head, tail, and guts of fish were removed. The remains were shoved into a hole at the centre of the table, from where they fell into the river. The fish was then sent to a washing tank. It was scraped, scrubbed, and the fins were removed. The slimer scraped off loose scales and the inside of the fish with a knife, and then scrubbed it with a brush. After a second washing, a cutter placed the fish in a manually operated gang knife. He cut the fish into lengths. The sections were then lifted by hand to chopping blocks and Chinese butchers split them into sizes for canning. The butchers slid the fish down an incline into a bucket. When the bucket was full, it was carried to the native women for canning. A little salt, deposited on a board, was placed in each can as it was filled. Women were paid by the number of tickets collected, receiving one per tray of cans filled. The number of cans per tray varied with the size of the can.

The trays were weighed and cleaned. Native women weighed the cans, while cleaners, usually native children, wiped them with pieces of old netting (the ratio appeared to be six children to one or two women). Chinese men put on the lids and fastened a piece of tin under the hole in each lid. The tops were crimped to the bodies, and then placed on a conveyor belt to a machine where the tops were soldered. Once through the machine, they were poled down a shoot to the "bathroom." Chinese workers placed the cans on coolers, on hand trucks (tables with wheels). They soldered the vent holes in the cans, and then the "bathroom crew" took over. This crew consisted of skilled Chinese workers who had made the cans earlier in the season. The cooler was hooked onto a crane and then run along to the first tester. Steam was turned on, the water boiled, and the coolers lowered into it. With the heat, the contents of the cans expanded. If there were any leaks, air would escape. These cans were lifted and sent to the leak stoppers. The cooler was then raised and lowered into the first boiler (three boilers, accommodating sixty-eight cases, were filled). These boilers (or retorts) were filled, sealed off, and steam was turned on them for seventy-five minutes. After cooking, the steam was turned off, and the first bath emptied.

After cooking, the coolers were sent to the testers. These men pricked holes in the tops of the cans to let out the steam. A few drops of flux were applied to seal the cans. They were then sent back into the boilers for a second cooking. Ladner (1979:107) provides a vivid description of the Fraser River operation. "The coolers were stacked on low-wheeled, steel-framed cars running on steel shod tracks. The cars were run into large steel retorts, set in a row and serially numbered, with a turntable in front of each." The tracks were then pulled out, placed on the turntable to remove the water, and pushed to the washers. The Windsor cooking operation appeared to be more manually intensive. Carmichael noted that, at Windsor, three native men washed 1,300 cases for $90. Each cooler was lifted and dipped into a caustic soda bath that removed the grease adhering to the cans. The cans were cooled for one day. Chinese labourers then sounded each can with a small, steel rod or large nail. They could tell by the noise which cans were defective or underweight. These were removed, repaired, or discarded.

When fishing ceased, the Chinese crew worked at casing the pack. Half the labourers made boxes, while the other half labelled and lacquered the cans. At Windsor, they were paid 37 cents per 1,000 cans. The cans were sounded once again for defects, and finally they were cased.

There was variance in the use of labour. For example, Ladner noted Indian women were almost never hired as fillers, whereas on the Skeena they were. At Wellington, women were usually employed as slimers, to wash the fish, while wipers were Chinese, many of them "mere lads" (Ladner, 1980:56). This work was boring and often done by children. Assignment of tasks to specific groups depended on the relative sizes of each labour force. However, tasks were assigned by racial distinctions, and there is no evidence, for example, that native women and Chinese men worked together on the same tasks in any one cannery.

Technology and Labour, 1905–1912

When labour was plentiful and cheap, machines were used to supplement the manual process. For example, gang knives were used to speed the cutting process, while retorts modernized the cooking process. "Prior to the introduction of butchering machines, the speed of the line depended on the capacity of the gangs to produce butchered fish, and there was little point in mechanizing other stages of the canning line, even where the technology was already developed" (Stacey, 1982:21). When labour became scarce and costly, efforts were made to mechanize the lines.

The iron butcher was introduced in the early 1900s and was sufficiently developed by 1907 to automatically clean the fish and supply two or three lines. This machine was the only one developed specifically for the salmon-canning industry. It became known derogatorily as the "Iron Chink," because it displaced skilled Chinese butchers. The butchers, whether human or mechanical, fed the canning lines. Mechanization increased the speed and regulated the quantities of fish being processed. As more machinery was introduced, more canning lines could run simultaneously, and overall output increased.

Although the iron butcher displaced skilled workers in one area, it, and other machines, obliged the canners to employ skilled workers for other purposes. Machines required maintenance, especially in the wet and dirty conditions of salmon canneries. In addition, since machines were operated for only a small part of each year, they had to be overhauled and prepared for each canning season. Cannery machinemen, most of them of European descent, replaced Chinese butchers as the elite labour force. Unlike other workers, these machinemen have always been hired directly onto company payrolls, usually on a permanent basis.

By 1912, the sanitary can and double seamer were added to the line. With the Smith butchering machine (a further refinement of the iron butcher), two operators performed the work of fifty-one expert Chinese butchers (Sessional Papers, 1906–1907, XLI, 9[22], Appendix 2, LXII). Stacey estimates that the adoption of the sanitary can and double seamer led to a 30 to 35 per cent reduction in the labour force. American Can Company introduced them in British Columbia and was able to attain a virtual monopoly over this type of can-making machinery (1982:23). It leased most of the machines on the lines to the canneries. Many plants ceased their local can-making operations, and American Can became the chief supplier. Local can-making operations continued in the north, because it was costly to ship empty and bulky cans to remote canneries. The invention of collapsed cans did not prove feasible, and preformed cans became standard in the province.

Thus a series of machines was introduced at various stages to overcome bottlenecks and labour shortages. These machines were usually fed by other machines interconnected by conveyor belts. Workers monitored the process and maintained, repaired, and adjusted the machinery. The machinemen, many holding engineering tickets, became ever more important. Yet until very recently, canning operations have employed large numbers of manual workers to wash the fish being fed into iron butchers as well as to fill cans. Native women provided the original labour, and, later, women from other ethnic groups joined them. The vast majority of these women have been recent immigrants, their job opportunities limited by their inability to speak English. They took the place of Chinese men as a reservoir of cheap labour.

Refrigeration Technologies and Labour

As refrigeration techniques were refined and cold storages were built, the fresh/frozen market was developed and expanded. In contrast to canneries, fresh fish plants have smaller, more permanent skilled labour forces. Women provided much of this labour, especially in filleting (unlike the Atlantic coast). In fresh-fish plants, distinctions between male and female jobs often became blurred, since many of the tasks overlapped and the total workforce was generally smaller. For example, both men and women fletched halibut. These overlaps became especially pronounced during World War II, when male labour grew scarce. However, the wage differentials remained.

Summary

Early canning companies tried to obtain a seasonal labour force at least cost. Contracted Chinese male labour and native women and children were the two groups initially exploited.

Trade union organization after World War II improved the wages and working conditions of cannery workers. Chinese contracts were replaced by negotiated agreements, and the Chinese were finally phased out of the industry (immigration had ceased after 1923 with the Exclusion Act, and the existing Chinese labour force grew elderly). However, native women continued to be employed. As plants were relocated from outlying streams to urban centres, canners made use of yet another cheap source of labour— new immigrant women who could speak little English. Although hourly wages are high, the seasonal nature of the work entails very low annual incomes.

NOTES

1. The various species of salmon return in cycle years (for example, sockeye return on four-year cycles) to specific rivers and streams. In the early period of salmon canning, canneries were established on sites close to major sources. The Fraser River was first identified as such a source, followed by the Skeena River in the north. Other locations were quickly exploited; for example, the Nass River and Rivers Inlet.

Canners could offset, to some extent, problems involved with cyclical runs by establishing plants on streams in scattered (strategic) locations.

2. Information on J. H. Todd & Sons, and on general events taking place in the industry in the 1950s, was taken from company files housed in the Special Collections Division, University of British Columbia Library.

3. As late as 1978, segregated housing was still used. A union observer noted in 1966 that housing at Cassiar, on the Skeena River, reflected "racial discrimination in its lowest form" (*The Fisherman*, vol. 29, no. 22:2). According to this observer, in the "white" bunkhouse each man had his own room and running water. Meanwhile, as many as ten men slept in a single unpartitioned room in the Chinese bunkhouse. The Chinese shared one toilet resembling turn-of-the-century "privies." In 1978, a complaint was filed with the Human Rights branch on behalf of the native workers at Cassiar, charging the company with "discrimination on the basis of race, colour, and ancestry" in its housing operations (*The Fisherman*, vol. 43, no. 24:1).

4. The following accounts are abridged from Alfred Carmichael, "Account of a seasons work at a Salmon Cannery," handwritten, Windsor Cannery, Aberdeen — Skeena, 1891 (UBC Special Collections) and T. Ellis Ladner, *Above the Sand Heads* (E. G. Ladner, 1929). Ladner gives a firsthand account of pioneering life in the area, which, in 1879, became the Municipality of Delta. Because of the costs involved in transporting workers from the south, there were proportionately more native people employed in northern canneries.

4

Competition Among
B.C. Fish-Processing Firms

Evelyn Pinkerton

INTRODUCTION

Raw fish markets on which fishers sell unprocessed products to buyers tend to be dominated by a few processing firms in many of the more developed countries. However, these few firms can effectively set prices only if they successfully avoid competition among themselves and prevent new competitors from entering the market. Their capacity to do so depends on several factors: the cost of entry, the processing technology most appropriate for particular fish, and the capacity of firms to lock fishers into a dependent or privileged relationship.

This chapter examines the behaviour of B.C. firms on three raw fish markets: salmon, roe herring, and halibut. One firm has been the major processor for several decades and four firms altogether supply the larger part of the most important domestic markets. The central issue here is whether and under what circumstances these dominant firms are able to control the raw fish markets and prices for different species of fish.[1]

THE RELATIVE IMPORTANCE OF SALMON,
ROE HERRING, AND HALIBUT

Salmon and roe herring provided about 85 per cent of the landed value of all fish production in B.C. in 1981 and 1982. Although the larger processors produce the entire range of fish products, competition has been largely confined to these two products. In earlier years, salmon was more important, providing between 60 and 83 per cent of the wholesale value of all fish production between 1968 and 1977 (*The Salmon Industry in British Columbia*, 1979). Since 1972, the value of roe herring has expanded and its wholesale value between 1973 and 1979 varied from 10 to 34 per cent of total fish production. Halibut is less important today, chiefly because of stock decline and the exclusion of Canadian fishers from traditional halibut grounds in Alaska by the two-hundred-mile limit. Halibut comprised one-fifth of the value of all landings in 1970; by 1981 and 1982, it represented only 1 to 2 per cent of the wholesale value of all B.C. fish products.

Table 4.1
Catches and Landed Value — Salmon and Herring

	Salmon				Herring	
	Metric tonnes	Value in $1,000	% value	Metric tonnes	Value in $1,000	% value
1966	76,589	38,654	88.3	139,550	5,107	11.7
1967	63,041	36,001	95.2	52,953	1,828	4.8
1968	83,709	44,887	99.5	2,891	231	0.5
1969	37,808	22,810	99.0	2,003	221	1.0
1970	72,487	45,076	99.4	3,865	290	0.6
1971	63,252	44,476	98.8	10,017	556	1.2
1972	76,831	50,341	94.9	39,021	2,726	5.1
1973[1]	86,861	100,216	90.1	50,960	10,923	9.9
1974	63,501	74,173	86.0	43,505	11,876	13.9
1975	36,384	46,913	77.2	53,293[2]	13,812	22.8
1976	57,462	91,942	79.7	78,832[2]	23,333	20.3
1977	65,581	109,176	78.7	73,485[2]	29,542	21.3
1978	70,603	158,686	74.9	63,408[2]	52,933	25.1
1979	61,214	160,533	55.8	37,480[2]	126,866	44.2
1980	53,871	117,003	81.6	16,414[2]	26,375	18.4
1981	74,530	154,532	81.1	29,486[2]	35,928	18.9
1982	62,597	165,193	82.3	27,658[2]	35,474	17.7
1983	71,765	111,100	67.7	39,155	52,902	32.2
1984	47,865	144,814	76.8	32,795	43,645	23.1

[1] Roe herring only, from 1973 forward.
[2] Includes spawn on kelp.
Source: Annual Statistical Review, Department of Fisheries and Oceans, various years.

Table 4.1 demonstrates the changing volume and landed value of salmon and herring production. Except for 1979 (when herring comprised almost 44 per cent of the value of all salmon and herring production), herring has not exceeded 25 per cent of the combined salmon and herring landed value. Herring has nonetheless played a role far more important than these figures indicate; it was a major focus of competition and contributed to altering temporarily the structure of the salmon and herring industries in the late 1970s.

DOMINANT FIRMS AND WHOLESALE/RETAIL MARKETS

The largest three or four B.C. processors are dominant on the Canadian domestic fish market. However, on most export markets, individual B.C. firms or groups of firms have little influence. Japan, France, and the United Kingdom import more frozen salmon from the U.S. than they do from

Canada. For a short period in the middle and late 1970s, B.C. was the largest supplier of herring roe to the Japanese market, creating the potential for a few large B.C. suppliers to gain market power.[2] In this case B.C. Packers and the Prince Rupert Fishermen's Co-op were the largest suppliers. The PRFCA claimed price leadership because of its higher quality, while B.C. Packers supplied a higher volume. Apparently there was no price collaboration between these two firms, and both firms lost part of the market to other B.C. competitors in the late 1970s. The export of herring roe remained highly competitive, even when B.C. processors were the most important players.

B.C. canned salmon, however, has had a long-term dominant position in certain countries — the U.K., Australia, and Belgium — where it accounted for over half of the canned salmon imports of these countries in 1982 (Fisheries Association, 1983). These three countries also bought 73 to 82 per cent of all the canned salmon exported from B.C. between 1976 and 1980. B.C. Packers has the largest canned production and longest established markets. It is possible that this firm and two or three other major firms control prices, but price control on the world market is less likely than on the domestic market.

The market which is unmistakably oligopolistic (i.e., a few sellers, here processors, sell most of the product) is the Canadian canned salmon market, where more than half of B.C.–produced canned salmon is sold. The existence of a 4 to 15 per cent tariff on canned salmon imports has restrained foreign competition in this market. A few major firms exercise market dominance. The top three canners sell 70 to 75 per cent of domestic canned salmon, while B.C. Packers alone sells over half of it (Schwindt, 1982). B.C. Packers is assisted in product distribution through the retail food chain of its parent company, Weston's, which was a major buyer from B.C. Packers even before it acquired voting control (Schwindt, 1982:51). In 1984 a parallel structure was established for the Canadian Fishing Company when it was purchased by Jim Pattison, owner of Overwaitea and Save-On-Foods retail chains, which can now assure the marketing of its "Goldseal" label.

B.C. Packers and Canadian Fish count on customer brand loyalty to help maintain their long-established share of this market. For example, of the seven labels used by B.C. Packers, the most popular (Cloverleaf, Paramount, Carnation, Bumble Bee) are sold by the Weston-owned stores and as well by competitors: A&P, Safeway, Dominion, and Steinberg. B.C. Packers reports that Weston's retail chain competitors may carry only one-quarter of B.C. Packers' forty Cloverleaf products, while offering a competing processor's entire product line. On occasion, a Weston competitor has been known to "delist" the B.C. Packers' Cloverleaf line after a controversy with Weston. To reobtain a listing (a commitment to carry a product line) can cost B.C. Packers up to $10,000. There are other ways a retailer can grant special advantages to one wholesaler, such as volume rebates, cash discounts, advertising, and cooperative plans. All can affect the profit margin of particular products while making them available to the consumer at the same price. Yet it is still likely that a firm with an established and popular label would retain the upper hand over the long run.

The major firms' oligopolistic position on the domestic canned salmon market contributes to their position on the raw salmon market because they are assured of sales and thus can withstand periods of price competition for the supply. However, it is probable that oligopoly on processed markets, at least originally, was a result rather than a cause of control in raw fish markets.[3]

Crutchfield and Pontecorvo (1969) argue that throughout the capitalist world raw fish markets are typically oligopsonistic, characterized by many sellers (fishers) facing a few large buyers (processors). However, they do not judge that any salmon oligopsonies in the U.S. or B.C. are involved in effective price collusion.[4] A high degree of concentration alone is not proof that the dominant firms control prices. The behaviour of firms, such as creating barriers to the entry of new firms and forming vertical linkages with fishers, is the critical factor.[5] It is through such arrangements that processing firms can inhibit price competition from other firms: simply purchasing the largest volume of raw fish in B.C. does not necessarily guarantee a firm or group of firms control over raw fish prices. The rest of this chapter is concerned with such behaviours.

DOMINANT FIRMS AND THE RAW SALMON MARKET

Two supply conditions have an important influence on the structure of the raw salmon market: high supply fluctuation and an extremely short harvestable time period. Salmon species differ markedly in size, colour, oil content, taste, firmness of flesh, and tolerance to different forms of processing. Each species has a different market, and annual fluctuations in the supply of each species require special management capabilities on the part of firms that wish to stock their markets regularly. A firm competing for a large share of the supply on a regular basis must carry over inventory from a high- to low-supply year, and use a technology that facilitates this. Canned salmon is easier and cheaper to hold in oversupply than are frozen or fresh salmon, so large canning firms are most able to adapt and take advantage of this supply condition.

The short harvestable time period also favours the major canners. The largest volume of salmon arrives in a three- to six-week period, when it must be processed with speed because of its perishability. Large-scale canning can be done more rapidly than large-scale freezing. Firms that

Table 4.2

Salmon Canning and Freezing, Number of Firms: Selected Years, 1951–1984

Year	Number of canners*	Canners who do not freeze	Freezers who do not can	Total canners and freezers	Total firms†
1951	12	4	5	17	15
1964	12	2	6	18	15
1971	9	3	11	20	16
1974	10	1	12	22	20
1976	11	0	15	26	21
1977	12	1	17	29	24
1978	12	0	20	32	26
1979	12	0	29	41	35
1980	11	0	23	34	26
1981	12	0	19	31	26
1982	11	0	20	31	29
1983	9	0	NA	NA	NA
1984	7	0	25[1]	33	NA

* Excludes custom canners, specialty firms.
† Only firms with a total production greater than $100,000 are included. All firms owned by a single entity are grouped together.
[1] Only firms producing frozen, dressed, head-off sockeye are included.
Source: Compiled from Department of Fisheries and Oceans, Marine Resources Branch, unpublished production statistics.

purchase large volumes of salmon during this time cannot freeze the majority of it quickly enough to avoid decomposition. Most must be canned because of the greater speed of canning technology and because at the height of the season, some fish will stay in the plant two or three days before processing, reducing its freezing quality. Thus large-scale canning is the most effective technology for dealing with the risks inherent in acquiring and marketing a large share of the raw salmon.

Canned salmon has traditionally been the mainstay of the entire B.C. fishing industry. Even after frozen salmon became important in the 1960s and 1970s, nearly the same number of salmon-canning firms, varying between nine and twelve, remained in operation (see Tables 4.2 and 4.3).[6] Indeed, while the identity of individual firms changed, the relative size and number of canners remained constant until 1984: about four large producers and eight small ones. As larger firms absorbed smaller competitors, small firms that had only frozen salmon operations moved into canning. In some cases, small firms grew to join or replace top producers. Individual firms' production data show that, despite the growing importance of frozen salmon, the top four canners maintained their dominant

Table 4.3
Total Salmon Production of Top Four Canning Firms as a Percentage of
the Value of B.C. Salmon Production: 1972–1985*

Year	% of production
1972	65.2
1973	69.3
1974	67.9
1976	67.6
1977	65.7
1978	56.2
1979	57.6
1980	62.2
1981	64.3
1982	57.0
1983	62.5
1984	53.7
1985†	68–70.5†

* All firms owned by a single entity are grouped together.
† Projections of Marine Resources Branch based on incomplete data on pack size.
Source: Compiled from unpublished production figures, Marine Resources Branch.

position in canning and their share of the canned salmon market. Table 4.2 also shows the stability of the number and size of canning firms compared to the relative instability of the number and size of freezing firms.

Another structural constant, significant to the discussion of price influence, is that the top four producers of canned salmon have normally been members of the Fisheries Association,[7] the body that bargains with the union (UFAWU) for shorework wages and for minimum prices for net-caught salmon. The Fisheries Association also lobbies for policies such as export regulations on fresh and frozen salmon, restrictions on the number of processors, sales promotion of canned salmon, and other matters relating to international negotiations, catch regulations, and habitat. Typically, the smaller six or eight canners are not members of the Fisheries Association, and feel that the major canners do not represent their interests. Canning for these smaller firms is a secondary concern after freezing, and they have had much higher frozen production during most of the 1970s. Thus it is less important for them to bargain for the minimum price of the lower-quality net-caught salmon. When frozen markets are high, canning is more an overflow of poorer quality salmon than a mainline operation for these smaller firms. They tend to be members of the B.C. Seafood Exporters' Association, a body concerned with promotion of the reputation and quality of exported B.C. fish products. Members of the Fisheries Association also hold memberships in B.C. Seafood Exporters' Association because frozen markets became so profitable that even the major canners began to participate in them more seriously.

The major canners in the Fisheries Association express their common interest by attempting to hold down the minimum negotiated price of net-caught salmon. In addition, they try to curtail competition among themselves during the season. Some industry observers doubt the companies' disciplinary abilities.[8] It is often claimed that negotiated minimum price does not affect prices because landed prices may be "bid up" during the season as processors compete for the fish (Shaffer, 1979). In the 1970s, this competition began with the small firms and then spread to the larger firms. However, a review of the relationship between minimum prices and average landed prices for gillnet, seine, and troll-caught sockeye between 1971 and 1982 shows that in most years (1971, 1972, 1973, 1975, 1976, 1977) average landed price exceeded minumim price by only 1 to 6 cents a pound. In three years, 1978, 1979, and 1980 average landed price exceeded minimum price by 18 to 30 cents. Throughout the decade, however, average landed price parallels and tracks minimum price, maintaining a relatively stable relationship to it. The same general pattern pertains even if troll prices are excluded. It appears that minimum price has an important influence on average landed price except in years of very buoyant frozen markets. While the Fisheries Association can inhibit competition among members under "normal" market conditions, their success in doing this varies inversely with the strength of the frozen salmon market. Thus price control can be less successful in the exceptional, high frozen market years.

Interviews also show that large firms can exercise control to the extent that they collectively set the size of fisher's bonuses and target "problem" fishers. When a firm offers a larger bonus or service to a skipper, small disciplinary actions are taken by other firms in the form of minute price rises or additional services to signal they are answering the competition. The response is usually for the competition to cease the offending behaviour, rather than to increase the price for all. Yet a small degree of nonprice competition among large firms is considered legitimate and takes the form of small differences in services provided.

What can be concluded from this about the importance of the number and relative size of firms, i.e., market structure to price competition on the raw salmon market? More firms and buyers participated in the market in the late 1970s though it stayed well within the definition of "oligopsonistic." However, the canner's ability to influence prices did change temporarily in the late 1970s. While concentration was still high at that point, discipline broke down and price competition was severe until the unusual export markets restabilized. In terms of the number and relative size of firms, an oligopsonistic structure therefore is a necessary but not a sufficient condition for behaviour such as inhibiting price competition.

The stability of the number of canning firms alongside a fluctuating number of freezing firms is often explained in terms of barriers to entry into canning. A barrier to entry is the cost disadvantage that a new entrant must bear relative to established firms.

There are at least three barriers to entry into canning. First is the absolute cost advantage that existing firms have over new entrants in equipment rental. New entrants must pay the escalating cost of the lease or purchase price of a canning line. In 1983 the purchase price of a high-speed line was $750,000; the lease price was up to $90,000 for the first two years, dropping to $14,000 per year or lower afterward. Longer-established firms have very low lease prices compared to new entrants, even though repair costs rise with the age of the line. Larger firms also benefit from volume discounts on cans and a warehousing discount on stored cans between January and June.

Second, the price advantage enjoyed by established firms through their brand names and markets constitutes a barrier to entry. This is one of the ways oligopoly on the domestic canned market contributes to the power of these firms on the raw salmon market.

But ultimately the major problem deterring the entry of a new or small canning firm is attracting a large enough supply of fish to benefit from low unit of production costs. Because they must start small, it is difficult for these firms to employ the strategies of the larger companies in attracting more supply, especially providing services and extending credit. Large firms enjoy economies of scale in canning and in affording services over time. The small firms need to keep their fixed costs low and are not equipped to take full advantage of the high supply years by competing in services. They are seldom able to provide in-season financing or repairs in exchange for deliveries. Instead, they compete by paying more for the fish; so freezing salmon is a more attractive alternative to canning.

In contrast to canning, there are low barriers to entry in freezing. It is possible to rent public cold storage at low cost, and the cost of small-scale freezing technology is modest. A small freezer firm can operate without high fixed costs and can tailor its overhead costs to the supply it can obtain in any particular year. It is economically rational for a small firm to compete in price for freezer-quality fish, rather than in services. This allows the greatest flexibility. Small freezer firms are quality-oriented or market-oriented in contrast to the quantity-oriented or supply-oriented canners: they compete by producing a higher priced product, rather than producing higher volume at lower cost. New freezer firms are attracted into the industry when frozen markets are high, not when supply is high. More established freezer firms stay in business because they have links with quality-conscious buyers who maintain buyer loyalty when frozen salmon prices are low and profit margins slim. Until the mid- to late 1970s frozen salmon prices were not consistently high enough to attract major competition into the raw salmon market. They were not high enough to lower the barriers to entry into the frozen market. The competition for raw salmon did not threaten the supplies of the major canners in any consistent way; the prices offered by the freezers were not high enough to attract away a significant portion of the supply.

This discussion emphasizes the polarization between a few large firms that dominate the raw salmon market and the more numerous small firms that compete on the margins of the industry. The supply of fish is limited, and neither large nor small firms wish to attract further entry nor bid up the price too much. Firms, when they have developed a long-term stake in the industry, find it convenient to cooperate in various ways, including holding raw fish prices low. The danger to human life and marine transport difficulties on the storm-bound coast reminded processors of how much they needed each other for emergency assistance. Small firms certainly valued the reliability at such times of the larger firms' capacity to provide ice, packing, and assistance in breakdowns to members of their fleet who were travelling or fishing in areas where the small firm had no processing facilities or services. In the case of cooperatives, the willingness of other firms to custom process or custom pack for them on an equal exchange basis was very important. Small firms also maintained good relations with larger canning companies because large firms sometimes bought the canned pack of a smaller firm and sold it under their own label, if supplies were short or customers' orders had to be filled. This was a considerable convenience for a small firm, which was normally an ad hoc seller in canned markets. Thus agreements between large and small firms, as well as mutual interests in restricting new firms, limited the competition between them.

Finally, small firms are often motivated to restrain excessive price competition because of the future career prospects of their managers: there is considerable circulation of management personnel among large and small firms. Frequently small firms are started by people with management expertise acquired in larger operations (see Chapter 13). Conversely, larger firms attempt to obtain new expertise by hiring people with an intimate knowledge of fish movements and fishing patterns on the coast acquired through independent operations. Given the vulnerability of small firms, managers know that they may some day be seeking employment from one of the larger firms, and so it is imperative to maintain good relations with them.

Under "normal" circumstances, aggressive price competition is disadvantageous, and firms maintain acceptable price differentials. Smaller firms follow the price leadership of the large firms by adjusting their prices to remain about five cents a pound higher. Their interests are best served when the leader's price is low. This avoids excessive price competition. The absolute cost of entry, the cost of establishing brands and markets, and the difficulty of attracting enough supply to lower unit costs keep the barriers to entry in canning high. Of these, the last is by far the most important factor. Attracting adequate supply, even by paying higher prices, is difficult for new or small canners because of pre-existing vertical linkages between processors and fishers, to which I now turn.

Assurance of supply may be at times the most important aspect of the vertical linkage for the canner, who is dependent on securing enough volume to lower production costs. Competition by processors in the provision of services to fishers is simultaneously the most direct method of acquiring supply[9] and of avoiding price competition. For example, the services provided, such as information on fish locations (given to favoured high producers), ice, net storage and transportation, and repairs on the fish grounds make it difficult for a fisher not to deliver to the servicing processor and are in themselves forms of enforcing the unwritten contract that a fisher so served has agreed to deliver the catch to that processor, even if another processor offers a higher price.

The provision of services as a form of competition is less quantifiable than price competition, and some companies attempt to provide these extra benefits as a less-obvious method of attracting fishers or rewarding highliners without direct price competition. These forms of competition among Fisheries Association members are secretive and minute compared to overt price competition. When a company offers a new and significant service to everyone in a bid to acquire more fishers, it is often detected by the competition and some form of retaliation occurs. As a former employee of the Anglo–British Columbia Company recalled: "ABC was offering (some service) to their fishermen, so Canadian Fish raised the price of fish one cent a pound. ABC then stopped their service and Canfisco dropped the price again." In more recent years, B.C. Packers and Canadian Fish appear to compete in the packing charge[10] they offer their seiners.

Offering special favours, such as skiff storage, to high producers is not only difficult to quantify; it also tends to be discussed less openly than fish prices by fishers who deliver to the same company. A company wishing to assure the loyalty of its best fishers without raising the price of all the fish will "throw in" as many noncash benefits as possible in order to offer the most attractive incentives to highliners. It is more difficult for other companies and fishers to keep track of these multiple small arrangements, especially as they may involve personal relationships of trust and patronage built up over the years.

The bonus or "charter" payment to the owner of the boat and net is another important form of nonprice competition. This is not to be confused with the seine skipper's bonus, which is paid by the processor to the skipper as a matching payment on top of the shares received by the skipper out of the traditional catch division. (Four-elevenths of the gross catch go to the boat and net owner after food and fuel have been deducted; seven-elevenths are divided evenly among all labour on the boat, including the skipper). The skipper's bonus is usually equivalent to one-half of a crew share, although some processors pay a flat 10 per cent of the gross catch. In either case, this bonus is important, but not as significant as the charter payment to the boat owner.

The charter, sometimes also called a bonus, is paid "under the table" to the owner of the boat and net, and is significant in competition between processors. Unless the boat is processor-owned, the owner of the boat and net is usually also the skipper.[11] The charter payment is a covert form of price competition: it is usually not shared with the crew,[12] amounts are not known by the crew (nor by other skippers), and it is not subject to price negotiation between the UFAWU and the Fisheries Association. Owners of large seine boats justify the charter as a legitimate return for the increased capital investment of seine boat owners that has developed since the original share system of granting four of the eleven shares to the owner of boat and net. They are less open about the exact amount of the charters and about who gets them.

Initially, this special "return to capital" on top of the share system was a flat rate by the day or season: one company's seasonal charter was $5,000. In the 1960s, when capital investments in vessels and repair costs rose, it was transformed to a percentage based on the gross catch; with some processors the percentage increased with the size of the catch and was based half on value, half on volume of production. Charter payments to seine skipper-owners varied over the 1970s from 30 per cent to 100 per cent.[13] In the 1980s, all charters were lower as markets tightened. Since skippers' capital investment was not lower at that time, the charter is apparently less a return to capital than an attempt to secure the loyalty of the skipper without paying more to the crew.

The exact amount of charters can be "one of the best-kept secrets in the industry." It is generally acknowedged that the size of the charter varies with the independence, productiveness, and indebtedness of the fisher (*The Salmon Industry in British Columbia*, 1979:178; Schwindt, 1982:18). Higher producers, whose loyalty the processor is anxious to maintain, are offered higher charters, especially if several members of a family also deliver to the same company. Conversely, there may be less need to offer a high charter to a skipper who is very indebted to the company, because he is already obligated to sell his fish to it. If the indebted fisher is a very high producer whose loyalty the processor wants to secure, however, he may still be offered a high charter or special services; alternatively, he may be encouraged to borrow even more. Holding a large debt is likely not only to make the boat owner feel even more obligated to deliver to the lending firm; it is also likely to make him even more dependent on the firm for ready cash and services. In a description of the social psychological mechanism of such a dependency, Langdon (1977) depicts the personality of Alaskan seine skippers who fall into such arrangements

— and often get stuck in them — as the charismatic, free-spending, high-living type who is worshipped as a local folk hero. Our interviews suggested that in B.C. as well, this type of indebted skipper tends to become addicted to such arrangements and to perpetuate them.

Such arrangements are of particular benefit to large firms, which are concerned with maintaining high productivity. Small firms may also lend to high producers, but not to the same extent as larger firms. Large firms not only offer a longer line of services; they also are able to underwrite the risk that a fisher might switch his deliveries to another company.

The charter varies with the market for fish; when different species of salmon are worth more, a processor creates incentives to target these fish without overt price competition. If a company wants more sockeye in a particular year, it may offer a 50 per cent charter on sockeye and only a 30 per cent charter on pinks. The company thus influences the skipper-owner to catch more sockeye, because he/she will receive a matching payment for 50 per cent of the sockeye delivered. The processor will not have to pay more for the entire load, only half of the load. With deliveries in which the boat owner receives a charter, price competition occurs covertly and usually over a portion of the catch rather than over the entire catch. If price competition were overt, it would have to be shared with the crew and would logically be offered on the entire catch. This would be equivalent to raising the negotiated "minimum price" of fish. But it is not necessary for the processor to bid higher for the fish, because all the company needs is the decision of the owner-skipper to deliver. One person only is rewarded.

Langdon (1977, 1982) has analysed a similar "bonus" system in Alaska as a method by which the processors secure the cooperation of the seine owner-skipper in extracting surplus value from the labour of the crew. Under expanding market conditions, it may appear appropriate to see the owner-skipper of a large seiner as a small "capitalist," a junior partner with the processor in extracting surplus from the crew. When markets fall, however, the same owner-skipper is more vulnerable than the crew because he has no control over the falling bonus, and may be unable to make bank payments. In a buyers' market, the processor may no longer be his "partner"; instead he may be in a position to require him to accept the brunt of price reductions. In the degree of control he has over the terms of exchange in the long term, then, the seine owner-skipper, excluded from the union, has a vulnerability and exposure perhaps equivalent to the crew's.

More importantly from the perspective of the crew, the owner-skipper shares the risk with them through the share system. Moreover, our interviews showed that many seine crewmen do not perceive the owner-skipper as an exploiter or "boss" supervising their labour, but as a coordinator in a relatively egalitarian social system predicated on a mutual dependence upon one another's skills. Their perception of the owner-skipper's role appears to vary with their own skill level, their experience in the industry, and their personal relationship with the owner-skipper.

The complexity of the owner-skipper's position lies chiefly in the fact that skipper-owners on seine, gillnet, and troll vessels alike have ideologically defined themselves as independent producers selling a product on a free, competitive market. Yet the market is not perfectly competitive in most years. To the extent that the seine owner-skipper receiving a high charter can pass on the noncompetitive conditions of the market to the crew, he/she acts as a capitalist. To the extent that he/she shares the charter with crew, as some owner-skippers do (Miller, 1978; see note 12), or otherwise is subject to pricing and market conditions beyond individual control, he/she may be more appropriately seen as either labour or a not-so-independent commodity producer (see also Part 2).

While the offer of high charters or bonuses is perhaps the favoured nonprice or limited price mechanism for securing fish from vessel-owning highliners, the financing and/or service contract is the oldest and most widespread form of securing supply and avoiding price competition. The contract may exist with or without a bonus. It offers security to the fisher and the processor simultaneously; it is an agreement on the part of the fisher to sell all the catch to the processor (in return for services and financing), and on the part of the processor to accept all the fish delivered by the fisher even in poor market years and during fisheries in which the processor might not want all the fish. The small canners seldom have such arrangements with fishers because of limits on their financing and services, and because they cannot guarantee the ability to handle almost unlimited volumes of fish as can the major canners. The majors operate canneries possessing between four and eleven lines, because "the cost of inadequate capacity at peaks is greater than the cost of maintaining excess capacity" (Schwindt, 1982). The "cost of inadequate capacity" includes the cost of losing fishers if not all their deliveries can be accepted in a peak year. One fisher declared he would have delivered to Nelson Brothers all his life, except for the quota system that the canners put on deliveries in the glut year of 1962 (to reduce the amount of fish they would have to accept from their fishers). Fearing the canners might again "betray their contract" with the glut of Nitinat chums in 1972, he joined a cooperative. Canning technology gives the majors the capacity to maintain vertical linkages with many fishers (in most years), by permitting them to deal with the tremendous supply fluctuations and time concentration in fish deliveries by large numbers of fishers.[14] The large canners provide the services because the volume of production in high-volume years lowers the average cost of services in the long run.

The contract system has remained relatively stable over the years and assured the major canners the supply they needed to develop and keep stable marketing arrangements. They could ignore the pockets of price competition from the smaller firms when it occurred, because as long as they secured enough volume to lower unit costs and to meet the average demand of their markets, they were little affected. In this way, vertical linkages also played into the relationship between raw salmon supply characteristics and the technological appropriateness of canning, which contributed to an oligopsonistic market for raw salmon.

In the late 1970s, the contract system was temporarily eroded, and the majors entered into price competition both with the freezer firms and with each other. Until then, when the frozen market reached unusual heights because of temporarily disrupted supply and market conditions,[15] fluctuations in frozen salmon markets had never really threatened the canners' supply nor their long-term vertical linkages with fishers.

Another new condition also threatened the contract system at this time. Because of the licence-limitation program implemented in 1969 and high expectations for increased productivity resulting from salmon enhancement and the two-hundred-mile limit, banks had become more willing to lend to fishers (McMullan, Chapter 6). Some of the fortunate fishers who purchased vessels in the early or middle years of the decade were able to pay off much of their debts before interest rates rose in the late 1970s or markets collapsed in the 1980s. If they did not expand further into larger boats, these fishers achieved greater independence. Although some independents still relied on the services of companies and valued the immediate loan with no questions asked, many changed their delivery pattern during the late 1970s and sold mostly to small companies, preferring to buy their services at a high price, or do their own repairs. The predictability of supply was reduced, and price competition resulted.

It is difficult to be precise about the timing and extent of this change. The system has always been able to tolerate a limited amount of price competition on a regular basis and a greater amount of price competition on an irregular basis. The competitive margin only becomes a threat to the major canners when it cuts into their supplies to the point of keeping their unit cost of production too high to realize profits. During the years of sustained high frozen markets in the 1970s, the percentage of production value in salmon captured by the four major canners declined steadily from 69 per cent in 1972 to 56 per cent in 1978, a significant drop of 14 percentage points (see Table 4.3). It seems plausible that the majors entered into price competition only when they feared that the supply essential for their profits was threatened. Since normal supply fluctuations make profit margins unpredictable anyway, it is difficult to separate fear from reality. At any rate, the majors clearly felt threatened by the late 1970s and responded by competing in price as well as services.

In the late 1970s the number of small firms offering immediate cash and higher prices for freezable salmon was increasing, and it was easier for fishers under contract with a major canner to sell part of their production to cash buyers, while continuing to use the services of the large company. The majors responded by demanding a moratorium on the issuance of new processing licences, successfully effected in 1979. But by 1980 the Japanese market for Canadian frozen salmon had declined as a result of Japan's success in enhancing chum salmon domestically. As well, Alaska's Bristol Bay sockeye fishery had expanded and was selling fish at lower prices.[16] Whereas only 22 per cent of the Bristol Bay pack had been frozen in 1976, in 1980 it was 49 per cent (Alaska Department of Fish and Game Statistics, 1980). In the 1980s Alaska producers also began setting much higher quality standards, with supporting legislation, in order to appeal to the Japanese market.[17]

The moratorium was removed in 1980, but the number of freezer firms continued to decline for two more years (see Table 4.2). The majors' share of the value of salmon production began to reverse its decline in 1980 and 1981: from the low point of 56 per cent in 1978, they recovered to 62 per cent in 1980 and 64 per cent in 1981. Although some "nonproductive" services (net storage, moorage) have been withdrawn by the major canners, there appears to be a general reversion to the oligopsonistic pricing and servicing of earlier times. There is less competition for the fish, and charters for seine skippers offered by the majors are reputed to fall between 9 and 17 per cent for different species in 1984. Some of the majors now own canneries in Bristol Bay and southeast Alaska, and/or import raw salmon from Alaska to supply their B.C. canneries. Importing Alaskan fish gives B.C. majors a better opportunity to keep their volume high and costs low. They can buy cheaper fish in Alaska and develop arrangements for a more secure supply through long-term arrangements with Alaskan fishers who are less militant trade unionists than their B.C. counterparts.

The failure in 1983 of Cassiar, one of the oldest majors, increased fishers' fear of not having a market for their fish in glut years unless they sold to B.C. Packers. Cassiar left its fishers unpaid for their season's deliveries. In addition, the failure of a firm creates a surplus of unattached fishers wanting contracts, and this decreases individual fisher's bargaining power over services, charters, and prices.

The failure of firms allows the surviving firms to acquire both attached fishers and plant facilities on favourable terms. This had led to even greater degrees of concentration and market power on the part of the existing majors. Most of Cassiar's seine fleet, its Fraser River plant, and its subsidiary, Royal Fisheries, were acquired by Ocean Fisheries. B.C. Packers acquired Cassiar's former production manager and so was able to attract independents who had delivered to Cassiar. At the same time, Canadian Fish acquired Prosperity Marine, a major shareholder in Quality Fish that was one of the smaller canning-freezing firms. The resulting loss of eight seiners to Canadian Fish may have lowered Quality Fish's chances of survival, as it was in receivership by the summer of 1985. By the end of the 1984 season, both Central Native Fishermen's Co-op and the Port Simpson Co-op had ceased to operate as independent firms.

Not surprisingly, by 1985 the top four major canners' share of salmon production had risen to an estimated 68 to 70 per cent, its level in the early 1970s. By this time, a counter trend also appeared. The major canners captured by far the largest share of production in years of greatest abundance, such as 1981 and 1985. However, in the years 1982 and 1984, when there were fewer fish overall and a higher percentage of troll-caught chinook, coho, and sockeye, the majors had a lower share, only 54 to 57 per cent (See Table 4.3). The dislocations of the 1970s resulted in the majors retaining and probably increasing their oligopsonistic position, although there is now greater competition at the margin than existed in the early 1970s, presumably related to the strength of frozen markets and the modest increase in independent fishers and small freezer firms.

This difference is reflected in landed price trends. In 1981 to 1983 the difference between minimum and average landed price for sockeye remained at ten to sixteen cents. Although this is higher than most years in the early and mid-1970s, it is far lower than the twenty-seven to thirty-cent differential of 1978–79. Again, degree of concentration or market structure tells only part of the story. Price behaviour completes it. The two together suggest that there is now a larger competitive margin in the raw salmon market, but that most fishers are unaffected by it.

Besides determining prices or guaranteeing supply, some forms of vertical linkage are structured as attempts to capture part of the profit from fishing. As Hilborn and Ledbetter (1983) suggest, fishers' catches appear to depend on skill and knowledge more than on investment in vessels or gear. Highly productive seine skippers are among those who do have profitable fishing ventures and are attractive to processors as joint-venture partners in boat ownership. B.C. Packers and Ocean Fisheries in particular have developed a number of such arrangements with seine skippers. If the processor is the 51 per cent boat owner, the company can pay half the "charter" to itself, but retain the interest of the fisher in maintaining the boat, sharing the risks, and having capital tied up. One such skipper estimated that he saved the company $20,000 a year in repair bills, because as skipper-partner he supplied his own labour free to lower the cost of repairs. As one processor put it, "We make money on boat ownership most of the time." Full processor ownership appears to be a less attractive option to processors than joint-venture ownership with skippers, because rental skippers with no stake in the boat have greater incentives to deliver to cash buyers and have no incentive to take good care of the boat. B.C. Packers claims that it was losing money on its fully owned gillnet fleet before selling it to the Northern Native Fishing Corporation.[18] The corporation hopes to avoid the pitfalls of this arrangement by another type of "joint venture" in which the corporation owns the licence and the fisher owns the boat (see Chapter 11).

Joint ventures appear to have evolved during the 1970s from two directions and for two different motives. B.C. Packers possessed a fully owned fleet of older seiners, some of them wooden vessels built in the 1920s. Repair bills and price competition increased at the same time, with the result that these boats generated little profit. When skippers became joint owners, however, B.C. Packers could liberate some of its own capital for price competition, free itself of some repair bills, and achieve a better guarantee of high production from the boat. As one joint-venture skipper indicated, "The boat is more dedicated in this situation. It fishes harder, and it's in better condition most of the time."

Ocean Fisheries developed joint ventures out of a different situation. They built a fleet of new seiners in conjunction with skippers they identified as highly productive and through a shipyard owned by Rivtow Straits Ltd., a transport company widely believed to have a part interest in Ocean Fisheries. Creating joint ventures was part of a strategy of building up new supply and avoiding large obligations in servicing. Apparently Ocean Fisheries also hoped to share in the profits made by fishers. These profits disappeared in the early 1980s, and joint-venture fishers, with too much already invested, were unable to extricate themselves. Possibly they took the brunt of the loss. Although the processors also lost potential fishing profits in poor years, at least they had guaranteed supplies and low maintenance costs. The skipper repaired the vessel at his own expense, since vessel maintenance signified control over the labour process rather than self-exploitation. More painfully expoitative for the skipper was his inability during poor markets to escape the obligation to supply, even when he was fishing at a loss. The processor realized long-term benefits by continuing to supply markets, even when profit margins were low. If the skipper were not tied to a joint venture at such a time, he would normally "let the fish rest" and seek temporary alternative employment.

Overall, vertical linkage in the form of joint ventures appears to be a risk-sharing arrangement that does not alter the noncompetitive structure of the raw salmon market and often makes it even less competitive. The exit of fishers is impeded, even when profits are low. Along with the contract system and the charter system, joint ventures have contributed to the maintenance of oligopsonistic structure and price behaviour.

SUMMARY

I have tried to show how the particular supply conditions of the raw salmon market work to the advantage of firms using large-scale canning technology. The risks inherent in securing salmon supply foster the development of vertical linkages between fishers and processors. As a consequence, the large canners who most benefit from the nature of the resource are also able to avoid price competition with other firms. Charter or bonus payments to skipper-owners, joint boat ownership, and service contracts are all forms of vertical linkage that large canners can use to guarantee supply, to avoid or reduce price competition, and to lower maintenance costs. Once established, large canners benefit from economies of scale in the provision of services and in canning, from the escalating costs of setting up new canning operations, and from the price advantage of established brands. But ultimately it is processors' vertical linkages with fishers that allow them to avoid price competition from new entrants.

The relationship between degree of concentration and degree of ability to influence prices on the raw salmon market has been demonstrated. Only exceptional world supply conditions permitted a temporarily competitive raw salmon market, but the reversion to old patterns since this episode permits us to conclude that under normal conditions, large salmon canneries can apply effective collusive pricing.

The task that remains is to test the notion that the supply nature and processing capabilities of salmon have an important influence on large firms' ability to influence prices. I do this by comparing salmon to roe herring and halibut. Although the raw markets for both these species are oligopsonistic in the sense of high buyer concentration, there is no evidence the large firms can influence the prices in these fisheries. I argue that this is primarily because of the nature of the fish.

CONCENTRATION, PRICE COMPETITION, AND THE RAW HERRING MARKET

Like salmon, the roe herring fishery is concentrated into a very short time. Openings come with no warning, except for updates on the percentage of roe from test fisheries, and may be less than one hour long. Also like salmon, the volume of herring fluctuates annually, although there is only one species. There is, also, only one market: Japan.

Unlike salmon, however, roe herring is not canned. Instead, it is brined or frozen and brined again before the roe is separated or "popped." There does not appear to be any technological advantage in having a large freezing or brining operation as opposed to a small one, so that the supply conditions of the resource do not give any competitive advantage to larger firms. As a result, large firms have no ability to influence prices, except in a very weak and sporadic fashion and partially as a spinoff of vertical linkages in salmon. Thus a fisher linked to B.C. Packers through sale of salmon might also deliver herring there. There was so much competition among processors for roe herring, however, that processors sometimes avoided using a fisher's herring deliveries to pay off salmon debts, fearing the fisher would then deliver herring elsewhere.

The number of firms processing roe herring (usually in addition to salmon) increased from twenty-two in 1974 to forty-two in 1979, and then declined to twenty-one in 1984. There have always been fewer firms processing roe herring than salmon. Table 4.4 shows that the value of roe herring production is much more evenly distributed than salmon. The small salmon canneries processed almost as much roe herring as did the large ones, and in 1979 the small freezer firms produced even more herring than the small canners. Thus there is a much lower level of concentration in roe herring production; the four large canners produced between 57 and 33 per cent of the value of all roe between 1974 and 1984 (Table 4.4).

Classifying roe herring production according to the frozen and canned salmon production of the same firms is a more relevant index of competition than classifying production simply by "the top four producers" of roe herring. These top four roe herring producers include the PRFCA and often a smaller canner, neither of whom is among the major salmon canners, and neither of whom belongs to an association excluding any of the smaller herring producers. The top herring producers could not be construed as an oligopsony in the sense of price influence, therefore, even though technically there is a high level of concentration among buyers. There is no oligopsonistic pricing associated with the high concentration, and no exclusive association membership held in common by the top producers.

There are also low barriers to entry in the processing of roe herring, since only freezing and brining technology are needed. Nor is large-scale investment required. For this reason, many Japanese firms, which had lost access to their own domestic roe herring fishery, and later the Chinese market (which was also temporarily overfished in the 1970s), turned to B.C. firms, eager to secure supply agreements. In the early 1970s, the fishery was developed by Japanese advisers working with the PRFCA, B.C. Packers, and Canadian Fish. Later, the Japanese became minority shareholders in other firms. In some cases, Japanese companies appear to have been the raison d'être of small firms processing mainly roe herring. They provided the capital to send cash buyers to compete on the grounds.[19] The existence of this type of firm was the principal reason for the moratorium on licences to new processors. Cash prices on the grounds were bid up so high that all firms had to enter full-scale price competition. New firms were attracted into the industry as ground prices steadily rose from 1977 to 1979. Because herring was a larger bonanza fishery than salmon in the late 1970s, it contributed to the erosion of entry barriers to salmon; firms could use profits from herring to enter the raw salmon market as well.

In general, fishers delivered salmon and herring to the same processor, and consequently, relationships based on the salmon fishery were "pol-

Table 4.4
Number of Firms Processing Roe Herring and Value of Herring Roe Production
by Top Four Salmon Canners as a Percentage of Total: 1974–1984

Year	Number of firms*	Herring roe production of top 4 salmon canners*
1974	22	51.1%
1976	27	57.3%
1977	24	55.9%
1978	30	51.0%
1979	42	39.0%
1980	23	(strike)
1981	29	41.3%
1982	23	33.1%
1983	24	41.6%
1984	21	53.9%

* All firms owned by a single entity are grouped together.
Source: Compiled from unpublished production figures, Marine Resources Branch.

luted'' by the herring fishery. Cash buyers offered such high prices for herring in the spring that some fishers entered their salmon season in the summer with less need for financing from processors and with a desire to sell salmon also to cash buyers. In some cases, the same Japanese-backed companies were paying premium prices for both herring and high-quality frozen salmon, giving fishers incentives to desert their old relationships. There were fewer herring than salmon licences, however, so many salmon fishers were unaffected. Out of a fleet of 497 seiners licensed in 1978, for example, only 230 or 46 per cent fished both herring and salmon. In 1982 some processors began separating the salmon and herring accounts of their fishers, in an apparent attempt to secure the herring deliveries of fishers who had large debts in salmon. They wished to assure the fishers that they would not deduct their salmon debts from their herring deliveries, because they were more eager to get the herring than to collect the salmon debts.

In sum, the market for roe herring illustrates the point that concentration is not always a good measure of price influence. In this case, there have been moderate levels of concentration in roe herring alongside fierce price competition. In addition, the maintenance of vertical linkages does not appear to have greatly affected price competition. Even indebtedness in salmon did not affect the competition processors exercised for a fisher's herring. If anything, the competitive market for roe herring increased the level of competition in the raw salmon market during the late 1970s.

THE RAW HALIBUT MARKET[20]

Halibut is unlike both salmon and roe herring in that it can be fished for a much longer season. The original year-round season has been shortened to three months or less for regulatory purposes, but it is still unneccessary to await the arrival or the spawning of the fish during open season, since fishers know where they are. Supplies have no natural fluctuation nor do they provide an incentive to harvest all the surplus in one year. With salmon, surplus not taken in one year is lost forever, because spawners ascending the river do not return to spawn again. In addition, when too many salmon return to a river, they may "overspawn": i.e., spawn on top of one another, destroying each other's eggs. Likewise, roe herring is lost for its commercial roe value once it has spawned.

In contrast, halibut is a long-lived demersal fish that may spend some eighty years at sea. Those not caught in one year may still be harvested in another. "The biology of the Pacific halibut dictates a fishery which exploits simultaneously a large number of year-classes [fish of many ages]. Consequently, short-term fluctuations in catch per unit effort and total landings are much less pronounced than in fisheries such as salmon and herring, where very wide excursions in catches are the rule rather than the exception" (University of British Columbia, *Economic Council of Canada*, 1980:8). Since halibut migration patterns are more limited, and fishers know where to catch them on their summer feeding banks, a halibut fisher has greater certainty of catch than a salmon fisher. Although halibut fishers have the same tendency to overcapitalize, they have fewer problems of indebtedness, and less need to be "carried" in poor years. This fishery is not characterized by the poor years and bonanza years of salmon. The greater risks and unpredictability in the salmon fishery create incentives for the contract system in both fishers and processors, but these conditions are absent in the halibut fishery.

More importantly, halibut does not have a sufficiently high oil content to be canned,[21] and thus offers no ready advantage to the large canners. In earlier years most halibut was sold fresh because its harvest was spread out over a long enough time period to be absorbed by Canadian and U.S. fresh markets. This was important to the early formation of an independent fleet. In 1930 only about 40 per cent of the product was frozen; this had increased to about 75 per cent by 1960, because of the shortening of the season (Crutchfield and Zellner, 1962).

The lower oil content of halibut also lessens its perishability, permitting some twenty days holding time on ice in vessels before halibut must be delivered. Fishers thus have more time to bargain for the best price in different ports. In the early days of the industry, many of the processed markets for halibut were local, and, because little processing was involved, the relationship between landed and final price (on processed markets) was more obvious than for canned salmon. The international character of the fishery from 1923 on (when the forerunner of the International Pacific Halibut Commission was instituted) meant that American and Canadian boats fished local waters and sold in each other's ports, contributing to the competitive price structure of raw halibut markets. The Seattle halibut exchange, in which an open auction system set the price, became competitive early in the history of the fishery, and Seattle fish buyers were instrumental in setting up independent halibut fishers. Halibut exchanges also operated in Vancouver and Prince Rupert. The purchases of the largest halibut buyer in Seattle declined from 26 per cent of landings in that port in 1931 to 15 per cent in 1955. The largest purchaser's share in Prince Rupert declined from 41 per cent in 1939 to 32 per cent in 1955 (Crutchfield and Zellner, 1962). Although there were only seven halibut buyers in Prince Rupert in 1955, and although the four largest buyers accounted for 90 per cent of the purchases, buying was highly competitive at that time. Canadian Fish, formerly dominant in the market, lost ground to PRFCA partly because of events during World War II.

During the war, halibut livers were sought for their high vitamin content, but the companies shared little of the profits from the livers with the fishers, who were largely unaware of their importance. The PRFCA was the exception, and most of the Canadian halibut fleet discovered this and joined this cooperative during the war.

The PRFCA has maintained its position in the 1970s and 1980s as one of the top two producers of halibut; it was rivalled by B.C. Packers in the late 1970s (Schwindt, 1982), and also by two other companies in the 1980s, who were not major canners. In the late 1970s and early 1980s, the top four producers accounted for 61 to 74 per cent of the value of halibut production. The high degree of concentration among these companies does not indicate price collusion, however. Except for the PRFCA, prices were negotiated either on the halibut exchange by open auction or individually between processor and fisher. Neither unions nor processors' associations have any impact on landed price. Even though concentrated, the market for raw halibut is clearly not collusive and is highly competitive.

CONCLUSION

In examining the markets for raw salmon, roe herring, and halibut it has been argued that only the first permits price collusion. All three markets show moderate to high levels of buyer concentration, but concentration is not necessarily related to noncompetitive pricing. All three markets also demonstrate high levels of price fluctuations because of highly variable world supply and demand characteristics. The characteristics that appears to distinguish the raw salmon market from the other two markets is the presence of large-scale canning technology, combined with the risks and uncertainty to both fishers and processors in acquiring enough or too much salmon on the high seas. The historical dependence of salmon fishers on processors for servicing, in-season financing, and purchase of the total catch continues even today, in contrast to halibut fishers, who became largely independent after World War I. Roe herring fishers require less processor assistance, because their short fishery requires little capital investment in addition to that already made for salmon fishing, and the season is too short to necessitate repairs in-season. Short-term debts incurred for start-up tend to be paid off immediately. Because the fishery developed in a condition of scarcity and supplies have been reduced by overexploitation, there has been no need for fishers to fear processor refusal of oversupply.

In the salmon fishery, however, vertical linkages between large-scale processors and fishers have persisted and constitute one of the most important aspects of the way both fishers and processors have dealt with the supply characteristics of salmon. The raw salmon market therefore continues its historical oligopsonistic pricing. There is no reason to suppose this will change, as long as the fishery remains an interception high-seas fishery in which fishers have to minimize risks and processors can find ways to restrict price competition among themselves and to reward some fishers at the expense of others.

NOTES

1. This analysis is based on interviews with some forty processing company personnel, some sixty fishers, about ten government personnel, numerous observers of the industry, and production records of processing firms made available to us by the Department of Fisheries and Oceans and the Marine Resources Branch of British Columbia.
2. All B.C. roe herring production fell to 36 to 40 per cent of Japanese imports in the 1980s because of renewed Chinese supply.
3. In an incisive analysis of early mergers among B.C. canners, Martin (1982) demonstrates that the drive toward concentration among canners was not motivated by a desire to achieve economies of scale in processing or to control processed markets, but to control raw fish prices. If the same conditions prevail today, one may surmise that oligopsonistic control is more important and attainable for processors than oligopolistic power. Bain (1956) demonstrated that the average size of the four largest firms in nineteen out of twenty industries studied exceeded technologically optimal plant size. In a similar vein Schwindt (1982) concludes that concentration levels in B.C. salmon canning far exceed those necessary to realize economies of scale on the plant or the enterprise level.
4. Crutchfield and Pontecorvo believe processor mergers in Alaska and Puget Sound in the 1950s, which increased the relative position of the four largest firms, "are clearly linked to the general decline in output

and . . . reduction of risk through diversification of sources of supply, rather than to any drive to control prices . . ." (p. 128). Similarly, exceptionally high processor concentration in Bristol Bay, Alaska, is explained by "natural" barriers to entry in the higher isolation costs and fewer sites (p. 108).

5. Although economists usually consider barriers to entry and vertical linkages with fishers as two other aspects of market structure (along with the number and relative size of firms), I have found it more useful in this discussion to consider them behaviours. Such a label facilitates making the distinction between concentration and collusive price behaviour or price influence, an important point in this chapter. In other ways, however, the industrial organization framework in applied micro-economics has been useful to this analysis (see Shaffer, 1979 and Schwindt, 1982).

6. Figures for canners in published sources are higher than this, because they include as "canners" the freezer firms who have some "custom canning" done for them by the large canners. I feel this to be an inappropriate designation, because the firm that has custom canning done may make no or insignificant profit on the custom work and certainly is not in the arena of competition with the other canners. Sport and specialty canners are also included inappropriately. Throughout this paper I use "value of production" figures instead of amount or value of raw fish purchased, because these were the only figures available on an individual firm level. Raw fish purchase is closely correlated with value of processed production, however.

7. In November 1984 the Fisheries Association amalgamated with three members of the B.C. Seafood Exporter's Association to become the Fisheries Council of B.C., concerned mainly with marketing and promotion. The Fish Processors Bargaining Association of B.C., made up of the four major canners and one minor firm, then formed to be the bargaining agent.

8. Thus Schwindt (1982) concludes that even a dominant-firm oligopsony does not imply price collusion on the B.C. raw salmon market. Dominant-firm oligopsony "runs the gamut from collusively arranged and rigorously enforced adherence to the leader's announced price, to the rather loose 'barometric' leadership model wherein the leader recognizes price rivalry in the market, initiates changes, and is generally followed, all without collusion or policing. . . . The sheer cost of policing a collusive price-fixing agreement would be prohibitive and this argues against the existence of such an agreement in the fishery today."

9. With the exception of the processing and marketing cooperative, which is the most direct form of acquiring supply, because delivery to the cooperative is a condition of membership.

10. In addition to the negotiated minimum price per pound offered for fish, processors offer seiners a few cents per pound for transporting or "packing" the fish directly to the plant. Because packing is not publicly negotiated, its exact amount has become one of the incentives that can be differentially used by various companies.

11. Often the skipper of a processor-owned vessel owns the net and receives a bonus on its use similar to the boat-and-net bonus received as part of the share system by a boat-owning skipper. This bonus has nothing to do with the "charter" discussed below.

12. Miller (1978) notes that some seine skippers in Sointula, originally founded as a Finnish utopian community, shared bonuses or charters with their crews. Conversely, Rohner (1967) notes the complaints of seine crews in a nearby Kwakiutl community on Gilford Island. "The bookkeeping ledgers are often not open to inspection by the crew members; they are maintained by the skipper. In this way he is able to include items which might be objected to by the members of the crew, and the crew seldom know how much is legitimately deducted from the gross earnings before their shares are allotted. Crew members often grumble because the skipper takes the boat to areas where they do not think it should be taken, using fuel, the cost of which must be deducted from all shares. Some skippers are accused of cheating their crews, particularly in managing the books."

13. Charters for owners of small boats were an extension of this system, but the charters they received were much smaller, about 10 to 20 per cent for gillnetters, and 4 to 5 per cent for trollers.

14. When asked why they canned instead of froze so much salmon, B.C. Packers replied that in many years, particularly high pink supply years, there was no other choice because of the speed requirement. Also, there would not be a market for frozen pink salmon in such quantity, especially since this fish does not freeze well unless it is processed very soon after catching.

15. In the late 1970s, the first stage of Japanese withdrawal from the two-hundred-mile limit fishing zones of other nations created a market for salmon in Japan that its budding aquaculture industry could not yet fill. B.C. was at first a main supplier to that market until Alaska salmon supplies, temporarily lower than B.C.'s in the mid-1970s, reached unprecedented heights in 1979–80. Because of different regulations and harvesting conditions, Alaska salmon could be produced more cheaply than B.C.'s (see also note 16).

16. In the late 1970s, Alaska's salmon catch began to increase dramatically; in 1983 Alaska salmon fishers landed 127 million fish, breaking a record that had stood for almost half a century. The 1983 bonanza also marked the first time in the history of the Alaska salmon fishery that more than one million fish have been landed in four consecutive

years. These record years followed exceptionally low catches in the early and mid-1970s, when Alaska's production fell below B.C.'s, and when B.C. was becoming established as a high-quality salmon producer on the Japanese market (Fisheries Association, 1983; *Pacific Fisheries Review, The Fishermen's News*, 1984).

17. In May 1982 the Alaska legislature passed Senate Bill 872, creating a voluntary "premium quality" program, and in June 1982 the Alaska Seafood Marketing Institute circulated widely a booklet entitled "Recommended Salmon Quality Guidelines for Fishing, Tendering, and Processing Operations" with the goal of developing "premium quality" product specifications for all forms.

18. B.C. Packers paid for the repairs of its fully owned gillnet fleet. By the late 1970s, the cost of repairs exceeded the returns from this fleet, partially because many of the vessels were old. Apparently this fleet was purchased by the corporation as much for the licences as for the boats, since fifty of the boats were already unusable the year after the purchase. One of the purchasers hoped to jettison the boats, but transfer some commercial fishing licences upriver to an inland fishery (N. Sterritt, speech to the Gitksan-Carrier Tribal Convention, 1982).

19. For an indication of the concern of Canadian processors over Japanese investment and the extent of Japanese investment, see Proverbs, 1978, 1980, and 1982.

20. Source material for this section includes University of British Columbia, Department of Economics, 1980; Crutchfield and Zellner, 1962; Bell, 1981; Schwindt, 1982; Canada Sessional Papers; B.C. Production Statistics, Marine Resources Branch; interviews by and conversations with Brian Hayward.

21. It is actually possible to can halibut, but the canned product has not proved an attractive or popular item, probably because of its low oil content.

5

The Production and Distribution of B.C. Salmon in the World Context

STEPHEN GARROD

British Columbia's fisheries have consistently generated about 30 per cent of the value of Canada's fish exports. The major export species has always been salmon, especially sockeye, which is high in oil and protein, easy to can, and excellent in flavour. In the 1970s herring roe became a second major export (Department of Fisheries and Oceans, Annual Statistical Review, vols. 9–14).

Though the commercial salmon fishery is less export driven than many other Canadian resource industries, world demand, prices, and competition have been significant influences on production. The herring roe fishery was entirely export oriented, and a single market, Japan, determined production and prices.

WORLD SALMON PRODUCTION AND MARKETS BEFORE 1950

Pacific salmon belong to the genus *Oncorhynchus* and range within the temperate regions that rim the North Pacific on the Asian and North American coasts. They are harvested and processed in the U.S., Canada, Japan, and the U.S.S.R. Sockeye, pink, and chum salmon are harvested in greater quantities than chinook and coho. While pinks are widely distributed, lower-value chum are more prevalent in Asian waters. Sockeye, the highest-value, high-volume fish, spawn largely in Alaska and B.C. (The only species similar to Pacific salmon is the Atlantic salmon, originating along the European and North American coasts of the North Atlantic. Though considered a premium fish, it has had very low levels of production over the past few decades.)

From the inception of the large-scale, commercial North Pacific salmon fishery a century ago, most salmon were processed in canned form, though the Japanese processed them in salted form for domestic consumption. Salmon canning expanded to Asia by 1910, and by the late 1930s, salmon stocks throughout the North Pacific were fully exploited. The United States, based on the enormously productive biomass of the Gulf of Alaska, was the dominant producer with 70 per cent or more of total production since 1900. It was the largest single market for canned salmon and a major exporter to the developed industrial countries, including Britain, Belgium, and Australia.

Japan, by exploiting its own salmon resources, through processing concessions in Siberia, and by developing a high-seas fishery based on Alaska-origin stocks, became a major canned salmon producer in the 1920s and the leading canned salmon exporter in the 1930s (*Pacific Fisherman Yearbook*: various years, 1915–1953).

SALMON PRODUCTION AND MARKETS, 1950–1970

In the postwar period, the pattern of salmon processing and exporting was largely re-established. However, the same four countries — Canada, Japan, U.S., and U.S.S.R. — were intensifying their exploitation of North Pacific salmon stocks, using new and improved technologies devised by the Japanese high seas fleet after 1955. At the same time, industrial expansion was degrading natural spawning grounds of some salmon stocks. North Pacific production, which had peaked at about 770,000 metric tonnes (mt) in the mid-1930s, was gradually declining (Table 5.1). By the early 1970s, Japan was harvesting about 34 per cent of the total catch; the U.S., 29 per cent; the U.S.S.R., 21 per cent; and Canada, 16 per cent. Total output had declined to a low of around 340,000 mt, about half the high mark of the 1930s (Fredin, 1980:59ff.).

Sockeye salmon production (Table 5.2) apparently had a more variable character.[1] Japanese high seas catches consistently declined, while U.S. and Canadian sockeye catches increased. The increase was the result, in part, of enhancement projects such as those initiated on the Fraser River by the International North Pacific Sockeye Commission. However, the Alaskan sockeye catch declined sharply in the early 1970s.

The marked decline of world salmon production from the peak years in the 1930s, combined with a slowly rising utilization of salmon in other than canned form, led to a gradual decrease in world canned salmon consumption. The established salmon utilization pattern remained fairly constant until after 1970, with the majority of the salmon harvest, particularly in Canada, the U.S., and the U.S.S.R. being canned. In Japan, salmon continued to be canned for the export market, while fresh or salted forms were reserved for domestic consumption.

By volume, on average, Canadian production for the period between the early 1950s and early 1970s was 70 per cent canned salmon and 30 per cent other (fresh, frozen, smoked, offal, roe, meal). U.S. production dropped from about 90 per cent canned at the beginning of the same period to about 70 per cent canned by the early 1970s. World production was, on

Table 5.1
Commercial North Pacific Salmon Harvest (All Species)

Period	Canada %	Japan %	U.S.A. %	U.S.S.R. %	Total
1955–59	12.6	36.4	24.9	26.1	488.8 mt[1]
1960–64	14.2	34.9	34.1	16.8	399.3
1965–69	15.0	34.1	34.3	16.6	403.5
1970	17.4	28.7	44.5	9.4	417.3
1971	14.5	32.7	34.9	17.9	434.1
1972	23.1	36.6	31.1	9.2	332.2
1973	21.2	34.2	25.4	19.1	406.6
1974	18.8	40.2	27.3	13.7	337.1
1970–74	18.8	24.1	33.0	14.1	385.5
1975	9.6	43.9	24.3	22.2	378.7
1976	14.5	32.5	35.4	17.6	399.8
1977	13.5	24.0	35.3	27.1	483.5
1978	15.5	22.7	44.5	17.3	455.2
1979	10.9	24.1	41.7	23.3	560.5
1975–79	12.8	28.6	36.9	21.7	455.5
1980	10.0	24.1	47.4	18.5	536.5
1981	12.3	24.3	47.6	15.8	641.6
1982	11.8	25.7	51.4	11.0	555.3
1983	11.3	25.2	44.7	18.9	662.7

[1]mt = metric tonnes.
Sources: See note 6 at end of chapter.

average, about 50 per cent canned by 1970 (*B.C. Salmon Industry: Survey of Economic Studies*, 1977:88). Part of the variation from year to year in proportion of salmon canned resulted from fluctuations in the supply of species that were primarily used for canning, such as sockeye and pinks.

World trade in salmon products was clearly dominated by canned salmon through the 1960s (Figure 5.1). Overall world trade totals were stable, with canned salmon exports slowly declining, and fresh and frozen exports slowly rising, up to the early 1970s.

In the immediate postwar period, U.S. consumption of canned salmon remained relatively high (U.S., Current Fishery Statistics 6129, 1973:25). As a result, U.S. canned salmon exports were at generally low levels from 1955 to 1975 (Figure 5.2). Indeed, in the early 1950s, with strong demand and low supply, the U.S. became a net importer of canned salmon, mostly Canadian pinks (U.S., Basic Economic Indicators, 1971:46). By the 1960s, U.S. consumption of canned salmon had begun to decline, and the U.S.

Table 5.2
Commercial North Pacific Sockeye Salmon Harvest (All Species)

Period	Canada %	Japan %	U.S.A. %	Total
1955–59	17.9	36.2	33.2	74.1 mt[1]
1960–64	12.4	35.5	45.7	70.7
1965–69	17.4	25.1	45.5	75.2
1970	10.7	16.9	68.2	106.2
1971	22.0	14.0	61.2	78.6
1972	21.9	14.0	61.2	43.3
1973	39.2	17.2	40.5	54.8
1974	41.4	15.6	40.8	52.4
1970–74	24.3	17.1	55.4	67.1
1975	14.8	19.9	61.4	38.6
1976	20.4	14.6	63.0	60.2
1977	25.7	6.6	67.7	67.8
1978	25.3	5.9	64.9	88.0
1979	13.8	4.8	78.7	105.2
1975–79	20.0	8.6	68.8	72.0
1980	7.4	5.5	83.3	103.6
1981	15.5	3.8	80.0	135.8
1982	22.9	3.1	71.7	131.6
1983	8.8	2.7	85.9	162.7
1984	10.2	2.7[2]	83.4	124.9[2]
1980–84	13.1	3.4	80.6	131.7

[1]mt = metric tonnes.
[2]Projected figures.
Sources: See note 6 at end of chapter.

generally ceased to import any major quantities and resumed exporting on a large scale.

Though Canada exported more canned salmon than the U.S. for most of this period, Japan was the largest exporter by far (Figure 5.2). Even in the British market, where Canadian canned salmon enjoyed a 10 per cent preferential tariff rate, Japan was the dominant supplier (Food and Agriculture Organization [FAO]: *Yearbook of Fishery Statistics; Fishery Commodities*, 1955–76). However, Canada had an established domestic market where price and profit were largely tariff protected. For the period 1955 to 1969, Canada exported over half of its total production of 432,600 mt. It is therefore unlikely that Canada could have had a greater share of the world trade in canned salmon.

Japan exported canned salmon products in the 1950s and 1960s, much as they had in the 1930s, to generate the foreign exchange needed

Figure 5.1
International Trade in Salmon

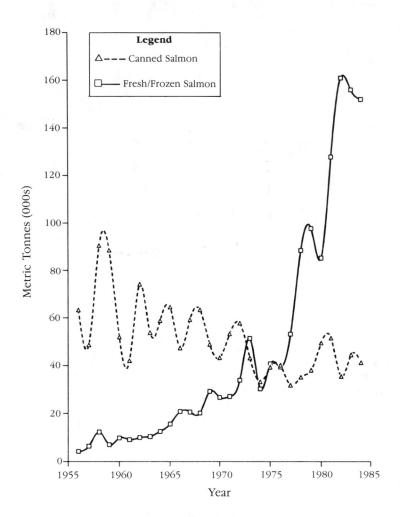

Source: See note 7 at end of chapter.

for various vital raw material imports. Until the mid-1960s, Japan ran a consistent foreign trade deficit, and its canned salmon was a proven money earner. Japan's weak international trade status at this time enabled it to compete with Canadian and U.S. salmon, since these countries had hard currencies while the Japanese yen was a floating-rate currency (Allen 1970:65).

Figure 5.2
World Canned Export Salmon Trade

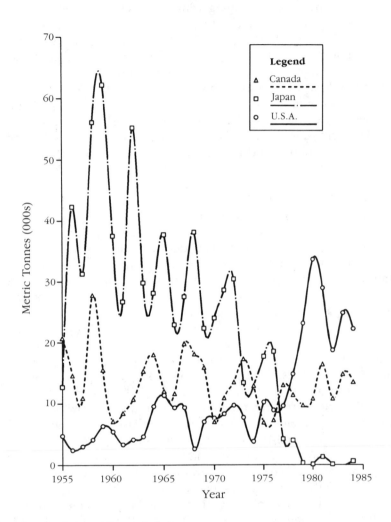

Source: See note 7 at end of chapter.

WORLD SALMON PRODUCTION AND MARKETS, 1970–1984

Though salmon production and distribution have been marked by great yearly variations in total volumes, species harvested, and locations of catch, there was an essential continuity in the world salmon industry in the period from about 1920 to 1970.[2] Total harvest was gradually decreasing because of a variety of factors including overexploitation of stocks. Japan was the pre-eminent exporter of canned salmon, and, through its high seas fleet, had become the leading salmon harvester. Canned salmon was the dominant product form and the leading export product, but as total production dropped and as freezing technology was more widely applied, canned salmon took a declining share of total production and trade.

This pattern of continuity has been sharply broken since 1970 through rapid and unpredictable changes in supplies, production locations, product forms, consumer tastes, and trade flows. These changes have had considerable consequence for Canadian fishers, fish-plant workers, and processing companies, and it is likely that the transformation of established patterns of production and marketing will continue for at least another fifteen years. These changes have occurred as a result of political decisions, most notably the two-hundred-mile limit imposed by most countries in 1977, which in the North Pacific excluded the Japanese high seas fishery from access to Alaska stocks and led to more restricted Japanese access to Soviet stocks. As well, ecological-biological developments, such as enhancement programs to restore degraded fisheries in all countries of the North Pacific, increased consciousness of the negative consequences of certain types of industrial development on fish habitat; the potential of fish farming/aquaculture and climatic factors have altered established production patterns.

World production of Pacific salmon, which hit its lowest point in the early 1970s, has almost doubled to over 660,000 mt since 1975 (Table 5.1). Most noticeably, U.S. production, which had fallen to about 100,000 mt in the early 1970s, has tripled to over 300,000 mt. As illustrated by Table 5.2, the growth of sockeye production is even more dramatic, with U.S. production increasing more than six-fold from its low point in 1975, and doubling or tripling the averages of the last thirty years. Canadian production has been relatively stable, though sockeye harvests have grown. Because of the two-hundred-mile limit, Japan lost access to Alaska-spawned sockeye and pinks and had their allowable catch of Soviet pinks cut in half after 1977. But their total harvest has remained stable because of the rapid growth of enhanced chum production. A hatchery program started in the early 1960s was intensified in the 1970s with the prospect of excluding Japanese fishers from traditional grounds. It has tripled chum production to 150,000 mt per year (Organization of Economic Cooperation and Development [OECD], Annual Reports, 1975–1984; INPFC, Annual Reports, 1976, 1978).

A major new addition to the world salmon supply has come from pen-raised, or farmed, Atlantic salmon (Figure 5.3). While production of wild-stock Atlantic salmon has remained at about 10,000 mt annually, farmed production has grown from less than 1,000 mt per year prior to 1975 to more than 30,000 mt in 1985.

Figure 5.3
World Atlantic Salmon Production

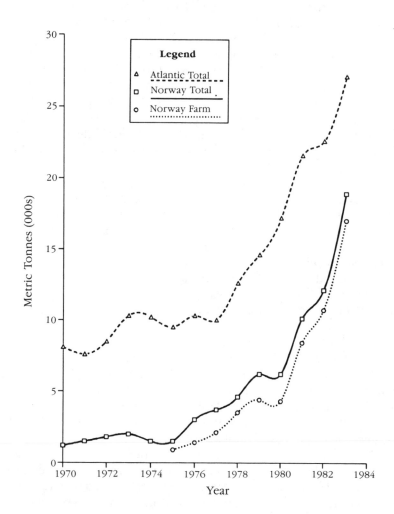

Source: See note 8 at end of chapter.

Norway is by far the largest producer, with predicted farmed produc-
tion of 30,000 mt in 1985, and a projected harvest of 90,000 mt by 1990
(*Fish Farming International*, March 1985; O. Lundby, interview data
1985). Farmed Atlantic salmon is also produced in Scotland, the Faroe
Islands, Greenland, the United States, Chile, Atlantic and Pacific Canada.
Efforts are underway in British Columbia, Washington, and Japan to farm

Pacific salmon species. British Columbia farmed salmon production has been projected at 15,000 mt by 1995 (*The Globe and Mail*, December 16, 1985:B4), and the U.S. Department of Commerce has projected world farmed salmon production of 100,000 mt by 1990 (*Fish Farming International*, 1985:10).

These changes in the supply sources and production volume have greatly changed world salmon trade. As illustrated in Figure 5.3, world salmon trade has tripled since the early 1970s to almost 200,000 mt per year, with virtually all the increase coming from the 500 per cent increment in trade of fresh, chilled, or frozen salmon. World trade in canned salmon declined slowly during the 1960s and 1970s as supply declined, but with the enlarged supply available in the 1980s, it has increased by 20 per cent over the previous five-year period. As shown in Figure 5.2, Japan went from being the world's leading canned salmon exporter to having almost no exports in the late 1970s. Indeed, Japan became the leading importer of frozen salmon in 1977, and its imports have continued to rise. Japan has taken about 100,000 mt of salmon a year since 1982 — half of all salmon products traded in the world market (UN, FAO, *Yearbook of Fishery Statistics; Fishery Commodities*).

The United States is the world's leading supplier of frozen salmon, exporting over 100,000 mt per year. Canadian exports have been stable at around 20,000 mt per year. Norway has dramatically increased its exports as it has only a limited domestic market for its greatly increased farmed-salmon production. From less than 1,000 mt per year, Norway now exports almost 20,000 mt and will soon surpass Canadian exports.

The Norwegian product is principally a fresh, chilled fish airshipped around the world for the "quality" restaurant market.[3] While Canada was a much larger exporter of canned salmon than the United States up to the late 1970s, the U.S. now has about two-thirds of the world canned-salmon trade because of increased Alaska production (Figure 5.2).

These major changes in consumption patterns have had considerable impact on the allocation of raw salmon to either canned or frozen product forms in Canada and the United States. In general, there has been a consistent trend away from canned-salmon production toward fresh and frozen production. In 1971, more than 80 per cent of production in Alaska was canned and less than 20 per cent frozen; by the mid-1980s, only 35 per cent was canned and 55 to 65 per cent was frozen. The change in sockeye production is even more dramatic. Traditionally the premier canning fish, over 90 per cent of Alaskan production was canned up to 1971. By the mid-1980s, only one-third of the sockeye was canned, the remaining two-thirds, frozen (*Seafood Business Report*, March/April 1985:70).

Because coho and chinook comprised a higher percentage of the catch in B.C. (traditionally consumed in fresh or frozen product form), fresh and frozen salmon in the early 1970s comprised 25 to 40 per cent of the wholesale value of the fishery, and canned salmon 50 to 70 per cent of the value. Other salmon products such as roe and meal comprised 5 to 10 per cent of the value. In four of the six years prior to 1984, the wholesale value of fresh and frozen salmon was greater than the value of canned salmon. It is in years of high volume, particularly of the traditional canning species, pinks and sockeyes, that canned-salmon production increases. This is due both to capacity and market constraints (B.C., Fish Production Statistics, 1984, Table 7:12).

International trade in canned and fresh/frozen salmon products has been affected by the general trend since 1947 to lower tariff duties under the General Agreement on Tariffs and Trade (GATT). For the first half of this century, Canada and the U.S. had matching tariffs averaging 25 per cent on imported canned salmon. Essentially, the tariff protected U.S. producers from Canadian competition, and Canadian producers appear to have wanted freer access to the large U.S. market. Because of declining production and strong demand, the U.S. tariff was cut to 15 per cent after 1951, and Canadian producers were able to export some pinks to U.S. markets (*Pacific Fisherman Yearbook*, 1952–1956). During GATT discussions in the 1960s, both Canada and the U.S. agreed to lower their tariffs. The tariffs at 15 per cent in 1967 declined to 7.5 percent in 1972. The Tokyo round of GATT talks began another set of reductions in tariffs, at a rate of 0.5 per cent a year beginning in 1980. In 1988, the tariff is scheduled to drop to 3 per cent in both Canada and the U.S. Canada and the U.S. have little or no tariff on fresh or frozen salmon. The United Kingdom used to give a 10 per cent preference to Canadian canned salmon, but since 1973 it has applied EEC tariffs of 5.9 per cent on canned and 2.5 per cent on fresh or frozen salmon. The Japanese tariff of 5 per cent on frozen salmon and 12 per cent on canned salmon has been stable since at least 1977. Both Japan and the EEC have tended to set numerous tariff and nontariff barriers in order to protect their agriculture and fisheries (E. Homma, interview data 1985; United Kingdom, 1985; Allen 1970:66, 97).

Relative currency values also have considerable effect on international trade. The increase in the value of the Japanese yen in the early 1970s tended to decrease their exports and increase their imports of salmon products (OECD, *Annual Reports*, 1973:11). Since 1980, most major currencies have declined relative to the Canadian and U.S. dollars. As a result, the prices of Canadian and American exports of salmon products have been inflated in local currencies. This may have limited their exports and helped Norwegian producers, whose currency has been devalued throughout this period. Canadian exporters, though possibly harmed by a strong Canadian currency relative to European currencies, have been helped by the value decline of the Canadian, compared with the American, dollar. Between 1978 and 1984, the relative decline of the Canadian dollar was 12 per cent. This may have been an important factor in allowing Canadian producers to effectively compete with U.S. products during this period (*Seafood Business Report*, March/April 1985:45).

THE BRITISH COLUMBIA FISHERY

A pessimistic view seemed to dominate the B.C. fishery in the early 1970s. Many felt that the resource was so threatened by the combined depredations of forestry, hydro dams, general industrial pollution, and overexploitation of the resource by American, Japanese, and Canadian fishers that the thriving fishing industry, so important along the B.C. coast, would soon be a memory. There was ample evidence to support this view: fishing was not highly regarded when compared to other industrial sectors by provincial policy makers. This was partly because it was under federal jurisdiction. High seas interceptions of Canadian-sourced salmon were not being controlled. The fishery in Washington and Oregon had been seriously damaged by industrialization and neglect, and Alaskan production appeared to be in a serious crisis. However, because of internal and international regulation, expanded rehabilitation and enhancement of salmon-spawning areas, and the amazing biological resilience of salmon, British Columbia and world salmon production has increased over the past decade, and continuing growth may occur. From Tables 5.1 and 5.2, it is clear British Columbia salmon production is showing a slight tendency to increase.[4] Though 1984 certainly was a low year, 1985 has been one of the largest volume years on record, with production in excess of 90,000 mt and high catches are predicted for 1986 and 1987. Sockeye production (Table 5.2) has shown a steady increase and is expected to increase substantially as the full effects of the Salmonid Enhancement Program are realized. Phase 1 of this program, completed in the early 1980s, is projected to increase the B.C. catch by over 20,000 mt by the early 1990s, and Phase 2, currently on hold, could add another 15,000 mt by 2000 (DFO-SEP, 1983, Appendix A). In the twenty-first century, total B.C. salmon production of existing and enhanced stocks could average in excess of 100,000 mt annually, 50 per cent higher than recent average production. This increased Canadian catch would be only part of an increased catch worldwide, as it appears likely that record catch levels will be reached in the 1990s when Japanese, Soviet, and American rehabilitation and enhancement projects increase production. It is conceivable, without using wildly optimistic projections, that total world Atlantic and Pacific salmon production could exceed 1,000,000 mt by the late 1990s.

While the increased British Columbia production figures auger well for the continued vitality of the commercial fishing industry, B.C.'s nearly 25 per cent share of world production in the early 1970s has been cut to between 10 to 12 per cent in the 1980s. In addition, the more than 20 per cent share in 1972 and 1973 was of a vastly smaller total world production, thereby giving British Columbia at the time a much more important role in supplying world markets. Indeed, as can be seen from Figure 5.2, B.C. was the world's leading canned salmon exporter in 1973 and 1974. These increased exports coincided with low supply of canned salmon on the world market, the upward movement of Japanese and European currencies, and increased Japanese consumer demand for salmon products. Export prices increased by 50 to 70 per cent (*Seafood Business Report*, March/ April 1985:70). The combined result of these sharply higher prices and high production levels was that the wholesale value of B.C. salmon products rose to the highest level on record (Figure 5.4).

Figure 5.4
Wholesale Values, B.C. Fisheries

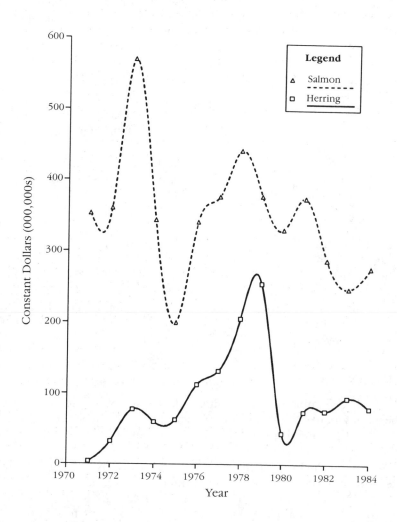

Source: See note 9 at end of chapter.

During the 1970s, B.C. also added herring roe, a major new fishery product, destined for export to Japan. Until the 1960s, herring had been a large-volume, low-value fishery principally processed for meal and oil. In the late 1960s, this fishery was restricted because of declining stocks. By the early 1970s, stocks had recovered and began to be exploited for their roe. The collapse of Chinese herring roe production allowed British Columbia to become the principal supplier to Japan. While salmon had traditionally accounted for more than 80 per cent of the wholesale value of the B.C. fishery, salmon produced just over 50 per cent of the value and herring roe almost 35 per cent by 1979 (Figure 5.4). This rapid increase in the total and relative value of the herring roe fishery took place while the actual volume of production was declining from the 1976 peak of about 80,000 mt to less than 40,000 mt in 1979.

The increased value received for herring roe in 1977, 1978, and 1979, combined with increased demand and prices for B.C. salmon, brought the wholesale value received by the B.C. fishing industry to record levels in 1978 and 1979. As discussed elsewhere in the book, different B.C. groups benefited unequally from this boom, and, as the Japanese became owners and financiers of processing companies, the issue of "foreign," i.e. Japanese, ownership was hotly contested.

The boom of the late 1970s just as suddenly became a crisis in 1980. Japanese consumers resisted the huge price increases on roe, and the market collapsed. In the face of proposed huge price cuts, union fishermen refused to fish for herring in 1980. As a result, production was curtailed, and with other fisheries and processors taking the price cut, the wholesale value of the herring fishery was cut to less than 20 per cent of its previous high. In 1980, production declined and price cuts were also imposed on the salmon fishery. The total wholesale value of the B.C. fishery declined by almost 40 per cent and, until 1985, remained less than two-thirds of the previous high. Herring roe has, on average, contributed less than 20 per cent of the total wholesale value, and, with the herring runs forecast for 1986 at the lowest since 1971, its value is likely to decline further.

SUMMARY

The British Columbia salmon fishing industry is not about to disappear. Rather, it has increased production, particularly of the higher value sockeye species, and production is predicted to continue to increase for at least the next decade.

World production has increased dramatically over the past fifteen years, and production is likely to continue to increase. The increase of world production has been proportionately greater than the increase of B.C. production, thereby making B.C. a less significant force on world markets and less of a price determiner.

Even if B.C. production of farmed salmon is able to follow the same growth curve as Norwegian farmed-salmon production, wild B.C. stock production will continue to be more important than farmed production in the foreseeable future.

It would appear that though the production choice between fresh/frozen and canned salmon may vary from year to year (depending on species and total volume harvested), and the fresh and frozen markets will continue to grow, the world canned-salmon market, whose bottom dropped in the late 1970s, will grow again as a result of increased supply and more competitive prices.[5]

NOTES

1. I say apparently because some analysts of the North Pacific salmon fisheries agree with the U.S. contention, put forward at every annual meeting of the International North Pacific Fisheries Commission, that Japanese high seas interceptions of Alaskan sockeye were depleting the runs. These analysts, who insisted on confidentiality, argue that the Japanese consistently underreported their high seas sockeye catches. These analysts suggest that it is no coincidence that Alaskan sockeye runs increased at the same time Japan fishing was restricted by the 200-mile limit, and further, they point to the hitherto unknown, and massive, Japanese sockeye market that emerged in the late 1970s. Other analysts, like C.R. Forrester, Executive Director of the INPFC Secretariat, agree with the long-standing Japanese contention that a series of extremely cold winters were responsible for the decline in Alaska salmon production, and that it was climatic change and improved management that restored the runs (INPFC, Annual Reports, 1959–1976).

2. In the period from 1940 to 1955, Japan had almost no canned salmon exports to its traditional European markets. From 1940 to 1945, the interimperialist war in the Pacific obviously disrupted normal trade patterns, though it is possible that Japanese canned salmon was exported to Axis allies. After 1945, the Japanese were excluded from Sakhalin Island and from their fishing concessions along the Russian coast. It was not until the development of the mothership fishery in the early 1950s that Japanese salmon production increased to prewar levels and large-scale exports resumed (INPFC, Bulletin no. 39, 1979:4–5; FAO, *Yearbook of Fishery Statistics and Fishery Commodities*, 1953–1960).

3. Because of concerns about the possibility of disease in Norwegian farmed salmon, they have been refused an import licence for Canada (D. Bevan, interview data, 1985). However, MOWI, the largest Norwegian producer, now has a licence to import into Canada (O. Lundby, interview data, 1985).

4. Note that B.C. production statistics must take into account social as well as biological factors. The production figures for 1975 are very low because of a major strike that year and the same applies for 1969.

5. The year 1982 was unusual in that canned salmon markets in Europe suffered because of a widespread bottulism scare based on defective Alaskan production. The market appears to have recovered from this setback.

6. For the years 1955 to 1976, all the data in Tables 5.1 and 5.2 are from International North Pacific Fisheries Commission *Bulletin*, volume 39, 1979. Additional detail for 1977 to 1984 is from the International North Pacific Fisheries Commission (INPFC) *Statistical Yearbook* or from UN, FAO *Yearbook of Fishery Statistics* and *Catches and Landings*.

7. Data in Figures 5.1, 5.2: FAO, *Yearbook of Fishery Statistics: International Trade* (1955, 1960–61); *Fishery Commodities* (1964–76); DFO, *Annual Statistical Review of Canadian Fisheries*, vols. 9–14; DFO, *Fish Product Exports of B.C.*; Japan, annual, *Abstract of Statistics on Agriculture, Forestry and Fisheries*; Japan, *Japan's Exports and Imports, Commodity by Country*; U.S., *Current Fishery Statistics* 1984. For early years, data on fresh, frozen, and chilled salmon may underestimate quantity since only frozen salmon appears to have been reported.

8. Data for Figure 5.3 do not include trout or salmonids. The major source is FAO, *Yearbook of Fishery Statistics* and *Catches and Landings*. Supplementary data are from *Fish Farming International*.

9. The wholesale values in Figure 5.4 are in constant 1981 Canadian dollars. All product types are included in salmon; herring is principally value for herring roe. The data sources are DFO, *Annual Statistical Review*, vols. 9–14 (1955–1981) and B.C., *Fisheries Production Statistics*.

6

State, Capital, and the B.C. Salmon-Fishing Industry

JOHN McMULLAN

This chapter analyses the structural contexts of the British Columbia fisheries at three policy periods: (1) an early regulatory phase in which the state created property rights, shaped oligopolistic relations in the industry, and established the social and technical relations of production, especially the position of petty capital producers; (2) a modernization stage in which the state, through wartime advisory boards, became directly and extensively involved in organizing the industry through capital assistance programs to processing firms and fishers, product marketing, and socializing the costs of fish production; and (3) a period of renewed licensing, regulation, and control since 1968, whereby state policy has been profoundly ambiguous and contradictory, resulting in crisis and conflict for processors and petty producers.

STATE, CAPITAL, AND SIMPLE COMMODITY PRODUCTION, 1880–1925

The pattern of the social relations of production in the B.C. salmon industry was formed under the influence of the processors in canning their salmon.[1] As noted in Chapter 3, canners faced problems of competition, overcapacity, and labour supply. British Columbia canners were in direct competition with American producers for markets and for the resource itself. Prices tended to be set on the world market by American production costs and by supply and demand estimates calculated in the U.S. Canadian canners relied on export markets, especially in the U.K. Their production costs were higher than American packers who could use low cost, capital-intensive fish-catching devices such as traps and drag seines. Processors along the Fraser River were in direct competition with U.S. firms for the harvesting of the resource. They faced a geographical limitation. Many of "their" fish were intercepted before they reached Canadian waters. Traps and purse seines captured large numbers of salmon, which frequently and seriously depleted fish stocks. Thus catches were irregular, incomes were lower and inconstant, and prices were unstable (Hayward, 1981a: Chapter 3).

Overcapacity was a second problem. As noted in Chapter 3, the fish supply was uncontrollable and highly variable, resulting in great seasonal variations in cannery operations. The capital accumulation process in salmon canning was also driven by a dynamic for renewed investments. Strong competition for supply meant the building of new, mechanized canneries. Gang knives, modern retorts, butchering machines, and double seamer cans were designed to overcome production bottlenecks and reduce or displace unwanted labour costs. Investments in sheds, canning lines, tenders, vessels, scows, nets, and buildings sufficient for runs of ample seasons (including days or weeks within seasons), meant idle capital in low seasons. By 1916, overcapacity was the major problem for Fraser River canners. According to Conley, "All the cannery machinery on the Fraser, running to capacity, could have packed the 48-day season's catch in 20 hours" (1985:204). Maintaining profits meant that new strategies were required to limit investment.

Control over supply was also crucial. As noted, fishing and canning operations required a large seasonal labour force. The introduction of more fishers led to overfishing, smaller catches per boat, pressure on the resource capacity, and paradoxically, frequent difficulties in recruiting fishers when prices were low (Ralston, 1968:41–44). At the same time, increases in the number of canneries resulted in increases in vessel numbers, operating expenses and processing capacity, and a proportionate decrease in the size of the pack per cannery. According to H. O. Bell-Irving, the established firms were being undercut:

> By 1910 . . . their capacity [was] in many instances already far larger than [was] necessary; year by year they have gradually seen their business cut down by new concerns. (Conley, 1985:20)

Indeed, packs were often reduced by half. Processing enterprises collectively coped with high production costs, competition, overcapacity, and labour supply problems by controlling the spread of canneries and the size and character of fish fleets.

State Strategies and Petty Capital Formation

The state was central to the collective strategies of capital. When overcapacity and production costs were high, canners struggled to control the pricing structure by combining to set fixed prices or by limiting the numbers of canneries and fishers by state controls and licences (see Chapter 3). Combinations, mergers, and price controls were not always successful. Some wayward canners did not cooperate, and there was stiff competition from American buyers who obtained supplies of fish in Canada. Informal arrangements and obligations in the form of loans and services were used to outcompete Canadian firms and ensure harvesting and deliveries (Reid,

1981:306–318). The state intervened to regulate these conflicts. Various prohibitions limited American competition. Canners agitated for export controls on salmon for canning and for other processing. They also pressured the state to create a system of "attached licences," especially in the north, in order to control fish prices (Gladstone, 1959; Hayward, 1981a).

A second area of state intervention was the promotion of limited entry for new canneries and the regulation of boat numbers. While conservation was the justification for these restrictions, they also bolstered the oligopsonistic position of established canneries by limiting competition and tempering supply. The most powerful state policy was the establishment of the boat-rating system in Northern District No. 2. The licensing of fishers and canneries was first recommended in 1882 and then implemented in 1888. It covered southern and northern regions, and by the time it ended in 1892, some 1,275 licences were issued as follows: Fraser River Canners 417; Fraser River Fishermen 270; Fraser River Fresh Fish Dealers 25; Fraser River Farmers 8; Northern Coast Canners 422; Northern Fishermen 132; seine licences 9 and 1 exclusive privilege to fish commercially in Nimkish River (Canada, Sessional Papers, 1892). This system was followed by an open policy on the Fraser River and a volunteer boat-rating system in 1902, 1905, and 1908 in northern waters. The volunteer system limited the number of boats a cannery could fish, and distributed Indian, Japanese, and "white" labour to each cannery (Hayward, 1981a: Chapter 3).

These voluntary arrangements were unstable. Bonuses, price increases, and new entrants were not controlled effectively. Older firms pressed further for a boat-rating structure as part of a *general* fishery policy, accompanied by an "attach licence" that obliged fishers to deliver to particular canneries rather than operate as independent fishers. This allowed canneries to exert temporary control over capacity and labour supply in the north. However, the system was short-lived. White settlers arriving to clear and cultivate land soon demanded work in logging and fishing, including the right to gillnet salmon. By 1913, independent licences were being allocated. This resulted in two distinct types of licences and, in turn, led to further changes in state policy. The regulation allotting boats to individual canneries was replaced with one that fixed the number of boats by area. Numbers of boats were determined by existing distributions to canneries in specific areas (Canada, Sessional Papers, 1917). This mixed system gave processors more control over fishers in the north than in the south, but it also restricted the mobility of many fishers because it enabled processors to control the pace of technological change.

A third area of state involvement consisted of measures to shape the technical relations of production. Salmon canners requested the use of trap nets and purse seines as methods of production and called for restrictions on the use of gas-engine boats for gillnetting in the north. The government had prohibited traps and drag seines since 1894, but after strikes in 1900 and 1902, canners lobbied the federal government for their return. They argued that traps and seines would allow them to control price demands from fishers and compete effectively with low-cost American producers. Against the protests of fishers, the state, in 1904, sanctioned the use of traps and seines. However, they remained a secondary method of production because technical limitations, costs of operation, and labour demands made them less preferable to gillnetting (Stacey, 1982:11).

Canners also sought limits on gillnetting operations. They attempted to stop fishing in the Fraser River above New Westminster, to limit the depth of gillnets, to set season closures, and to close completely the Fraser River as a conservation measure in 1906, 1908, and 1910. They also lobbied for embargoes on the export of raw salmon, preventing American competition for B.C. fish (Conley, 1985:208).

A major technological change in salmon fishing was the introduction of motorboats for gillnetting. Small gasoline motors were first used on the Fraser in 1906 and then in the north in 1909 and 1910. In 1911, two northern canneries employed motorboats on a large-scale basis, and this created a crisis, pitting canners against simple commodity producers and simple commodity producers against wage-labour fishers. Fishers with motorboats could roam farther, make more drifts, and increase their catch at the expense of the "sailboat fleet." If one cannery supplied its fishers with gas engines, then all thought them necessary in order to hold on to their share of the catch (Stacey, 1982:26–28). This technological change was regarded as threatening by most canners, increasing their problems of production costs, overfishing, and overcapacity. On the other side, the simple commodity-producer fishers wished to extend their right to fish to northern waters. They repeatedly opposed the prohibition of motorboats, arguing that they could not afford to allow their boats and capital outlay to lie idle or be limited by species and district restrictions.

Canners in the north generally favoured a ban on motorboats. Their employees, mostly share fishers, also favoured a ban because they feared that added costs to canners would undercut prices, and thus threaten their very survival in the industry. This alliance restricted petty commodity relations by persuading the state to forbid gas-engine boats for gillnetting from 1911 to 1923 (Hayward, 1981a; Conley, 1985:221–223).

In the south, simple commodity production was the solution to many of the canners' problems, since the fishers owned the equipment, took many of the risks of fishing and selling their catch, and provided a stable and productive seasonal-fishing labour force, especially on the Fraser River. They fished yearly for the entire salmon season and were present when runs were small and when it was hard to attract fishers from other occupations. Since they were fishing for themselves, they were also intensive and productive fishers.

   In the sparsely colonized northern region, there was no year-round small-boat fishery. The needed fishing labour force was provided by the "shares system." The cannery supplied the boat, net, and other gear and received one-third of the proceeds of the catch, the rest being divided between the net man and the boat puller. As a form of piece wage, this system encouraged intensive fishing efforts by fishers in a situation where direct labour surveillance and control were impossible, and where compatible wage-labour forms in agriculture and forestry coexisted (Conley, 1985).
   The state played a role mediating between large capital and small capital and wage labour. This is most evident in its regulations to exclude certain groups of fishers. While there were conflicts between simple commodity producers trying to limit or contain wage-labour fishers and vice versa, the most serious state-sanctioned exclusionary policy was cut along racial lines. Complaints against Asiatic fishers were voiced frequently by white and native Indian fishers. Asiatics were accused of illegal fishing practices, improper licensing arrangements, and lacking a critical distance from the canners, as evidenced in their perceived unwillingness to support strikes (Ward, 1978). Political pressure for their exclusion from specific fisheries and rivers, and for their reduction in numbers as a percentage of the fishing fleet began in earnest in 1914. For the next six years, boards of trade, labour councils, merchants associations, town governments and the Fishermen's Protective Association lobbied for a preferential licence system for whites and Indians, a limitation on Japanese fishermen on the Fraser River, and a ban on licences to Japanese fishermen above the New Westminster Bridge. The situation for Japanese fishermen was further compounded with the return of demobilized veterans who also pressed for special fishing privileges. By 1919, returning soldiers were guaranteed 30 per cent of gill-net licences in the north, and with white and Indian fishermen they continued to demand restrictions on Asiatics. By 1921, the question was no longer whether Asiatic licences would be reduced, but how many and when. The state monitored this process closely. In 1922, they reduced by one-third "the number of salmon trolling licences to other than resident white British subject and Canadian Indians" (Canada, Sessional Papers, 1924). In that same year the Fisheries Commission recommended a 40 per cent reduction in the number of "Oriental licences." The unlicensed hand-line cod fishery was licensed in 1923 to enforce the 40 per cent reduction rule and to decrease Asiatic participation in the fishery (B.C. Fisheries Commission, 1922). Further reductions and area discriminations were made in the years following so that by 1925, Asiatics held only 24 per cent of salmon gill-net licences, 10 per cent of salmon trolling licences, and their employment in seining and dry salting herring was reduced to 50 per cent of what it had been in 1924 (Department of Fisheries Annual Department Reports, 1925). The thrust of state policy was to completely displace Asiatics by white fishers with great dispatch

and little disruption. This was stopped in 1928 and 1929 when Supreme Court and Privy Council judgments overruled the practice.

The point to be grasped about this exclusionary policy is that it went against many of the canners' interests. They opposed the elimination of "Oriental licences" but to little effect. The state directly bolstered the claims of the small-commodity producers, in this instance filtering out the number of fishers by race, and ensuring that the white and native Indian groups were maintained. Indeed, the 1922 commission, which sanctioned the eliminations, also ended the prohibition on motorboats in the north, so white, simple-commodity producers won a dual victory. They were able to extend their mobility and fishing privilege, curtail the spread of wage-labour fishers, and exclude their most effective ethnic competitors from the salmon industry (B.C. Fisheries Commission, 1922; Conley, 1985).

In sum, the role of the state was pivotal to the early formation of the commercial fishery. State regulation was central to the definition of property rights and to the fashioning of an oligopsonistic structure in the industry. Within this dominant form, the state was active in shaping the social and technical relations of production, encouraging, at different times, wage labour and the "shares system," and petty capitalist production. Ultimately, however, the state promoted the interests of petty capital against wage labour and fishers of minority groups, both of whom were direct resource competitors to white simple-commodity producers. By 1925 the social relations of production were well formed: a thriving petty capitalist sector of harvesters were the dominant class of fishers, with wage labour increasingly concentrated inside the cannery factories.[2]

THE MODERNIZATION PERIOD

The outbreak of World War II and the usurpation of extraordinary powers by the federal state created a major change in the role of the state and in the structure of capital in the fishing industry. From 1925 to 1937, major processing companies such as B.C. Packers suffered losses. Soft markets and weak profits led to plant closures. In 1925 the total number of B.C. canneries was sixty-five, by 1940 there were only thirty-eight (Muszynski, 1984:94). The impact of the Depression started to ebb by 1937, and a minor recovery occurred that was further bolstered by considerable state involvement between 1940 and 1946. Federal war purchasing and demands for increased production and efficiency focussed attention on the west coast salmon fishery.

War Boards, Fish Markets, and Prices

The War Measures Act centralized the organization of the state under a small war committee of eight members. Prominent people such as C. D.

Howe, minister of munitions and supply, came to exercise immense powers, requisitioning, directing, and restructuring industries in accord with the announced needs of war production. By the end of the war, the Canadian state had spent a total of $28 billion on the war effort, established twenty-eight new Crown corporations, and employed 1.1 million people in war-related industries (Newman, 1975:316–333).

This development catapulted Canada from a minor industrial nation to the fourth largest in the world in five years. It was controlled by the Canadian capitalist class who served directly on the War Supply Board and in the Department of Munitions and Supply. Wartime advisory boards were also established in many sectors of the economy to give representation to men with practical business experience. Teams of specialists were recruited to form a galaxy of industrialists, and they carried out policy with little government interference (James, 1983:280). The powers of the War Measures Act and the Munitions and Supply Act were enormous. Government contracts were seldom awarded through tender, and there was a mad scramble for them. In the four-month period ending October 31, 1939, the total number of contracts placed by the Defence Purchasing Board was $41.3 million, and the next two months added $19.7 million. The value of war contracts for 1939 and 1940 was over $500 million. Following from the input of Munitions and Supply, total production rose in 1941 to $1.17 billion and then more than doubled in 1942 to $2.46 billion (Granatstein, 1975:187). Throughout 1940 factories were being created on an almost weekly basis. As Roberts (1957:87) noted:

> He [C. D. Howe] built huge plants with government funds and turned them over to industrial corporations to manage. He urged other corporations into tremendous expansion programs by guaranteeing fast write-offs for capital invested. What Howe started in 1940 was an industrial revolution . . . .

Between 1940 and 1945, depreciation privileges totalling $514 million were granted and over $300 million were spent on machinery and equipment. For the war years, new business investments in buildings, structures, machinery, and equipment exceeded $4.5 billion, of which $3.5 billion was directly related to the war effort.[3] According to Newman (1975:335) 50 per cent of the new investment during the war years was a direct result of state tax credits and allowances.

The Department of Fisheries came under the authority of the War Committee, two committees of Economic Defence, the Food Production and Marketing Committee, the Price Control and Labour Committee, and a Wartime Fisheries Advisory Board. The department wasted no time in reorienting its policy to meet the new situation.[4] The Canadian state took an active role in promoting new markets for west coast fish products. Through their boards and committees, they sold two-thirds of the 1941 canned salmon pack and 1.6 million cases of canned herring directly to the British Ministry of Food (B.C. Packers, Annual Report, 1941). The war years brought an increase in demand for Vitamin A oil, and this spurred on new markets in dogfish liver oil production. Most of the oil was sold under state contract, chiefly to Great Britain, where it was used as a dietary supplement. In 1942, the British Ministry of Food purchased the entire provincial canned salmon and herring packs, including the sockeye pack, which was the largest since 1913. British plants operated day and night and processed in three days an amount equal to that received in the six previous weeks (B.C. Packers, Annual Report, 1942). By the end of 1942, B.C. Packers had no inventory of canned salmon, "a situation unique in the history of our company" (B.C. Packers, Annual Report, 1943). In 1943, the Canadian government purchased the canned salmon pack, but released 200,000 cases for the domestic market, an amount of fish equal to about one-third of the prewar Canadian market demand. That same year the entire canned herring pack was purchased by the government and sold overseas. All fish meals and oils were also requisitioned to assist domestic food production for the war effort (B.C. Packers, Annual Report, 1943). In 1944, the sockeye runs were low, but pinks were unexpectedly high. The Canadian state bought 80 per cent of the canned salmon pack and 90 per cent of the canned herring pack. They also contracted for a large quantity of the dogfish liver oil produced and purchased 88 per cent of all fish oil and 100 per cent of the fish meal produced (B.C. Packers, Annual Report, 1944).

In 1945, the salmon pack was small. The state granted canners a subsidy to compensate for expected low earnings and to offset their price-fixing policy, which, since 1942, had set prices below the principal U.S. market. Increased volume, guaranteed markets, higher production, and subsidies meant higher profits for major firms that were even further inflated by an increase in U.S. demand for Canadian fresh/frozen fish, since there was a dearth of available Canadian canned salmon. In 1945, 60 per cent of the canned salmon pack and the entire canned herring pack were absorbed by government agencies. Furthermore, 75 per cent of the fish oil produced was used to fill government orders. In 1946, 50 per cent of the canned salmon pack was requisitioned, as was 100 per cent of the canned herring, fish meal, and fish oil production. Ready guaranteed markets at fixed prices were available for the 1947 canned herring and salmon packs, as the British Ministry of Food continued to buy fish through the intermediary of the Canadian government (B.C. Packers, Annual Report, 1947).

State Assistance, Modernization, and the Corporate Fishery

The role of the state in guaranteeing markets and prices had a direct impact on the corporate fishery and on the petty capitalist sector. A process of modernization was unleashed. In 1942, a federal program of assistance to vessel building was implemented for the trawlers, draggers, and seiners. On the west coast, subsidies were paid at the rate of $165 per gross ton to help in the construction of new packer-seiner vessels measuring between seventy-two and seventy-eight feet in length. Provision was also made for special depreciation allowances to owners so that they could save taxes until they had written off the costs of the vessel. Between 1942 and 1946 (when the subsidy was in effect), twenty large packer-seiner vessels were built to a total subsidy value of $333,000.[5] Such fleet expansion programs were supplemented by state sponsorship for upgrading and re-equipping small-scale fishers. In Atlantic Canada, Provincial Fishermen's Loan Boards, with federal assistance, launched subsidy programs for independently owned, medium-sized draggers and longliners. On the west coast, fleet developments were constant but not as dramatic. The B.C. fishing industry did not have pride of place as in Nova Scotia or New Brunswick. It was a distant fourth in economic importance to forestry, mining, and agriculture. The provincial state did not use the available federal transfer payments to underwrite direct producers in the fisheries. The west coast fleet was not chronically inefficient or undercapitalized. In general, modernization had proceeded more rapidly in Pacific Canada.

The trend toward amalgamation of plants and business organizations, noticed earlier, was intensified in the interwar years and was accompanied by advances in motorization and improved technical relations in harvesting the resource. It was during the war and postwar years that "the fisheries in other regions began to close the gap formerly separating them from those of the Pacific Coast" (Royal Commission on the Commercial Fisheries of Canada, 1956:8). State intervention was the driving force behind this process of catching up. They "pump-primed" the Atlantic fishery with capital investment loans and subsidies to underwrite plant redevelopment and fleet expansion. As Barrett notes (1981:19–20), "Between 1947 and 1960, 125 longliners and 34 draggers were built in Nova Scotia with such assistance." Offshore expansion was encouraged so that the number of "vessels over 50 tons in the Northwest Atlantic increased 165 per cent, from 211 in 1959 to 558 in 1968." Tonnage over the same period increased by 320 per cent.

On the Pacific coast, there was a steady momentum to fleet modernization, but it was less state supported. Comparatively speaking, there already was higher large-scale investment of capital, higher-priced products, and a more concentrated and mechanized fish-processing sector on the west coast. By 1947, the capital equipment of the primary fishing industry was valued at $26,801,000, including vessels and boats worth $22,666,000 (Department of Fisheries, *Employment in Canadian Fisheries*, vol. 1, 1949:13). The estimated total number of vessels in 1945 was 8,596 vessels: 38 draggers; 26 gasoline vessels (ten to twenty tons); 78 diesel vessels (forty tons and over); 219 diesel vessels (twenty to forty tons); 228 diesel vessels (ten to twenty tons); 990 sail and row-boats; 6,738 small gasoline and diesel boats; and 279 carrying boats and fish packers. Approximately 75 per cent of the fishers were engaged in the small-boat fleets (Ibid., vol. 2. 1949:A.29). A year later, the total number of vessels increased to 9,413 with most of the additions occurring in the gasoline- and diesel-boat category. The number of fishers employed in the primary sector also increased from 9,609 in 1939 to 13,292 in 1945 (B.C. Packers, Annual Reports, 1939, 1945). By 1948, overcapacity was a growing concern, and a degree of political pressure was mounting for the enactment of licence limitation on a fishery-specific basis (Sinclair, 1978:22).

The major thrust of modernization occurred in the processing sector. Between 1930 and 1950, the salmon-processing industry experienced depression and then marked recovery. Annual production fluctuated around a slightly declining trend, but the value of the landings dropped sharply from $8 million in 1930 to $3.3 million in 1931, reflecting the depressed prices. Then it recovered to an annual average of $5.5 million for the years 1932 to 1940. Commencing in 1941, the demand created by the guaranteed markets of the World War II period moved the salmon catch value up to a ten-year annual average of $10.3 million, exceeding for the first time in 1950 the $20 million level (Ministry of the Environment, *Commercial Salmon Fisheries of British Columbia*, 1958).

Up until 1920, canned salmon was traded primarily on U.S., Canadian, and British markets. New markets and producers emerged in the post–World War I period. From 1920 to 1940, B.C. canners faced stiff competition from the burgeoning Japanese industry. Japanese production rose from nil in 1909 to 800,000 cases by 1919, and to 1.7 million cases by 1928. In the 1930s, their production was approximately 50 per cent higher than that of British Columbia, all for world export. Japanese canned salmon undercut B.C. canners in the U.K. market, especially at times when free trade operated, and they made major inroads into the new European markets, particularly France, Belgium, and Italy. Indeed, between 1930 and 1934, Japanese sales to the French market far exceeded those of B.C. The abrogation of government-negotiated trade quotas and productions such as the Franco-Canadian Trade Agreement spelled trouble for Canadian canners in Europe. By the end of the 1930s, B.C. canners had oriented their supply to the price- and market-protected areas of the U.K. and its empire, particularly Australia. Their share of exports going to the United States varied; in 1932, 1933, 1935, and 1938 there were large volumes of low-priced canned salmon exported to meet shortages in U.S. domestic production; in other years, much less. Exports to the U.S. were heavily restricted by tariffs, and it was only when the domestic U.S. price moved higher than the Canadian export price, plus 25 per cent, that exports were profitable.

Not surprisingly, B.C. companies were eager to develop their domestic market. From 1920 to 1929, the domestic salmon market accounted for about 22 per cent of the ten-year annual production average, well below the pre–World War I rate of 42 per cent. During the Depression, the domestic market jumped to an annual average of 35 per cent of salmon production. Throughout the war years, B.C. canners tried to maintain their domestic market position. They petitioned the government for supplies for this market, and they advertised their national brands even when they had few products to sell. In the years from 1943 to 1946, the domestic market dropped to an annual average below 20 per cent for salmon production. It recovered after 1947 to the 45 to 50 per cent range.

State guaranteed markets, price controls, depreciation packages, investment incentives, new market potentials, good salmon fish runs, and increased production resulted in a modernization program for the capital sector. Utilizing government subsidies, the major firms built new plants and expanded and modified their facilities. They re-equipped their operations and technology to accommodate increased quantity and diversity of production. Fresh fish plants and cold storage facilities were added to existing physical plants. Reduction plants were built alongside canneries or designed as new combined cannery and reduction operations. For example, B.C. Packers began expanding and retooling in 1939. They developed a policy of product and income modification that led, in 1940, to the purchase of Edmunds and Walter and to a major involvement in the fresh, frozen, and cured sectors of the industry. In that same year, they constructed and redesigned reduction plants at Alert Bay, Namu, Steveston, Pacific, Port Edward, and Deep Bay for producing fish oil and fishmeal (B.C. Packers, Annual Report, 1940). In 1941, they added more fresh fish plants to their holdings and they acquired interests in the Prince Rupert fresh fish and cold storage business. The introduction of Clover Leaf canned herring in tomato sauce meant the installation of at least eight new machines at the Steveston plant to gut and clean this new product. A separate herring cannery was built at the Imperial plant, in an integrated design, so that it could also can salmon. The salmon cannery had ten lines using vacuum suction machines. Six to eight thousand cases (forty-eight pounds per case) could be processed daily. By 1946, the Imperial plant in Steveston was the largest in the British Commonwealth (B.C. Packers, Annual Report, 1946). The processing capacity of B.C. Packers was bolstered by their own spectacular growth in cold storage operations. In 1945, they obtained the controlling interest in the Canadian Fish and Cold Storage Company and became the dominant owner of the largest refrigeration storage operation in the British Empire (*The Fisherman*, vol. VII, no. 5, February 1945). By 1945, the capital equipment of all fish-processing establishments in B.C. was valued at $18,935,000, while that of the Atlantic coast including Quebec, was valued at only $14,917,000 (Department of Fisheries, *Employment in Canadian Fisheries*, vol. 1, 1949).

The most significant development in the fishing industry, however, was the restructuring and consolidation of existing Canadian capital into increasingly monopolistic relations under the control of one firm — B.C. Packers. In a manner similar to the consolidation of National Sea Products in Atlantic Canada, B.C. Packers engaged in amalgamations and buy-outs. Centralization resulted in the closure of many plants as well as in the enlargement, modification, and construction of others. The overall number of B.C. canneries declined from forty-three in 1935, to thirty-eight in 1940, and then to twenty-seven in 1948. Most of these closures were on Rivers Inlet, The Nass River and in northern outlying districts (Muszynski, 1984:94). While the size of the labour force was remarkably stable in the years from 1935 to 1947, with 17,989 being employed in 1936 and 17,934 remaining employed in 1947, there was an increase of over one thousand employed in the primary sector and a decrease of over eleven hundred employed in the secondary sector of the industry (Muszynski, 1984:95). More importantly, by 1945, B.C. Packers alone accounted for over ten thousand fishers and plant employees, which was 52 per cent of the entire fishing labour force in the province (B.C. Packers, Annual Report, 1945). Net income for B.C. Packers rose from $6,190,953 in 1940 to $13,910,667 by 1945. Near the end of that year, the company expanded its holdings further by purchasing companies in the United States. The income from these new subsidiaries increased their net income to $23,547,246 in 1946. In general, net profits also rose from a decade low of $206,852 in 1941 to a decade high of $1,578,898 in 1948 (B.C. Packers, Annual Report, 1941; 1948).

Much of this restructuring, increased production outlay, profit maximization, and new-boat construction was assisted by state subsidies, many of them awarded under government orders in council. This did not stop with the end of the war. Economic and currency crises after 1947 led to a collapse of the U.K., European, and sterling bloc markets (i.e., Australia and New Zealand especially). Processing firms had to increase their reliance on the domestic market. They were assisted in this transition by the Fisheries Prices Support Board, which was set up in July 1947 to monitor fisheries adjustment from wartime to peacetime conditions. As part of its mandate, the board had authority to buy quality products and to dispose of them by sale, "or to pay the producer of a fisheries product the difference between a price prescribed by the board, and the average price the product had in fact commanded." They could not directly fix prices (Department of Fisheries, *Employment in Canadian Fisheries*, vol. 1, 1949:44). On the west coast, the Fisheries Prices Support Board assisted in the reduction of canning activities by purchasing, in 1948, a proportion of the canned herring pack. They also purchased part of the east coast pack at a total cost of about $1,250,000 and donated most of the fish to charitable and relief agencies (*Employment in Canadian Fisheries*, vol. 1, 1949:44). The Fisheries Prices Support Board also provided a direct subsidy of $300,000 to west coast processors for a marketing program to sell canned salmon to the Canadian consumer (B.C. Packers, Annual Report, 1949). In the period from 1950 to 1954, the domestic market expanded to take 50 per cent of the average annual production of 1.6 million cases of salmon.

Fisheries, Economics, and the Treadmill of Expansion

A number of documents released during the 1950s and 1960s reveal the policy concerns and consequences of modernization in the west coast industry. There were concerns expressed about excessive capacity and licence controls, but the publication of H. Scott Gordon's "An economic approach to the optimum utilization of fishery resource," by the Fisheries Research Board gave a clear rationale for further modernization. This work, prepared for the Royal Commission on Canada's Economic Prospects, pioneered the theoretical analysis for the free exploitation of a sea fishery. The central proposition was that the optimum level of development in the primary fishery occurred when there was an uneasy balance between biology and economics. The biological "maximum sustainable yield" was reached when older and larger fish were removed and the rate of growth and reproduction were increased. The accumulated stock had to be "fished up" so that there was no overabundance of the spawning stock, and natural productivity could be maintained. This was a warning not to fish too little. On the other hand, overfishing could result in a peculiar law of diminishing returns. Catches could increase but size and weight would be fewer and smaller as the age structure of the stock declined (Pearse, 1982:10–11). At bottom, Gordon realized the tendency for the reproductive ability of a stock to become ever more reliant upon a diminishing number of spawning stock, but he concluded optimistically that large broods "do not appear to depend on large numbers of adult spawners" (Gordon, 1954:126). He downplayed the conservationist concern about the depletion of the resource and argued in support of the theory that the fish population is generally unaffected by the activity of man.

> The optimum intensity of fishing is that which maximizes the difference between costs and returns, and this normally is somewhat less than the level of intensity that maximizes the yield in physical terms. (Gordon, 1954:129–130)

But the free access to fish and the spread of fishing units and capacity make this balance unstable. Since fishers aim to maximize their incomes, they will fish the stocks until the average cost of production equals the market price for fish. The effect is the dissipation of potential economic rents and reduction in fish stock so that the fishery is no longer economically viable. This is the "tragedy of the commons." To overcome the problem of overexploitation, Gordon argued that a more rationalized, capital-intensive, vertically integrated fishery was required. More cost-saving and cost-efficient measures were required, and in turn this would galvanize the operation of the natural balancing principle (Royal Commission on the Commercial Fisheries of Canada, 1956:107–111). This could be achieved in a number of ways. Since complete ownership was not to be given to any one corporation, the state had to regulate exploitation as "sole manager" of the resource for other than biological conservation. This meant establishing controls on the modernization process, especially in the salmon fishery where the costs of exploitation were comparatively low and where the population in a free-entry situation could be drawn close to extinction. The "free entry" equilibrium had to be modified for the B.C. situation. As Fraser observed, "The government was early aware of the conservation problem posed by this particular species [salmon] — hence the earliest attempts at license limitation and increasingly stringent gear and area restrictions" (1977:14).

On the east coast, modernization of the fishing fleet went ahead full steam. The entire process was viewed as progress and a testimony to federal-provincial cooperation. While not as spectacular, fleet modernization on the Pacific coast did entail a new capital investment emphasis. Major concern turned toward diesel-powered vessels of ten to forty tons or more. Between 1946 and 1953, the number of such vessels increased by over 260, while the value rose from $6.8 to $22.2 million, reflecting an increase of eighty-seven in the number of vessels over forty tons (Royal Commission, 1956:112). The investment per person in primary fishing operations doubled in the ten years following the war and represented an increase in the size of fishing craft, in power equipment, and electronic, navigational, and fish-locating devices.

There was also a parallel consolidation and mechanization of the fish-processing sector. As Table 6.1 shows, more than one-half the plants in Canada produced less than $50,000 worth of products each in 1954. But, there were a number of large establishments, each of which processed a large output of fish products. In 1954, forty of these large plants each produced fish products valued at more than one million dollars. Together they accounted for almost 66 per cent of the yearly output. Over half of them were located in the Pacific area. In this region alone, twenty-one firms representing 25 per cent of all B.C. fish establishments handled about 83 per cent of the gross value of sales. The number of establishments operating in the Atlantic region was 504 with an average value of sales per establishment of $163,500, whereas in the Pacific region there was a concentration of eighty-two firms with an average value of sales of $866,700.

Capital investment in the processing industry for all of Canada in 1954 was estimated at $150 million. A regional breakdown according to value of fish products marketed in 1953 reveals the following division:

Table 6.1
Distribution of Numbers and Gross Sales of Fish-Processing Establishments
Classified by Value of Production, with Percentages,
Atlantic and Pacific Regions of Canada,[1] 1954

(a)

Value of production (dollars)	Number of establishments			
	Atlantic region		Pacific region	
	no.	%	no.	%
Under 10,000	119	23.6	5	6.1
10,000 to 24,999	73	14.5	9	11.0
25,000 to 49,999	81	16.0	12	14.6
50,000 to 99,999	77	15.3	10	12.2
100,000 to 199,999	73	14.5	6	7.3
200,000 to 499,999	46	9.1	10	12.2
500,000 to 999,999	16	3.2	9	11.0
1,000,000 to 4,999,999	19	3.8	19	23.2
5,000,000 and over	—	—	2	2.4
TOTAL	504	100.0	82	100.0

(b)

Value of production (dollars)	Gross value of sales			
	Atlantic region		Pacific region	
	(thousand dollars)	%	(thousand dollars)	%
Under 10,000	573.6	0.7	21.6	0.0
10,000 to 24,999	1,156.5	1.4	163.4	0.2
25,000 to 49,999	3,035.2	3.7	435.9	0.6
50,000 to 99,999	5,563.4	6.8	700.1	1.0
100,000 to 199,999	10,338.3	12.5	726.6	1.0
200,000 to 499,999	14,481.0	18.0	3,365.2	4.7
500,000 to 999,999	11,202.4	13.6	6,498.1	9.2
1,000,000 to 4,999,999	35,677.6	43.3	59,157.8	83.3
5,000,000 and over	—	—	—	—
TOTAL	82,387.9	100.0	71,068.7	100.0
Average value of sales per establishment	163.5		866.7	

[1] Includes a few firms engaged in marketing but not processing.
Source: Royal Commission on Canada's Economic Prospects, The Commercial Fisheries of Canada, Appendix C, 1956.

Maritimes and Quebec $62 million; Pacific coast $61 million; inland fisheries $16 million; and Newfoundland $11 million. A later survey in 1956 showed total investment in fish processing in B.C. as approximately $66 million. Of this investment, almost two-thirds was in fixed assets, mostly processing machinery and equipment (Royal Commission on the Commercial Fisheries of Canada, 1956:116).

Modernization of the fleet and of the processing sector continued unabated throughout the 1950s and 1960s. Fleet development expanded excessively to the point of threatening some of the stocks and undermining the level and stability of economic returns. Temporary economic gains stimulated fishers to expand their vessels' fishing capacity and enhance their catch. New entrants were brought into the fishery, with the result that excess fishing units were chasing the same number of fish (or less). Redundant capacity also raised the capital, labour, and production costs of harvesting and so diminished the net benefits of fishing. Modernization was profoundly connected to technology. Pacific vessel owners in the late 1950s and 1960s adopted innovations to increase the speed of their boats and their hold capacity, to reduce running time, and to construct larger vessels capable of offshore operations in order to intercept fish before competitors. Indeed, between 1956 and 1965, the number of fishing vessels increased from 7,650 to 9,593. As Table 6.2 shows, tonnage and length of boats increased generally, but most important was the increase of 583 vessels over ten tons and 812 vessels over twenty feet in length.

Corporate Consolidations

The fortunes of B.C. Packers also reflect, in good measure, this overall modernization trend. In the 1950s, they, along with Canadian Fish Company, were the strongest competitors in the industry. These companies were able to design, construct, and adapt multiline and multiproduct plants concentrated for canning, but capable of diverting the product to the fresh/frozen market if demand was strong. The expansion of processing techniques allowed for considerable horizontal diversification, and elimination of rivals permitted increased vertical integration. In 1954, B.C. Packers and Canadian Fish Company jointly absorbed J. H. Todd and Sons Ltd. In 1969, they once again took over the assets of a medium-sized competitor, ABC, and in that same year B.C. Packers announced control of Nelson Brothers. These mid-sized firms were unable to capitalize their plants, machinery, and equipment effectively. Nor were they able to retain their market volume. The two dominant firms reduced competition and increased the administrative expertise needed to run their expanding operations.

The structural weakness in B.C. Packers' and Canadian Fish Company's position in the industry was their vulnerability in forward-market linkages. As early as 1950, George Weston Ltd. was engaged in a large, $150 million acquisitions program, part of which involved capturing major food suppliers as a strategy to outcompete other food retail conglomerates. In 1962, Weston secretly purchased a controlling interest in B.C. Packers for

Table 6.2
Number of Fishing Vessels by Net Tonnage and Length Classes and Number
of Fishermen in the British Columbia Commercial Fishery, 1953–1965

		Number of fishing vessels by				
		Net tonnage class		Length in feet class		
Licence year	Number of fishing vessels	10 +	< 10	20 +	< 20	Number of fishermen
1953	8,155	702	7,453	6,243	1,912	12,008
1954	8,106	729	7,377	6,066	2,040	12,680
1955	7,841	759	7,082	5,873	1,968	11,860
1956	7,650	767	6,883	5,795	1,855	10,853
1957	7,580	816	6,764	5,904	1,676	12,016
1958	7,884	829	7,055	5,879	2,005	14,266
1959	8,188	860	7,328	5,966	2,222	14,463
1960	8,321	873	7,448	6,048	2,273	14,191
1961	8,504	960	7,544	6,178	2,326	15,660
1962	8,768	980	7,788	6,525	2,243	15,060
1963	9,454	1,161	8,293	7,194	2,265	15,374
1964	9,240	1,453	7,787	7,055	2,185	13,300
1965	9,593	1,285	8,308			13,000

Source: Abridged from Sol Sinclair, *A Licensing and Fee System for the Coastal Fisheries of British Columbia*, Vancouver: Department of Fisheries and Oceans, 1978, p. 78.

an estimated $6 million. By 1974, Weston's had obtained a solid equity interest of 81 per cent for an estimated value of $10 million (UFAWU, 1984:13). The decision "to sell to Weston" was influenced by existing business connections. B.C. Packers had for many years sold the Clover Leaf brand through the Loblaws grocery chain, which in turn, was part of the Weston empire. Loblaws was B.C. Packers' major customer in Canada and Britain and the purchase guaranteed economy-of-scale advantages, increased market control, a better financial portfolio, and secure retailing assistance and capacity. Between 1962 and 1974, it resulted in $20 million profits for Weston, and it marked the beginning of B.C. Packers' journey to monopoly domination (see Muszynski, Chapter 3; Pinkerton, Chapter 4).

THE RETURN TO REGULATION

The modernization process of the war and postwar settlement period resulted in new policy developments. Two issues quickly became important: there was too much capacity in the fleet and there was a growing problem of overfishing the stocks. The first concern led to a new regulatory policy involving licensing and surveillance of fishers; the second resulted in a program of fish conservation and enhancement. Ironically, the state's structured relationship with the processing sector and with petty capital producers undercut the intentions of control and catch enhancement. State policy pulled in different directions and was decidedly contradictory.

The Limits of State Licensing

Just as the large and petty capital sectors were undergoing modernization, so the state was undergoing substantial transformation. For the first time since the Sanford Evans Commission of 1917, the government was discussing the free entry basis of the salmon fishery. While conservation had been a concern, there really were few controls imposed on canneries or fishers in the forty years following the Sanford Evans report. In 1919, a recommendation by the Sloan Commission for government ownership of the fishing industry was rejected. The Duff Royal Commission of 1923 changed social and technical relations in the industry, permitted the full use of motorboats, removed restrictions on cannery licences, reduced and controlled licences issued to fishers other than Caucasians and Indians, and opened the way for "free competition" for the resource. Specific adjustments on gear size and type, fishing areas, and special licences for interest groups were made, but this did not amount to a systematic program. Sinclair notes that until 1958 "the salmon fishery continued to function essentially on a free entry basis controlled only for conservation purposes" (1978:22).

The work of Scott Gordon, the Royal Commission on Canada's Economic Prospects, and the Fisheries Research Board underscored not only the need for capital-intensive, maximum economic yield, but also argued increasingly for a new, strong, and expanded state presence in the industry. The dominant theme was to control free entry by means of state property limitation. Throughout the 1950s, proposals were circulated and debated but with no consensus. Finally in 1958, the federal state commissioned a study of the limited entry problem. The Sinclair Report, published in 1960, rejected the idea of transforming the resource into "sole ownership" on "legal, political or social grounds quite apart from any possible abuses that may arise out of the monopoly position" (Sinclair, 1960:101). Instead, it opted for a position "in which the government assumes ownership but permits its use by private firms" (Ibid., 102). It also resisted a tax system on catch or fishers. Overcapacity and generally depressed earnings made it "hardly appropriate or politically acceptable to attempt to correct a situation by adding a high tax to already burdensome costs of operation" (Ibid., 106). Sinclair proposed a system of restricted vessel licences and levies on the catch to halt overinvestment. Implementation would be gradual, with a five-year moratorium on new licences, followed by a permanent transferability under competitive bid.

Political action on these recommendations was delayed for eight years. Debate, research, and planning did continue, and in 1966, the federal state introduced a licensing system that attempted to identify every fishing vessel on the west coast. Then in 1968 the Davis Plan for the salmon fishery was announced.

The Davis Plan was designed to eliminate overcapitalization and excess labour usage. It was intended to reduce the costs of production and create an economic surplus. Fishers' income should increase and the state should take a return for the costs of resource use and management. The plan was to have a minimum dislocation effect on existing capital and labour. It involved four phases. The first phase was to freeze the fleet by licensing only those who could demonstrate a significant dependence on the salmon fishery. The second was to gradually reduce fleet size by "buying out" and returning excess vessels. The third was to improve vessel standards and product quality. The final phase, never implemented, was to introduce economically optimum regulations to improve fishing effort for the reduced fleet.[6]

With few changes, the first phase followed Sinclair's recommendations. Fleet stability was achieved by creating a specific salmon licence instead of the general fishing licence previously available. The licence was tied to a vessel, granted on proof of level of participation in the salmon fishery and made renewable annually.[7] Licence fees were doubled, but still stayed nominal. Class "A" vessels could be replaced, if retired, and licence transfers upon vessel sale were permitted. An appeal committee was struck to consider "special circumstance fishermen" who were nevertheless "bona fide."[8]

The second phase dealt with reducing the fleet by substantially increasing licence fees, phasing out the "B" licensed fleet, funding a "buy-back" program, and adjusting original regulations. For the 1970 season, fees for category "A" vessels were increased from their low level of $10 to either $100 or $200, depending on vessel ton size. Fees were doubled the following year for "A" vessels, and the "B" fleet was set to be retired over a ten-year period. In order to speed up attrition, "B" licence fees were kept at $10 and a buy-back program was inaugurated. The proceeds of the higher licence fees were banked in a fund and used to buy out a portion of the "A" fleet. Prices were negotiated and the scheme was voluntary. When a boat was purchased, the state resold it at public auction with the proviso that it never be used in any B.C. commercial fishery. The monies from these sales were rechannelled into the buy-back fund for future purchases. Between 1971 and 1973, 350 "A" licence vessels had been retired by this method (Fraser, 1977:30).

Also, under phase two, the problem of growing company control of the fleet resurfaced. Earlier in 1969, a limit was put on company ownership of "A" vessels, but with the shrinking fleet, processors' proportional share of the fleet was increasing. Companies were advised by the minister to make parallel reductions in their own fleets, although this was never formally written into the regulations (Fraser, 1977:32).

Finally, phase two addressed the special place of native Indian fishers. The increased fees and stringent licence eligibility rules meant that many Indian vessels had been classified as "B" vessels and were subject to be phased out in ten years. In order to maintain the native presence, special provisions were established. An "A-I" licence was created with a $10 fee. However, such boats were ineligible for a buy-back purchase. These vessels, however, could be sold to either native Indians or whites, but if sold to the latter, the vessel reverted to "B" status unless all exempted licence fees were fully paid (Pearse, 1982:102).

The third phase of the program was designed to improve vessel standards, safety, and product. When the provisions were finally implemented in 1973, they stressed greater hold size, better insulation from heat, cleaner, nontoxic surfaces in fish holds, and improved refrigeration requirements (Pearse, 1982:79; Fraser, 1977:33).

The Davis Plan was controversial, and in response to pressures, the state made a number of concessions that weakened the scheme. They relaxed the requirements of licence eligibility, allowing additional vessels. They did not, at the outset, control for net tonnage increases when boats were replaced. There was no impediment to increasing the productive capacity of the fleet through larger unit investments. They did not prevent the practice of "pyramiding," whereby the tonnage licence from a smaller boat or boats was combined to meet the tonnage licence capacity of a larger vessel. The scavening of tonnage licences led to the redistribution of the fleet from small boats into the larger, more productive seine vessels. Finally, the state-run, buy-back operation was short-lived, underfunded, and wholly ineffective in purchasing and retiring fishing capacity (Hayward, 1981b; Pearse, 1982, Chapter 9).

Most important, there was a fundamental error in the basic assumption that vessel numbers could be used to control the intensity of fishing power. In retrospect, the plan succeeded in downsizing the fleet, but it did not remove excessive capacity. After fifteen years of restrictive licensing, already excessive capacity in 1968 had doubled or trebled by 1982 (Pearse, 1982:78)

Contradictions of State

Administrative lapses, loopholes, and misconceptions about licensing were compounded further by a second set of state initiatives: major capital-assistance programs for small producers. The capital costs incurred in buying inflated tonnage, new technology, larger replacement boats, and expanded quality hold capacity were not supplied by processing companies or cooperatives. Many of these firms were reconsidering their financial and merchant capital arrangements with the fishers. They were expanding their own operating capital. They withdrew their capital for major loans, repairs, and equipment and resisted providing fishers with the instruments of production. They sold their holdings and rental fleets and concentrated capital in plant reorganization and product markets.

The banks entered when industrial capital exited. They had high deposits and were eager for new investment markets. Banks were encouraged and aided by state policies. Despite licensing regulations, the state created a contradictory program; they spurred on further capital growth by expanding their independent business programs for fishers and by increasing capital subsidy programs to shipbuilders and the processing sector. With state guarantees for loans and subsidies, chartered banks had a degree of protection should defaults become a problem (McMullan, 1984:65–88).

The difficulties and failures of licence limitation contributed to the treadmill of overcapacity and established a state-property value to the right of fish. Licences were endowed with a marketable commodity value subject to inflationary pressures. This occurred when large salmon runs combined with high prices to make the fishing industry a lucrative investment. By the end of 1973, licences were sold for $4,000 to $5,000 a ton (Hayward, 1981b:45). The escalation of these licence "rights" for sale was an enticement to major lending institutions. Bank capital started to flow into the industry at about the time when the proportion of the percentage average rate of return taken up by the cost of licences was increasing from 2.7 per cent in 1972 to 10..5 per cent in 1973 to a decade high of 24 per cent in 1974 (Hayward, 1981b:47).

The state-supported investment incentives and expansions in fishing technology were bolstered by developments in the economy of fishing. Two points are important. A constant succession of good production years from 1973 to 1979 with few real sustained low-income returns, combined with increasingly high prices for herring, created a generally optimistic investment mood in the industry. Fishers engaged in high levels of investment and carried large debt loads. Also, the expansion of markets, especially fresh/frozen salmon and herring to Japan and the ground-fishery sector to the United States spurred on large amounts of new investment. Vessel capitalization was extensive. The estimated market value of the salmon fleet alone increased by 270 per cent in ten years from 1968 to 1977 (Fraser, 1979:276–77).

The composition of the fleet was altered. The number of seine vessels fishing for salmon only increased from 286 in 1969 to 316 in 1980 and the number of combination roe herring/salmon vessels increased from 83 to 216 in the same years (Pearse, 1982:100). Transfers, conversions, and new additions accounted for much of the restructuration. Gillnetters and trollers declined, while combination vessels increased as gear was added to expand fishing opportunities. Approximately 30 per cent of the roe herring seine fleet was built in the 1972 to 1977 period. In those years, seine vessels participating in the fishery doubled and the average roe herring seine income quadrupled to $76,000 for a one-month season (Sinclair, 1978:143).

In this context of economic growth, chartered banks and the federal state made large amounts of capital available to fishers. Table 6.3 shows that the major purpose was to provide financial capital and guarantees for capital costs on vessel purchases, improvements, and reconstruction, especially through the Fisheries Improvements Loans Act.[9] Table 6.4 indicates that from April 1, 1976, to March 31, 1979, the numbers, value of loans, and their average size increased dramatically. In the fiscal year 1976–1977, 854 loans were made, the average size was $14,461, and the total value was $12,349,366. By 1978–1979, the number of loans granted had increased to 1,490; the average loan size had reached $19,005, and the total amount had more than doubled to $28,317,786. Table 6.5 reveals that $133,780,146, over two-thirds of the dollar value of loans since the inception of the program in 1955, were granted in the seven years from 1974 to 1981. Repayments were met by borrowers right up until the end of the 1974 fiscal year; then debt started to accumulate. The outstanding balance of loans payable to lenders at the end of period 5, June 30, 1977,

Table 6.3
Government Financing Utilized by B.C. Fishermen
by Purpose*

Purpose	Years	Amounts ($,000)	Per cent
Capital costs			
(a) Fisheries Improvement Loan Act	1976–1977	9,564	
(b) Fish-chilling assistance	1973–1976	932	
(c) Federal Business Development Bank	1976–1977	1,658	
Total		12,156	99.0
Labour	1977–1978	8	0.4
Educational upgrading	1977–1978	12	0.6
Total		12,176	100.0

* This tabulation does not include the Fishing Vessel Construction Assistance Program or the Fishing Vessel Insurance Program, the former because it was annulled in 1975, the latter because it is an indirect assistance.
Source: *Financing in the B.C. Fishing Industry*, Foodwest Resource Consultants, 1979, Chapter 3, pp. 23–29.

Table 6.4
Loans Made and Claims Paid under the Fisheries Improvement Loans Act
from Inception to March 31, 1981

Fiscal year	Loans made		Average size of loan	Claims paid		Recoveries of claims paid
	No.	Amount ($,000)		No.	Amount ($,000)	
December 12, 1955 to March 31, 1957	305	335,196	1,099	—	—	—
April 1, 1957 to March 31, 1958	136	149,960	1,103	—	—	—
April 1, 1958 to March 31, 1959	144	177,040	1,229	1	278	—
April 1, 1959 to March 31, 1960	127	170,382	1,342	—	—	—
April 1, 1960 to March 31, 1961	142	217,296	1,530	—	—	—
April 1, 1961 to March 31, 1962	179	283,808	1,586	2	338	—
April 1, 1962 to March 31, 1963	144	214,816	1,492	—	—	—
April 1, 1963 to March 31, 1964	202	368,229	1,823	—	—	—
April 1, 1964 to March 31, 1965	238	459,648	1,931	—	—	—
April 1, 1965 to March 31, 1966	341	1,097,962	3,220	3	1,894	—
April 1, 1966 to March 31, 1967	348	1,420,539	4,082	2	1,433	46
April 1, 1967 to March 31, 1968	280	1,273,924	4,550	1	600	—
April 1, 1968 to March 31, 1969	202	1,103,491	5,463	2	2,347	—
April 1, 1969 to March 31, 1970	360	2,579,005	7,164	3	6,510	3,891
April 1, 1970 to March 31, 1971	467	3,299,641	7,066	2	8,000	—
April 1, 1971 to March 31, 1972	692	5,296,814	7,654	—	—	—
April 1, 1972 to March 31, 1973	910	7,070,356	7,769	4	10,532	750
April 1, 1973 to March 31, 1974	1,152	10,762,552	9,342	4	16,613	878
April 1, 1974 to March 31, 1975	933	12,420,268	13,312	3	10,785	3,409
April 1, 1975 to March 31, 1976	920	12,839,670	13,956	11	52,236	—
April 1, 1976 to March 31, 1977	854	12,349,366	14,461	16	45,591	5,603
April 1, 1977 to March 31, 1978	927	18,830,481	20,313	42	401,979	8,437
April 1, 1978 to March 31, 1979	1,490	28,317,786	19,005	28	298,437	1,272
April 1, 1979 to March 31, 1980	1,408	27,805,493	19,748	29	228,302	4,474
April 1, 1980 to March 31, 1981	1,158	23,370,203	20,182	22	221,491	3,546
Total	14,059	172,213,926	12,249	175	1,307,366	32,306

Source: Fisheries Improvement Loans Act; Annual Report 1980–1981.

Table 6.5
Loans and Repayments to Lenders under the Fisheries Improvement Loans Act
from Inception to March 31, 1981

Period	Loans made	Repayments	Balance of loans payable to lenders
	($,000)	($,000)	($,000)
Period 1: December 12, 1955, to June 30, 1965	2,650,865	2,650,865	NIL
Period 2: July 1, 1965, to June 30, 1970	8,025,341	8,010,995	14,346
Period 3: July 1, 1970, to June 30, 1971	3,798,934	3,773,484	25,450
Period 4: July 1, 1971, to June 30, 1974	23,958,640	22,549,647	1,408,993
Period 5: July 1, 1974, to June 30, 1977	38,738,792	30,958,585	7,780,207
Period 6: July 1, 1977, to June 30, 1980	76,601,802	32,652,844	43,948,958
Period 7: July 1, 1980, to June 30, 1983 (As at March 31, 1981)	18,439,552	3,405,102	15,034,450
Total	172,213,926	104,001,522	68,212,404

Source: Fisheries Improvement Loans Act; Annual Report 1980–1981.

was $7,780,207; by June 30, 1980, the amount had escalated to $43,948,958, and as of March 31, 1981, the debt load was $68,212,404, of which $54,569,920 was owed in the B.C. region.

Tables 6.6 and 6.7 provide a profile of bank involvements. Approximately $98 million of the $123 million, that is 80 per cent of total loans, went to British Columbia fishers; chartered banks accounted for $94 million and credit unions held $4 million. Two banks accounted for almost two-thirds of all loans. From 1975 to 1981, the Royal Bank made 1,013 loans for a total amount of $30,137,049, a 31 per cent share of the market. The Canadian Imperial Bank of Commerce made fewer loans, 888, but had larger average amounts per loan for a total amount of $30,798,236, a 31.5 per cent share of the market. The Toronto Dominion Bank had a smaller investment interest. They made 426 loans for a value of $16,445,364, a 16.8 per cent share of the market. The top three banks made about four-fifths of all loans for a total value of $77,380,649. Four other chartered

Table 6.6
Loans to B.C. Region and by Lender, 1975–1981

Year	1975–76		1976–77		1977–78		1978–79		1979–80		1980–81		Since 1975–1981	
Lender	No.	Amount ($,000)	No.	Amount ($,000)	No.	Amount ($,000)	No.	Amount ($,000)	No.	Amount ($,000)	No.	Amount ($,000)	Total no.	Total amount ($,000)
Bank of Montreal	29	634,200	52	1,448,615	62	1,842,370	65	2,111,460	63	2,429,143	47	1,492,738	318	9,958,526
Bank of Nova Scotia	12	285,787	30	719,667	36	992,732	30	854,525	40	1,157,623	18	477,828	166	4,488,162
Royal Bank of Canada	275	5,820,586	154	3,175,505	109	3,412,751	213	6,744,636	165	6,635,106	97	4,348,465	1,013	30,137,049
Toronto Dominion Bank	57	1,252,558	51	1,481,995	97	3,569,855	95	4,243,100	73	3,699,520	53	2,198,336	426	16,445,364
Canadian Imperial Bank of Commerce	62	1,405,143	85	2,431,992	149	4,591,169	227	7,823,670	181	7,674,051	184	6,872,211	888	30,798,236
Bank of British Columbia	5	91,000	4	117,800	17	613,680	19	681,450	10	446,250	4	58,700	59	2,008,880
Total Chartered Banks	440	9,489,274	376	9,375,574	470	15,022,557	652	22,508,841*	532	22,041,693	403	15,448,278	2,873	93,886,217
Credit Unions	—	—	5	188,700	26	1,283,900	14	1,006,200	9	936,725	12	564,110	76	3,979,635
Total Other Lenders	—	—	5	188,700	26	1,283,900	14	1,006,200	19	936,725	12	564,110	76	3,979,635
Total	440	9,489,274	381	9,564,274	496	16,306,457	666	23,515,041	551	22,978,418	415	16,012,388	2,949	97,865,852

* This figure includes three loans made by the Provincial Bank of Canada in the total amount of $50,000.
Source: Compiled from Fisheries Improvement Loans Act; Annual Reports 1975–1981.

Table 6.7
Profile of Bank Involvements with Regard to F.I.L. Program, 1975–1981,
B.C. Region

Banks	Loans made		Average size of loan ($,000)	% of market
	No.	Amount ($,000)		
Royal Bank	1,013	30,137,049	29,750	31.0
Canadian Imperial Bank of Commerce	888	30,798,236	34,683	31.5
Toronto Dominion	426	16,445,364	38,604	16.8
Other Banks	543	15,455,560	28,463	15.8
Credit Unions	76	3,979,635	52,364	4.7

Source: Compiled from Fisheries Improvement Loans Act;
Annual Reports 1975–1981.

banks and three credit unions shared the remaining 619 loans valued at $19,435,195. Credit unions held few loans but had the largest average size per loan at $52,364.

Tables 6.8 and 6.9 show that the purchase of vessels accounted for almost two-thirds of the loans (1,742) for a total amount of $73,912,201. There were 173 loans assigned for building and constructing boats totaling $5,730,896 and 448 loans for fishing equipment purchases amounting to $3,584,026. Repairs and overhauls to vessels and engines resulted in 254 loans worth $2,948,647, and 151 loans were made to purchase new engines, the total value of which was $1,496,293.

In addition to the Fisheries Improvement Loan program, state financing of fishermen also occurred through the Department of Indian Affairs and the Fisheries and Marine Services. Between 1968 and 1979, a special Indian Fishermen's Assistance Program (I.F.A.P.) provided loans, grants, and training subsidies amounting to $17,030,887, of which $15 million was applied toward the purchase of new or used vessels, rebuilding, and the purchase of fishing gear and equipment (Foodwest Resource Consultants, 1979:69). A second temporary emergency plan allocated another $4.1 million in direct loans, grants, and loan guarantees to assist fishers, most of which were used for repairs, refits, and repayments on bank and company debt (Regional Indian Fishermen's Assistance Board, 1982:49, 61–4). Similarly, the Federal Business Development Bank operated a special loans program supplying fishers with capital for boat purchases, fishing equipment, and operating costs. In 1976–1977, $1,658,500 was allocated for the purchase of vessels. By 1981–1982, the total amount available was

Table 6.8
Loans for Various Improvement Purposes and for Building and Construction, 1975–1981,
B.C. Region

Years	Purchase of boats/vessels		Purchase of engines		Repair/ overhaul		Building and construction		Total	
	No.	Amount ($,000)	No.	Amount ($,000)	No.	Amount ($,000)	No.	Amount ($,000)	No.	Amount ($,000)
1975–1976	232	7,085,709	23	188,830	42	401,803	57	1,341,555	354	9,017,897
1976–1977	234	7,477,180	16	121,367	36	219,922	50	1,392,850	336	9,211,319
1977–1978	311	12,150,425	28	264,575	37	446,722	66	2,996,491	442	15,858,213
1978–1979	393	16,831,235	27	232,548	48	715,447	—	—	468	17,779,230
1979–1980	334	18,414,047	33	389,240	40	609,298	—	—	407	19,412,585
1980–1981	238	11,853,610	24	299,733	51	555,455	—	—	313	12,708,798
Total	1,742	73,812,201	151	1,496,293	254	2,948,647	173	5,730,896	2,320	83,988,042

Source: Compiled from Fisheries Improvement Loans Act; Annual Reports 1975–1981

Table 6.9
Loans for Fishing Equipment, 1975–1981, B.C. Region

Years	Nets and traps		Radio/electronic equipment		Vehicles		Other equipment		Total	
	No.	Amount ($,000)	No.	Amount ($,000)	No.	Amount ($,000)	No.	Amount ($,000)	No.	Amount ($,000)
1975–1976	18	667,052	15	38,702	4	13,425	49	291,915	86	471,377
1976–1977	7	43,060	10	58,340	2	30,760	26	220,795	45	352,955
1977–1978	10	102,490	10	32,738	6	37,210	28	275,806	54	448,244
1978–1979	19	227,793	27	121,231	10	52,074	45	538,543	101	939,641
1979–1980	14	107,780	20	87,139	9	50,874	38	451,225	81	697,018
1980–1981	10	58,594	79	163,820	8	29,290	44	423,087	81	674,791
Totals	78	667,052	101	501,970	39	213,633	230	2,201,371	448	3,584,026

Source: Compiled from Fisheries Improvement Loans Act; Annual Reports 1975–1981

reduced to $450,000, but as of July 1982, 116 loans remained outstanding for a total debt load of $4 million (Pearse, 1982:61).

Subsidy programs to Canadian shipyards, under the auspices of the Department of Industry, Trade, and Commerce, were a further incentive for increased capitalization on vessels. Under the present plan, shipyards receive a 9 per cent subsidy on the agreed-upon cost of building or remodelling boats greater than seventy-five feet in length, thus promoting the shift into larger, more technologically sophisticated troll and seine vessels. Between 1979 and 1982, a total subsidy of $5.7 million was paid on thirty-two vessels, and this in the years after the vessel construction boom (Pearse, 1982:160, 169). The Income Tax Act also provided considerable incentives for high-income fishers to invest and reinvest in fishing production. Tax credits of 10 per cent were allowed from tax payable (not taxable income) for new investments designated as equipment on new fishing vessels. Additionally, the act allowed fishing vessel depreciation at an annual rate of 15 per cent, but new vessels were permitted an accelerated rate of 33.3 per cent, claimable on a direct-line basis. This represents an enormous tax shelter on an amount of income equal to the value of the boat in only a three-year time period. This encouraged expanded fishing capacity (Pearse, 1982:161).

Finally the federal state sponsored a Small Business Development Bond Program as a hardship measure for fishers in the industry. Initially intended to encourage capital expenditure and purchase of real property (i.e., buildings, equipment, upgrading, and improvements) for incorporated companies, the bonds were extended, in 1981, to include firms in financial difficulty. Provisions were added to include individuals and partnerships. Under this new program, the banks converted fishers' debts into bonds. The lender received the interest paid by the borrower free of tax and was able to accord a lower rate of interest to the fishers than would normally be the case, usually 2 or 3 per cent above half the bank prime rate. But fishers were unable to use the interest paid under the bond as a deductible expense for tax purposes. Increasingly in 1982 and 1983, these bonds were sold to insolvent fishers as a temporary means of refinancing their continuation in the industry. By 1983, some five hundred British Columbia fishers held business bonds amounting to an estimated $70 million (Reid, 1982).

In sum, between 1969 and 1982, the capital value of licences went from zero to $145 million, and the capital value of vessels escalated by almost 500 per cent from 91 million to $432 million; this despite a reduction of 1500 vessels in the overall fleet. The debt to equity ratio across the fleet is about 50/50, much of this increase has been borne by individual producers borrowing back capital. Of a total capital value of $580 million, about $300 million represents debt: $55 million is outstanding Fishermen's Improvement Loans (F.I.L.), $55 million is on credit union ledgers, and $200 million is carried by the major chartered banks.[10]

Since 1968 state intervention has resulted in uneven, ad hoc, and problematic social policy for fishers. Through licensing and capital assistance programs, the state has transformed property rights and relations, reduced the size of the fleet, and strengthened the formal position of petty commodity producers in the fishery. But it contributed to increased and redundant overcapacity and overcapitalization, created an overregulated control apparatus, restricted the "privilege to fish," and facilitated the *real subordination* of petty producers to financial and industrial capital.

The role of the state has also been contradictory vis-à-vis small and middle-sized processors. During the 1970s, there were twenty financial assistance programs geared to the processing industry: eight were for capital costs; four each for export and marketing and new product development; two for manpower assistance and one each for working capital and counselling assistance. The funds expended allow for a tentative assessment. Table 6.10 shows that between 1973 and 1979 (where information is available) the provincial and federal state allocated $4,373,000, to B.C. processing firms, of which 97 per cent was for capital cost expenditures. Much of this money was used in constructing new and improved marketing facilities, expanding existing capacity, diversifying product processing, purchasing new equipment, and upgrading services (Foodwest Resource Consultants, 1979:31–78). Interventions on behalf of cooperatives and small companies through Agriculture and Rural Development Agreement, Federal Business Development Bank, Department of Fisheries and Oceans, and B.C. Development Corporation marginally increased the financial strength of the fresh/frozen and salmon roe segments of the processing industry. They also contributed more capacity to already excessive processing ability. Reducing overfishing and overcapacity were undercut by a contrary set of initiatives designed to expand capital growth and keep processors in the industry.

State interventions on behalf of middle and small firms were insufficient to prevent their demise. Oakland Fisheries went into receivership just after expanding its processing facilities, courtesy of a Federal Business Development loan. Tofino Fisheries received a similar loan as well as a special ARDA grant, and then shortly went bankrupt. Most telling is the financial assistance provided for the Pacific North Coast Cooperative to construct a new cannery. Between 1974 and 1979, state aid exceeded $7 million and the operation is still unable to support itself (Shaffer, 1979: 135–140). This policy of financing capital-cost requirements without providing working-capital assistance meant that many companies acquired assets and expanded their fixed capital. But they could not meet their operating costs. When prices, markets, or production declined, the pattern was familiar: debt, bankruptcy, takeover, and/or failure.

Table 6.10
State Financing Utilized by B.C. Processors
by Purpose (Selected Years)

Year	Category	Amount	Per cent
	Capital costs	(,000)	
1973–1979	Fish-chilling assistance	1,144	
1978–1979	Regional Development Incentives Act	518	
	Small Business Loans Act	—	
1977–1978	Assistance to small enterprises	78	
1977–1979	Agriculture and Rural Development Subsidy Agreement	1,055	
1977–1978	Low-interest loans assistance	225	
1974–1978	B.C. Development Corporation	1,238	
	Subtotal	4,258	97.0
	Working capital and capital costs	—	
1977–1978	Manpower assistance industrial training program	6	0.5
	Export and marketing		
1973–1978	Program for export market development	30	
	Promotional projects program	—	
	Export development corporation	—	
1977–1979	Market development assistance program	79	
	Subtotal	109	2.5
	Efficiency and new product development	—	—
	Counselling	—	—
	TOTAL	4,373	100.00

Source: F.R.C., Financing in the B.C. Fishing Industry, Vancouver, M.R.B., 1979:77.

The State and Monopsony Relations

State intervention has been more protective of the major corporate firms, especially B.C. Packers. There have been two major rounds of corporate consolidation since the Davis Plan. Each involved takeovers, plant eliminations, usually in outlying communities, and new constructions for fresh fish and cold storage capacity and/or for increased canning capacity.

In 1969, Anglo–British Columbia Packing Company folded and sold its assets to B.C. Packers and Canadian Fish Company. The state assisted in the transaction by purchasing ABC's rental fleet to support a native presence in the fishery. Immediately the Klemtu (B.C. Packers) and Butedale (Canadian Fish) plants were shut down. Both companies then closed their J. H. Todd operations, removing facilities from Sooke, Rivers Inlet, and the Skeena. In 1971, B.C. Packers reduced their Namu cannery and turned it into a cold storage plant. They withdrew 150 boats from their company fleet

(125 gillnetters, 25 seiners), and converted the licences into new, higher capitalized vessels. A year later Canadian Fish Company closed its large North Pacific Cannery in Prince Rupert. Between 1968 and 1972, nine canneries were shut down, and the two major firms established between them a near monopoly position in the industry (UFAWU, 1984:13, 28).

The second phase of consolidation came in 1980, after the collapse of the roe herring market. B.C. Packers acquired all the northern holdings of its competitor, Canadian Fish Company. The bankruptcy of New England Fish Company of Seattle prompted its subsidiary, Canadian Fish Company, to sell off many of its B.C. assets.[11] For $15 million (U.S.) B.C. Packers took over the major assets, including the large Oceanside cannery, the rental fleet of 122 gillnetters, and the Atlin, Tofina, Phoenix, Brittannia, and Scotch Pond operations. Since then, B.C. Packers has closed the Port Edward operation, the Tofino plants, and the Seal Cove storage and fillet plant. It reduced Namu to a small service operation and sold Coast Oyster, the St. Mungo plant, and its Los Angeles holdings. It also obtained a net gain of $3 million by disposing of its interest in United Oilseed Products (UFAWU, 1984:14).

As B.C. Packers was eliminating unprofitable operations and moving out redundant assets and personnel, it embarked on a heavy capital investment program to maximize productivity. In 1981, it invested $9.8 million to expand its Prince Rupert plant (formerly C.F.C. Oceanside) and $1.3 million to improve the reduction plant operation in Steveston. A year later, it purchased Petersburgh Processors in Alaska for $9.4 million and closed down the Namu station (B.C. Packers, Annual Report, 1981).

The move toward monopsony in the canning field was initially financed internally, with a subsequent increase in company debt for 1981. But the role of the state was very important. Federal assistance to B.C. Packers–Weston, in the years 1981 to 1983, exceeded the $15 million spent to take over Canadian Fish Company. In 1981, $12.7 million was advanced to B.C. Packers to purchase its gillnet fleet for native Indian fishers. Ostensibly a job-creation program for some two hundred fishers, the sale allowed B.C. Packers to sell a costly archaic fleet and use the profits for plant restructuration. Ironically, much of the capital was used at the Oceanside cannery and eliminated native jobs in the process. Indeed, the Northern Native Fishing Corporation, set up to administer the fleet on behalf of the northern tribal councils, quickly ran into trouble. Many of the members could not pay the $1,000 licence rental, and in order to fish, they borrowed directly from B.C. Packers, subordinating themselves to the company through debt obligations.

A year later, a major land transaction between the federal state and B.C. Packers resulted in a $9.8 million gain for the company. The sale of Paramount Pound and associated properties in South Richmond will likely lead to a ten-year housing development program, of which there may be even further returns to the company. In all, state subsidies and purchases of unwanted assets amounted to over $20 million.[12] Still on the agenda is the long delayed sale of one hundred B.C. Packers seine boats, which probably hinges on a state buy-back or purchase for a third party, and the possible sale of Port Edward and Namu operations to the federal government (*Vancouver Sun*, May 11, 1983).

Going Out with the Tide:
Crisis, Bankruptcy, and Debt

B.C. Packers emerged as the monopsony power in the industry precisely at the time when small and middle-sized firms were failing. The withdrawal of Japanese equity and portfolio investment, an important feature of the B.C. industry from 1972 to 1979, combined with poor markets, rising costs, and high bank interest charges led to a substantial "shake-out" of firms from the industry. Proverbs lists eight firms that had foreign investment in 1979 that were out of business by 1982. Most importantly, Oakland Fisheries Ltd., owned by Marubeni and Hako Fishing, was closed in 1982 and has not reopened. Production was shifted to Cassiar, but in 1983 it and its subsidiary Royal Fisheries were forced into receivership by Marubeni and the Royal Bank of Canada for debts in excess of $20 million. Norpac Fisheries, 30 per cent owned by Japanese interests, went out of business in 1980 and its assets were purchased by J. S. McMillan and Island Cash Buyers. Three other firms, Northwestern Fisheries, Pacific Rim Maricul-ture, and Tradewind Seafoods also went into receivership (Proverbs, 1978, 1980; UFAWU, 1984). Millerd Fisheries remains in grave trouble. It did not do business in 1979 or 1980; in 1981 it received $300,000 from Nippon Suison to reopen; two years later it was placed in receivership owing $5 million to creditors, including $1.9 million to Nippon Suison and $1.8 million to the Royal Bank of Canada. In 1984, it reorganized and now operates on a very reduced basis (Proverbs, 1982; *Vancouver Sun*, March 4, 5, 1983). Two other firms, Quality Fish and Central Native Fishermen's Co-operative, operated under conditions of a bank-ordered merger and with the assistance of a $270,000 government pack subsidy (*Vancouver Sun*, September 1, 1983; UFAWU 1984:4–6). They went into receivership in 1985.

Fishers also faced a serious crisis condition by the end of the 1970s. Rising salmon prices peaked in 1979 and then fell off for the period from 1980 to 1983. Export sales slumped with the result that there were excessive inventories of Pacific canned salmon held by B.C. processors.[13] The roe herring fishery, which had a "bonanza" year in 1979 because of strong Japanese demand and high prices, went into a serious decline. In 1979, the total landed value reached $125 million, gillnet vessels averaged gross earnings of $50,000, and seine vessels made, on average, about $268,000 (subject to fluctuations around these averages). In 1980, a weak market, a collapse of roe herring prices (down to 52 cents from 134 cents per pound in 1979), and a fishers' strike drove down landed values by over $100 million. Landings went up in 1981, 1982, and 1983, but prices stabilized at bottom levels (Pearse, 1982:101).

Halibut, an old, valuable fishery, continued to deteriorate. Large catches by foreign trawl fleets, exclusion from the Alaska fishery because of the two-hundred-mile limit, low catch-sharing agreements for Canadian fishers (in 1982, the quota was 5.4 million pounds compared to more than 30 million pounds in 1967), and an increase in fleet size because of generous state landings qualifications and licence appeal regulations resulted in a decrease in the landed value of halibut from $16 million in 1978 to about $6 million in 1981 (Pearse, 1982:122–123). The low price has remained for the last four years.

The groundfish fishery, which doubled its landings between 1971 and 1978, levelled off in 1979 and declined slightly in 1980, 1981, and 1982. Like the halibut fishery, it expanded well beyond its capacity and is now depressed. Pearse (1982:129–130) notes:

> This fishery is under extreme financial stress. The markets for groundfish species have declined. The vessels involved in trawling consume exceptionally large amounts of fuel, so that escalating fuel costs have had a particularly heavy impact on this fleet. This has been aggravated by vessel subsidies that biased construction toward larger, less fuel efficient vessels.

The decline in markets, prices, and earnings was matched by rising costs of production, soaring interest rates, and currency instability. Equipment, repairs, labour, and fuel costs escalated in the years from 1978 to 1982. Fuel prices increased by an astounding 140 per cent between 1978 and 1981 (McKay, 1982). Currency changes had an impact. Devaluations of the British pound, French franc, Australian dollar, and Japanese yen made the purchase of B.C. fish products increasingly expensive. Interest rates also reached unprecedented levels in 1979, 1980, and 1981, hovering around 20 per cent.

By the beginning of the 1980 fishing season, fishers were caught in a severe economic "squeeze." The substantial investments in vessels, gear, and equipment could not be paid from returned earnings. From 1980 to 1984, fishers have faced growing indebtedness, defaults on loans, boat arrests, repossessions, seizures, and forced sales of vessels, gear, and equipment. By the end of the 1980 fishing season, it was found that 46 per cent of 1,266 fishing vessels contacted were unable to make full loan payments, compared to the usual 3 per cent of earlier years. One-quarter of the fishers in arrears (12 per cent of the total) were making no payments at all on their vessel loans, compared to the usual rate of 0.5 per cent. The number of vessels seized or put up for sale at lender requests was 2 per cent of the total, compared to the usual 0.5 per cent.[14]

The situation worsened in 1981. A survey in the lower mainland area (Vancouver, Richmond, and Steveston), found that half of 350 fishing accounts were in arrears. A follow-up study of 187 fishing accounts found that 61 per cent were in arrears or had been refinanced because interest and principal payments could not be met. Three per cent of the financed vessels were seized. In the Prince Rupert–North Coast region, 320 accounts were surveyed and 90 per cent were in arrears. The follow-up study found that 219 of 290 accounts (76 per cent) were in arrears or had been refinanced. Between twenty-five and thirty fishing vessels were seized or repossessed.[15] One lending institution holding accounts for four hundred fishers estimated that they had $36 million in loans out on 207 vessels. The average debt on trollers (107) was $137,298.00; on gillnetters (67) $69,586.05; on seine vessels (14) $589,187.41; on bottom fish trawlers (7) $810,739.01; on halibut vessels (9) $278,938.97; and on C licence and herring-only boats $37,260.00.[16] The debt situation of the west coast troll fleet was also desperate. Of 175 accounts surveyed in Victoria, Port Alberni, and Nanaimo, 80 per cent were in arrears or were refinanced, many for the second consecutive year. Approximately twenty vessels were repossessed.

A third study in 1982 revealed that the economic condition for many fishers had deteriorated further. The three major lending institutions, holding loans on 3,100 vessel owners, reported that 35 to 45 per cent were nonproductive. On vessel-only loans, the two major banks reported that of 451 fishers unable to make full payments, approximately 220 made no payments at all. The average size of the loan in arrears (based on 363 vessels) was $80,000. From July to December of 1982 these banks seized a minimum of eighty-one vessels — twenty-one seiners, nineteen gillnetters, and forty-one trollers, of which thirty have been resold.[17]

A third lending institution financing 260 vessel owners reported that 80 per cent of their vessel loans were in arrears. The amount of interest outstanding increased from $1.5 million to $4.5 million in one year. The average amount outstanding was $173,000. One-quarter of those financed were told to leave the industry. Four boats were seized. By the summer of 1984, lending institutions had impounded some 150 fishing vessels for financial default.

Consolidations, high indebtedness, bankruptcies, and economic failures have contributed to further centralization of economic development. The economic base of coastal communities has eroded. Large sections of the B.C. coast are now abandoned villages. As corporate capital merged and restructured, plants were closed, and services to fishers declined. Risks and costs were shifted to the producer at the same time that jobs disappeared. The number of jobs lost as a result of B.C. Packers' recent restructuring is more than one thousand. The bankruptcies of Cassiar, Royal Fish, and Oakland have put another eleven hundred shoreworkers and fishers out of work. Between 55 and 65 per cent of the shore workforce are now based in either Prince Rupert or Vancouver (UFAWU, 1984:5, 30).

B.C. Packers is the giant of the industry. It controls the north coast and they own at least twenty of the fifty-four canning lines; the rest are divided among a dozen firms (UFAWU, 1984:15). Table 6.11 shows that the events of 1980 to 1983 resulted in total additions to fixed assets of $54.7 million, to receivables investments of $9.1 million, and to proceeds on disposal of assets of $37.5 million. The rates of return, however, have been poor. B.C. Packers' net surplus after tax as a percentage of equity was −4.5 per cent a year between 1980 and 1982. Surpluses have been taken by the banks in the short term. According to union estimates, lenders took 98.8 per cent of the surplus value produced in 1980; they accounted for 93.3 per cent of total surplus in 1981; their share was a staggering 230 per cent in 1982; and in 1983, they absorbed 89 per cent of all surplus produced (*The Fisherman*, April 8, 1983).[18]

The real source of B.C. Packers' domination is its vertical integration at the wholesale level. Weston, through Loblaws, has a powerful position in Canadian and international food distribution. Loblaws cooperates with Provigo in the Foodwide of Canada Buying Group. Together, they accounted for $9.1 billion in sales in 1982. Provigo, in turn, is linked through M. Loeb to the I.G.A. – Safeway Buying Group ($6 billion in sales in 1982), giving B.C. Packers direct or indirect access to some thirty domestic wholesalers and retail chains (*The Globe and Mail*, March 13, 1983). In addition to assured domestic and foreign markets in the United States and Britain, the Weston connection provided B.C. Packers with central cash-management services for finance and investment purposes, guaranteed display and product promotion, and reduced costs in purchases of goods and services from member companies.

The remaining assets of Canadian Fish Company have also been restructured at the wholesale level. The recent purchase of the company, by Jim Pattison, has vertically integrated the company into a large Canadian grocery-buying group — United Grocery Wholesalers. Pattison bought Canadian Fish Company to supply his Overwaitea retail food chain. In turn, Overwaitea cooperates with Woodward stores, Alberta Grocers Wholesales, A&P, Federated Co-operatives, and six other wholesale chains giving Canadian Fish Company direct or indirect access to eleven domestic outlets (*The Globe and Mail*, March 13, 1983).

The position of the remaining small-sized firms and cooperatives was uncertain. The processing firms that are tied to the remaining Japanese trading companies did have direct vertically linkages to Japanese retail chains, but the withdrawal of Japanese interests and the threat of state controls on foreign investment seem to have fractured this wholesale linkage (Shaffer, 1979:64, 168–188). In 1984, firms like the Central Native Fisherman's Co-operative (with Marubeni capital) and the Pacific North Coast Native Co-operative were unable to market their salmon packs and sold them directly to B.C. Packers and Canadian Fish Company (UFAWU,

Table 6.11

Chronology of B.C. Packers Consolidation and Control, 1980–1983

Year	Event	Known cash involved	Additions to fixed assets	Additions to receivables, investments	Proceeds on disposal of assets
1980	B.C. Packers takes over most of Canadian Fish Company holdings.	$15 million (U.S.)			
	B.C. Packers sells interest in United Oilseed Products.	$ 8 million, net gain $3 million			
			$28.5 million	$2.1 million	$12.5 million
1981	B.C. Packers invests in Prince Rupert — Oceanside Plant.	$ 9.4 million			
	B.C. Packers improvement in Steveston — Reduction Plant.	$ 1.3 million			
	B.C. Packers closes Port Edward operation.				
	B.C. Packers sells St. Mungo Plant.	?			
	B.C. Packers sell gillnet fleet to D.I.A. for tribal councils.	$12.7 million			
			$ 9.4 million	$2.7 million	$ 3.9 million
1982	B.C. Packers closes Seal Cove and Tofino operations.	?			
	B.C. Packers sells Coast Oyster.				
	B.C. Packers purchases Petersburgh Processors in Alaska. Namu operation reduced.	$ 9.4 million			
	B.C. Packers lays off office staff of one hundred.				
			$16.8 million	$4.3 million	$11.3 million
1983	B.C. Packers sells Paramount and related properties to federal government.	$ 9.8 million			
	B.C. Packers sells Los Angeles Plant.	?			
TOTAL			$54.7 million	$9.1 million	$37.5 million

Source: UFAWU, *The Future of the B.C. Fishing Industry*, vol. 2, Vancouver (1984:15–17).

1984:18, 19). The implications are clear. The small-volume, locally advertised brand-name suppliers are on the verge of extinction. Through vertical integration, food distribution chains are dictating the terms of business to fish processors.

Crises of State

State policies in licensing, promoting petty commodity producers, assisting small processing firms, and bolstering the dominant position of B.C. Packers had one single effect: a rapacious harvesting of the resource resulting in the erosion and destruction of the fish stocks. Fleet capitalization and overcapacity resulted in overfishing, and in turn, this led to a regulatory and management crisis for the federal Department of Fisheries and Oceans. Technological changes generated acute problems. The efficiency of the new fishing vessels outmatched the research knowledge and data collection ability of the state. The lag in scientific-biological information led to a new regulatory policy. An emphasis was placed on increased information gathering, new classifications, surveillance, and enforcement. Two things became clear. First, resource management did not go far enough in securing ample catches. Enhancement programs were needed. Second, producers and their catches had to be monitored more carefully.

Natural salmon stocks have been declining by 1.5 per cent per year in the aggregate. They are currently at low levels. The valuable chinook and coho stocks are close to extinction. Every year pink stocks fall rapidly, and chum stocks are in a major decline. Plans for increased escapements of fish have not relieved the pressures on the stocks. Some species are at danger levels from which rehabilitation may be impossible. The rebuilding of chum stocks, for example, will require the complete closure of commercial fishing (D.F.O., Pacific Fisheries Policy Options, 1984:2–5). Habitat degradation has exacerbated already delicate situations. Dams, dyking, dredging, landfills, logging, and mining have destroyed spawning and estuarial habitats and contributed heavily to stock declines and low productive capacity (UFAWU 1984: Appendix 2). New fishing gear and indiscriminate fishing in terms of species have also fed a growing "by-catch" problem.

These deleterious effects, among others, prompted the state to build up existing stocks. Conservation ideas were not novel. They have been a persistent theme in state fisheries ideology. From 1882 to 1937, the state promoted the value of hatcheries and fishways, but many of their projects were ad hoc, unsuccessful, and short-lived. In the twenty-five years following World War II, they constructed fishways and spawning channels in particular. Then in the 1970s they re-experimented more successfully with new hatcheries.

The major salmonid enhancement effort began in 1975. The conservationist thrust to "preserve, rehabilitate, and enhance natural salmonid stocks," however, was subjugated to definite economic and social ends. Between 1975 and 1977, the federal Cabinet authorized $6 million to develop a comprehensive plan. In 1977 they initiated the Salmonid Enhancement Program (SEP) aimed at doubling the stocks of salmon and anadromous trout. Planned as a joint project with the B.C. government, phase 1 lasted seven years and cost $157.5 million: $150 million of federal state capital and $7.5 million of provincial monies. Fully $80 million went into capital spending, most for major hatchery and fishway construction. Between 1977 and 1983, ten major, six middle-sized, and thirty small hatcheries were built, as well as three major fishways, one spawning channel and thirteen distributory channels (SEP, Annual Report, 1983).

Fish production capacity increased to 43.5 million pounds of salmon annually, about 85 per cent of the projected target. This had spinoff benefits. More discrete harvesting practices were developed and better stock research, management, and upgrading resulted (Pearse, 1982:50–51). The major benefits, however, accrued to the locally based building materials, engineering, and construction firms, who built many of the major production projects, and to foreign firms who supplied the major machinery, technology, and equipment. Not surprisingly, employment in constructing and operating enhancement facilities exceeding the target figure by more than 33 per cent. Major engineering projects accounted for over half the costs and three-quarters of the gains in net income (SEP, Annual Report, 1983).

The increases of SEP production also favoured large corporate capital in the fishing sector. Increased catch size did lead to some new entries and competition in the fresh/frozen market, but overall it did not challenge the monopsony structure. As noted, existing firms expanded their own collecting, packing, and processing facilities to inhibit entry and maintain concentration. Moreover, the species emphasis of enhancement was skewed in favour of the major processing firms. Figure 6.1 reveals that about 75 per cent of the total increases were in sockeye and chum salmon. Pink, coho, and chinook species were underemphasized by SEP. This species mix favoured the net-caught, traditionally canned fish over the troll and fresh/frozen ones. The more concentrated, vertically tied, canning-oriented segment of the industry was reinforced by the enhancement program. They exerted a greater domination over raw salmon purchases, and their buying concentration increased overall (Shaffer, 1979:150).

Economic returns to fishers were low. The benefit cost ratio of 1.3:1 was short of the projected 1.5:1. Net national income gains were about one-quarter of the original amount expected. Benefits to target areas were poor, about 40 per cent of the original target, and continuing employment

Figure 6.1
SEP Production Capacity by Species:
Phase 1 Targets and Performance to Date*

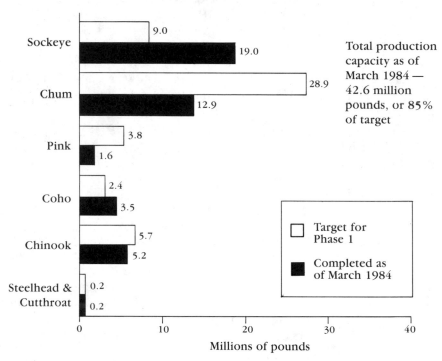

Total production capacity as of March 1984 — 42.6 million pounds, or 85% of target

Sockeye 9.0 / 19.0
Chum 28.9 / 12.9
Pink 3.8 / 1.6
Coho 2.4 / 3.5
Chinook 5.7 / 5.2
Steelhead & Cutthroat 0.2 / 0.2

☐ Target for Phase 1
■ Completed as of March 1984

Millions of pounds

* Production capacity measures the capacity of facilities that have been constructed since 1977 (SEP facilities only) and are now operational. The capacity is stated in terms of expected future adult catches, using current facility capacities, current expected egg to adult survival rates of enhancement techniques, and projected future Canadian catch rates.

Source: Salmonid Enhancement Program, Annual Report 1983, Government of Canada, Department of Fisheries and Oceans, p. 78.

for native people was below 50 per cent of the expected total (Pearse, 1982:50).

Ostensibly a state policy designed to maximize the social interest, SEP ended up facilitating the restructuring of capital in the industry. More fish and the species mix with enhancement did not lead to a conservationist situation or even to reduced levels of concentration and more industry competition. The objectives of SEP fit nicely with the strengths of large existing firms who preempted the entry of other firms by expanding and upgrading and by limit pricing, which transferred much of the gains in rents to the harvesting sector. More state capital = more fish = more capacity = more concentration seems to be the leitmotif of SEP.

Enhancement efforts went hand in hand with a new regulatory resolve that reconsidered free entry and the "right to fish." This did not upset the corporate sector. As noted by Marchak (Chapter 1), processing capital endorsed licence restriction because it reduced the number of fishers without affecting the processor's control of market supply. But regulating the Pacific commercial fisheries did draw the ire of petty capital producers, especially when the Department of Fisheries and Oceans, in its struggle to conserve the resource, commenced new regulation on the fishing grounds.

Wide, finely meshed, strong control nets were needed to monitor vessel licensing, gear restrictions, fish quality standards, the manipulation of open and closed areas and fishing times, and the protection of the fish habitat. These technical and complex provisions had to be enforced on thousands of commercial fishers working in a vast area, as well as on sports and native fishers. The entire complex grew to be confusing and cumbersome. Rules were applied differently according to separate contexts. Restrictions proliferated. Internal procedures were tightened and dispersed down through the system. In the end, the "great Canadian fish chase" involved increasing numbers of experts, bureaucrats, and enforcement officers apprehending, separating, processing, and returning fishers to the sea, newly tagged, ready to chase the appropriate fish in the required location, at the right time. The "privilege to fish" was radically refashioned as bureaucratic imperatives changed.

Internally, however, the Department of Fisheries and Oceans was in some disorder. Coherent policies were absent. Administrative changes and disruptions were frequent. Staff turnovers were regular. Throughout the 1970s, strains between Ottawa and the Pacific region were severe. As Pearse (1982:235) noted:

> Aggravating the general vagueness of policy is the widespread perception that administration itself lacks consistency and vigour and that policy decisions are pliable in the face of lobbying and other pressures.

The administration and enforcement of policy was suspect. Constant modifications and outright reversals undermined the substantive claims of the state. Symbolically they performed a "scarecrow function," targeting, declaring, and warning, but they were weak in practical enforcement, in statistical information, in the consulting process, and in policy documentation. This led to a lack of response to urgent problems and needs. Internecine disputes, ambiguous and burdensome procedures, and contradictory rules and decisions contributed to loose administration and ineffective state action.

Institutional instability was twofold. Internally, the responsibility for fisheries was negotiated and renegotiated considerably and politically. From 1971 to 1978 the department was one of a number of disparate agencies in the conglomerate of the Department of Environment. It suffered for attention, focus, consistency, and qualified personnel at senior levels. Externally, there were administrative encroachments and jurisdictional disputes, with a corresponding dilution of regional influence and ability. Absorption by Ottawa led to successive waves of change and "undoings" of change. Qualified personnel retired, resigned, or were fired as management regimes came and went. In these seven years, there were three directors general of the Pacific region, each of whom made major organizational changes during his tenure (Pearse, 1982:234). The instability continues to the present.

This centrifugal tendency to integrate the Pacific region into the orbit of Ottawa operated in concert with a centripetal tendency within the Pacific region proper. In the 1960s, there was a trend toward centralization in Vancouver. Professional biologists, engineers, and economists were recruited to manage the medley of technical problems from a central headquarters. In 1970, this trend was reversed by the decentralization of fisheries management into northern and southern regions. Then a renewed drift back toward centralization occurred, followed by a current decentralization thrust, especially for fisheries management and habitat protection. A third region was created, and separate structures and jurisdictions crisscrossed in a bewildering pattern. Area managers were only partially responsible for activities in their regions. Vancouver office had authority in some regions on some matters, but not on others. There was a continual shuffling and shifting of policy priorities, research, and implementation responsibilities between departmental units (Pearse, 1982:235).

In 1978, a separate Department of Fisheries and Oceans was created, in effect cancelling the earlier decision to consolidate fisheries with forestry, wildlife, meteorology and water, and environmental protection. With this new structure came a new series of management regimes. But the sole focus on fishing activities did not result in effective policy. Administrative lapses and ruptures persisted, except now the state was confronted with a crisis situation.

Overcapacity, overexploitation, fisheries jurisdiction disputes, the Indian fishery problem were not responded to in a decisive manner. Hesitation and confusion were prominent. The state had an agenda for reform and regulation, but it was vague and only weakly presented. It could be undone by the vociferous. Processing firms and associations, state experts and administrators, biologists, unions, and professional associations of fishers lobbied for their particular interests and pressured for partisan policy agendas. The state was an arena of social conflict, but its personnel were weak participants in policy disputes. Political autonomy was not profoundly obvious. A frequent pattern was for the department to address a problem by creating new priorities, new regulations and rules, and then promptly undo or compromise them. Licensing, boat replacement procedures, and fish escapement targets, for examples, were routinely established, undermined, and reset. These interventions were politically insensitive and economically irritating. Directed primarily at petty producers, they fostered intense conflicts. The ad hoc and paternalistic style of policy, decision making, and administration bolstered the view, among fishers, that the state was uncaring and incompetent.

The consequences are telling. Already complicated laws, regulations, and rules have been further compounded so that the expanded myriad of regulations is a complex and unworkable basis for administration and enforcement. The state can seldom enforce its cumbersome collection of closed areas, seasons, and times. It is relatively ineffective in monitoring gear restrictions and the conditions governing harvesting and sale of the resource. The use of informers, bounties, special undercover units, and citizens' fish patrols have been uncertain aids to state control. One hundred and twenty-five officers are no match for a defiant fleet of thousands. Ideologically, the state is perceived as a negative power: defining, surveying, limiting, and disciplining. User groups in the industry resent, mistrust, and contest many decisions. They ignore new regulations, sell fish "over the side," confront fisheries officers, withdraw from government committees, take legal action in defence of their perceived rights, occupy fisheries docks, buildings and offices, and continue to wage political campaigns against the Department of Fisheries and Oceans. In a word, a profound polarization now characterizes the relations between the state and the fishers.

NOTES

1. There were other uses to which salmon were put — cured, frozen, fresh, salted, and dried—but these were decidedly secondary compared to canning.
2. This is not to suggest that there were no wage-labour fishers. There were, but they had become increasingly marginal and the petty capitalist cycle of reproduction of labour power was the dominant one.
3. It is important to note the overall scale of state involvement. Up to V-J Day, August 15, 1945, the Dominion government spent more than $700 million on industrial plant expansion, of which more than 75 per cent comprised wholly owned Crown companies and the remainder war equipment and installation added to private industrial plants. In addition, more than $500 million worth of private war investment was encouraged through special tax credits and special allowances for depreciation and depletion, and about a similar amount for tooling

costs and related expenditures was allowed to be charged to current expenses. Measures providing tax relief represented wartime variants of special depreciation provisions. In addition, some $800 million were spent on defence construction and about $23.6 billion on war goods of all kinds, including munitions, military equipment, and military pay, or a total of $28 billion. This is five times the gross national product in 1939 (James, 1983:299).

4. Policy tended to centre around four main programs: large capital contributions and allowances for plant development; subsidies and depreciation allowances for vessel construction; local educational/ vocational activities for fishers; and resource conservation.

5. The Subsidy Program for Construction of Fishing Vessels in British Columbia breaks down as follows:

Fiscal year	Expenditures
1942–43	$ 61,000
1943–44	122,000
1944–45	116,000
1945–46	34,000
Total	$333,000

The subsidy value was at a rate of $165 per gross ton. But the real value of twenty packer-seiners was between $960,000 and $1 million.

6. For a more detailed description of this often-discussed program see G. Alex Fraser, "Licence Limitation in the British Columbia Salmon Fishery," Vancouver, 1977; and P. H. Pearse and J. E. Wilen, "Impact of Canada's Pacific Salmon Fleet Control Program, *Journal of Fisheries Research Board of Canada*, vol. 36, no. 7, 1979, pp. 764–9.

7. Vessels that landed in excess of 10,000 pounds of pink or chum salmon (or other salmon species) qualified for Class "A" licences. Those with smaller production output qualified for Class 'B' licences.

8. Usually the term "special circumstances" referred to those fishers who were ill and unable to fish in the 1967 and 1968 seasons or those who were in the process of having their salmon fishing vessel constructed.

9. The Fisheries Improvement Loans Act guarantees loans of up to $150,000 from banks for fishers for the purchase, repair, or land improvement of vessels. The program started in 1955 and is now under the authority of the Department of Fisheries and Oceans.

10. The estimate (probably conservative) is based on data compiled from bank managers, DFO personnel, and credit union managers. It does not include company loans to fishers, and it does not consider indebtedness to other state agencies or the special positioning of

native financing. It refers exclusively to fishers and does not include the debt load of processing firms.

11. By most accounts Canadian Fish Company was solvent and competitive. The sale was directly related to the bankruptcy in the American market.

12. In 1982, Weston was ranked as Canada's fifth largest corporation with sales of $7.8 billion and a return on invested capital before tax of 14.6 per cent.

13. Canada's share of foreign salmon markets underwent stiff competition and displacement. In terms of the frozen markets, the U.S. has 83 per cent of the Japanese volume compared to 39 per cent for Canada; 54 per cent of the French volume compared to 39 per cent for Canada; and 53 per cent of the United Kingdom volume compared to 23.7 per cent for Canada. The canned salmon markets are as follows: Canada's share by volume of the United Kingdom market is 51.7 per cent as compared to 47 per cent for the U.S.; 63.6 per cent of the Belgium market as compared to 18 per cent for the U.S.; and 30.7 per cent of the Australian market as compared to 53.1 per cent for the U.S. In addition, Japan and the U.S. have increased their domestic supplies, impeding Canadian performance in their markets, while the latter has expanded its involvement in Canada. For example, in 1981, 11,001 tons or 12.1 per cent of exported American frozen salmon and 235,946 full cases or 18 per cent of exported American canned salmon came into Canada. In that same year, the B.C. canned pink salmon pack totalled 1,105,915 cases, by far the largest since 1962; 321,000 cases sold domestically while 101,006 cases of U.S. pinks were imported. The unsold B.C. inventory at the end of 1981 was 604,000 cases. The effect was to lower earnings in the industry. For 1980 the total value of salmon declined $133 million from $187 million in 1979. See Fisheries Association of British Columbia, "A Status Report on the Commercial Fishing Industry of B.C.," Vancouver, 1983.

14. I am grateful to the Department of Fisheries and Oceans Canada, Pacific Region, for providing me with access to this information. The data are a summary of their findings.

15. The two surveys were conducted in December 1981 and May 1982. While not complete, they nevertheless identify the major trends. The first survey was of 800 fishing accounts covering the financial institutions in Prince Rupert, Victoria, Port Alberni, Nanaimo, Vancouver, Richmond, and Steveston. The second survey was of 750 accounts, and the sample was somewhat different because of personnel changes at some of the financial institutions. Many of the data in this paragraph are summarized from their reports.

16. This was obtained in an interview with a financial manager of a major lending institution.

17. These figures should not be taken as representative of the total numbers in each category. One of the banks refused to release its demand loan portfolio to me, so the numbers are certainly higher. It must also be stressed that these figures are a minimum number. The three other chartered banks in the industry made some vessel seizures and sales. But their numbers are low.
18. In 1982, B.C. Packers declared a loss, but its poor performance must be qualified by considering the larger Weston affiliation. In 1982, the year of the record loss, Weston reported a 15 per cent return on equity and net earnings in excess of $68 million.

7

"Because Fish Swim" and Other Causes of International Conflict

PATRICIA MARCHAK

Former Prime Minister Trudeau's quip in response to the question of why the federal government retained control of the fisheries might equally have been given to explain why the United Nations devoted thousands of hours to deliberations on the Law of the Sea. Salmon travel from the west coast spawning streams as far as the tip of the Alleutian chain and they mingle with Asian-sourced salmon, so that it is impossible to determine their origin until they return to spawn. Halibut do not range such a distance, but they do navigate international boundaries between Canada and the United States. Groundfish, if caught far out at sea, are gone before Canadian inshore fishers can put out their lines. Of the major B.C. fish stocks, only herring have no wanderlust.

The habits of fish are part of the reason that so many negotiators of so many countries have spent years fashioning the law of the sea. There are other reasons. The same habitat shared by swimming fish teems with international ships, contains oil and mineral beds, provides routes between military bases, and carries pollutants from shipwrecks and shore-based industries. It is the subject of endless jurisdictional disputes for many reasons besides national rights over anadromous fish. At a subnational level, as well, there are numerous vested interests that affect the bargaining stances of national governments and that are not always about fish and fishers. Complex as all this is — and too vast to deal with in a single chapter — it is pointless to discuss the law of the sea with reference solely to fish. The objective of this chapter is to situate bilateral and multilateral fisheries questions within that larger context, though specific issues will not be discussed in exhaustive detail. Except for a brief reference to early bilateral disputes between Canada and the United States, the focus will be on the post–World War II era.

OVERVIEW

Bilateral negotiations between Canada and the United States began (with Britain negotiating for Canada) in the 1850s. Many of the contended issues have never been resolved, and repeat performances recur at irregular junctures. Canadian negotiators have complained frequently that American governments are relatively unconcerned with issues of conservation, and that conservationist measures taken in Canada, especially with respect to those species that migrate south and into the high seas, are frustrated by the lack of complementary measures in the United States. They have also complained that it is impossible to arrive at a negotiated settlement because so many diverse interests and different governments or government agencies participate in the U.S. decision-making process.

On negotiations between these countries, the two major pre-1950 agreements affecting the west coast established the International Pacific Halibut Commission (1923, originally named the International Fisheries Commission), and the International Pacific Salmon Fisheries Commission (1937). The central unresolved disputes are over salmon spawned in the Fraser River and the ownership of the entire Dixon Entrance–Hecate Strait region.

International law of the sea was sparse until the late 1950s. Several conventions were signed in 1958, respecting the territorial sea and contiguous zone, the continental shelf, the high seas and conservation of living resources. To these, Canada was a signatory only to the Continental Shelf Convention. However, Canada was very much involved — as a junior negotiating partner to the United States — in negotiations with the Japanese at the end of World War II. These negotiations, which had long-term fisheries impacts on the west coast, had, as well, many implications beyond fisheries.

Canada's role in shaping international law grew during the following two decades, as Canadian negotiators became leaders in pushing first for the twelve-mile and then the two-hundred-mile limits of national jurisdictions for coastal states. The United Nations Law of the Sea (UNCLOS) Conferences of the 1960s and 1970s dealt with these proposals and simultaneously dealt with numerous conflicting issues related most particularly to sea-bed mining. The two-hundred-mile limit was eventually proclaimed by Canada and the United States and then by other nations in the mid-1970s. However, in 1981, the U.S. government withdrew from UNCLOS, and two decades of agreements, together with numerous bilateral agreements, were thrown in doubt; its 1982 return to the negotiations failed to result in new agreements. Bilateral agreements between Canada and the United States are likewise tenuous, their status uncertain from year to year. Conservation and stock replenishment programs continue to be frustrated for lack of international agreement, and fishing issues continue to be overshadowed — even where the issues on the agenda are ostensibly about them — by the interests of mining companies, shipping, and American military objectives.

EARLY BILATERAL CANADA–U.S. NEGOTIATIONS

The United States imposed import duties on Canadian fish in 1854, contrary to Reciprocity Treaty provisions negotiated between Britain, the U.S., and pre-Confederation governments. After Confederation, Canadian governments apparently still had difficulty acting in their own interests, as evidenced by the 1869 defence of a new law: "To absolutely prohibit fishing by United States fishermen in colonial waters" (Sessional Papers, vol. II, no. 4:11).[1]

These early negotiations referred to Atlantic coastal fisheries, and it was not until 1871 that British Columbia was affected by them. The Washington Treaty (1871) included articles that allowed Americans the right to unrestricted sea fishing and British subjects the same rights in U.S. eastern seas, together with reciprocal tariffs and trade arrangements. However, United States customs officers argued that this treaty did not apply to B.C., since it joined Confederation two months following the signing of the treaty. The treaty was, in any event, terminated by the United States in 1883.

A joint commission established in 1892 resulted in a new agreement including major B.C. rivers and lower mainland fishing regions. However, new conflicts erupted and continued throughout the remainder of the century. One issue was the permissive use of traps and purse seines in the United States, although these were prohibited in Canada. American seine fishers were denied entry to Canadian waters in 1903 because they used illegal gear (Sessional Papers, 1905). Resentment of American nonconservationist practices and demands for an International Fisheries Commission and uniform regulations were repeatedly expressed by Canadians. An initial treaty was devised in 1908. The Canadian government amended the Fisheries Act to conform with this, but the U.S. Congress refused to approve it or enact appropriate legislation.

In 1914 the United States did remove duties on fresh and unprocessed fish and requested an extension of privileges for their vessels in Canadian waters; at the same time, the U.S. persisted in refusing Canadian vessels entry to U.S. ports from the fishing grounds, although since 1897 Canada had allowed American fishers the privilege of entry and bonded shipment of catch from Canadian ports. Apart from the effects of this on Canadians, free entry to Canada created unrest in Seattle and Ketchikan ports because they were by-passed by their own fishers. While the war was in progress, special orders in council in both countries allowed free entry, provisions, and sales, but by 1917, the U.S. Congress entertained a bill to prevent Pacific fish being shipped into the U.S. from Canada. Defeated on the first occasion, the bill came up again the following year when American legislators called for a new Joint Commission. A tentative agreement was reached in 1918, but remaining issues regarding the Fraser River salmon fishery required further negotiation.

The Fraser River salmon fishery was then, as it is today, a major contested issue between the two countries. Fish spawned there intermingle with fish spawned in the U.S. waters in the Strait of Juan de Fuca, both before travelling far out to sea and on return to their spawning grounds.

Overfishing has threatened the stocks throughout the century, and they have been threatened by landslides connected to railroad construction in the Fraser Canyon as well. Canadian objections to the American use of seines and American objections to the failure of Canadian governments to clear the Fraser Canyon obstructed efforts on both sides of the border to reduce overfishing. A draft treaty of 1919 was rejected in the U.S. Senate in deference to opposition from the State of Washington (Sessional Papers, 1923), and further drafts were drawn up over the next decade. But it was not until 1937, with a treaty establishing the International Pacific Salmon Fisheries Commission (IPSFC) that some agreement was achieved (Johnston, 1965:384–90).

The treaty assigns half of the Fraser's pink and sockeye salmon stocks to each country and bans net fishing on the high seas and designated areas of the territorial waters. The commission was charged with the task of building fish ladders in the Canyon, undertaking research and improving spawning beds (Logan, 1974:50). Other salmon species are not included in the treaty and continue to be under respective national jurisdictions (Pearse, 1982:37–38).

Pacific halibut fisheries negotiations were facilitated earlier (1923) by the establishment of a joint International Fisheries Commission (renamed the International Pacific Halibut Commission in 1953) to recommend regulations to both governments for management in view of acknowledged overfishing and stock decline.[2] That agreement had weaknesses, but with some experience, the commission succeeded in reducing fishing effort, and the stocks did, eventually, increase. At the time of enactment, two-thirds of all landings were made by American vessels, mainly at Prince Rupert. Licences were not required, since nearly all operations were conducted in extraterritorial waters. The 1923 and 1924 seasons provided large catches, and fishers intensified their efforts during the open season. The development of refrigeration techniques allowed fish buyers the opportunity of saving the surplus for sale during the close seasons enforced by the commission after 1923. Thus the regulations failed to reduce overfishing immediately, and over the next several years recommendations to establish quotas, longer close seasons, and licensing were advanced (Sessional Papers, 1923–27). Over time, the commission acquired power to establish catch quotas by area, to regulate gear, and to close nursery areas. To deal with excess capacity, the season was progressively shortened (Pearse 1982:122). These measures appeared to succeed in conserving the stocks, and the situation was stable until the late 1960s when catches sharply decreased. Pearse attributed this partly to environmental changes, "but mainly to incidental catches of halibut by foreign high seas trawl fleets" (1982:122). These were fleets of Japanese and U.S.S.R. origin.

MULTILATERAL NEGOTIATIONS, 1945–1980

Immediately following World War II, and with an eye to possible deposits of oil and manganese nodules,[3] the United States proclaimed ownership of its continental shelf (Truman proclamation on coastal fisheries, 1945), and a number of Latin American states laid claim to twelve and, in some cases, two hundred miles of territorial seas. Canada at this stage did not state similar claims, but with the entry of Newfoundland into Confederation in 1949, fisheries issues gained prominence in Canada and pressures mounted within the country for stronger protections. Debates of the period relating to Canada's participation in the Northwest Atlantic Fisheries Commission (NAFC, established in 1949 by sixteen countries), bilateral negotiations on both coasts, and early international preparations for Law of the Sea Conferences indicated a changing and more expansionist stance in Canadian attitudes (Johnson, 1977:59–60).

North Pacific Fisheries Convention (1953)

The American fishing industry formed lobbies immediately following World War II to press for restrictions on Japanese fishing rights as part of the peace negotiations. Early drafts of the peace treaties were particularly punitive, but a more moderate stance emerged within the American Defence Department as American policy became more concerned with establishing Japan as a Pacific ally in the cold-war arena. The fisheries were thus not treated alone and in their own terms, but rather became part of the American postwar strategy. Because fish provided the staple in the Japanese diet, compromise between the demands of American and Canadian fishing interests and the political aims of the American government had to be achieved (Langdon, 1983, Chapter 4). In the negotiations Canada was a secondary power.

At the invitation of the Canadian government, the major interest groups in the B.C. fishery met in 1951. They issued a joint statement demanding that Japanese fishers be excluded from the west coast fisheries. However, the U.S. persuaded the Canadian government to sign the 1951 Peace Treaty, which provided only that Japan would enter further negotiations respecting conservation and development of the fisheries.

A fisheries treaty was proposd by the U.S., placing limitations on the Japanese for the following five years and providing reciprocal fishing rights for Americans and Canadians on the west coast. This proposal was strongly opposed by the United Fishermen and Allied Workers Union (UFAWU), which proposed instead a bilateral treaty with Japan and restrictions on American rights in B.C. waters (*Pacific Fisherman*, Nov. 13, 1951, pp. 1, 6; cited in Langdon, 1983:59).

According to Langdon's interpretation of these events (1983), the American proposal included no reference to freedom to fish the high seas; the Japanese counterdraft asserted this principle subject to conservation restrictions. The compromise was "voluntary abstention" in place of "waiving of rights," with a protocol covering the species and areas where such abstention would apply. At the last minute — when many delegates had departed — the Japanese negotiators proposed a demarcation line for abstention, arguing that it would reduce later friction. The remaining delegates agreed, finally, to a line at longitude 175 degrees west; the North Americans were uninformed about the extent to which salmon migrated into mid-ocean. They subsequently realized that North American–sourced salmon mingled with Asian-sourced fish west of that line, and the line was later changed to their advantage (Johnson and Langdon, 1976:10).

The Peace Treaty was signed in 1952, and the fisheries treaty was ratified by the Canadian government in 1953. The outcome for Canada was protection of salmon, halibut, and herring fisheries from which Japanese fishers were excluded for a period of ten years, and the right to fish in American waters except for Bristol Bay (for which there was voluntary abstention). In return, Canada granted reciprocal rights to American fishers in Pacific waters. Japanese fishers were severely restricted, and Japanese merchants and investors began seeking out alternative means of obtaining the staple of the Japanese diet. Their success may be measured by the data shown in Chapter 5: from 1955 to 1969, Japan was the leading harvester and exporter of salmon.

The International Commission for North Pacific Fisheries (INPFC) was established in connection with these fisheries treaties. Its task is to ensure that noncoastal states refrain from exploiting fishstocks already captured at maximum sustainable yield.

International Law of the Sea Conferences, 1958, 1960

In the first two conferences sponsored by the United Nations to develop international agreements on fishing rights, national positions emerged that tended, not surprisingly, to reflect the divergent interests of nations according to geographical location, extent of fishing capacities, and shipping interests. Those with shipping interests, known as the maritime group, generally opposed an extension of territorial limits beyond the traditional three miles because they feared that such national jurisdictions would limit their navigational freedoms. This group included the U.S.S.R. and the United States, Japan, and European states bordering the sea. Ranged against them were numerous Third World countries, including Caribbean states, Mexico, India, Libya, and many new nations. Fish were important food sources for them. As well, they feared the potential predations of imperialist, developed nations once the seabed resources, other than fish, were fully utilizable by new technologies. Some of these countries, together with Canada and a few other developed nations with no major shipping interests, but with fisheries and coastlines, were known as the "coastal states group." They were generally more concerned with preservation or expansion of their fisheries jurisdictions. These early designations and positions blurred as offshore oil and mineral exploration altered national understandings of what lay beneath the ocean floor.

In the opinion of the International Law Commission when preparing for the first UNCLOS, no extension of boundaries beyond twelve miles would be acceptable. The Canadian government proposed a fishing zone (carrying no other ownership rights) for nine miles beyond the traditional three-mile zone, and apparently perceived itself to be defending Third World interests as well as its own in this proposal (Johnson, 1977:61–62). However, Britain and United States, which had both defended narrower boundaries to this point, suddenly proposed a six-mile territorial sea limit with an additional six-mile fishing zone beyond it. Four proposals eventually reached the Plenary Session in 1958, but none passed.

Four conventions did pass in 1960, including the Continental Shelf agreement, which Canada signed, but none of these resolved problems of the west coast fisheries. The reasons for not passing the other conventions were never fully explained. Johnson (1977) speculates that Canada was concerned with the possibility that a freedom of fishing clause in the High Seas Convention might be construed as implying that no state could, later, unilaterally declare an extension of fishing rights. With the failure of the 1960 conference, Canada's efforts moved toward bilateral negotiations. Meanwhile, several coastal states unilaterally introduced twelve-mile fishing zones (Iceland in 1959; Norway in 1960. Several Latin American states had already done this prior to the 1958 convention).

Twelve-Mile Limits

During the 1960s, Canada repeatedly expressed concern about dwindling stocks in the Atlantic, apparently the result of overfishing by offshore freezer-trawlers. The International Commission for the Northwest Atlantic Fisheries established quotas, but Russian and Japanese fleets were, in the opinion of Canadian observers, ignoring them.

In the west coast fisheries, the major dispute concerned the baselines determining the boundaries at sea. The Canadian position was that these should be drawn from the northern tip of Vancouver Island to the southern tip of the Queen Charlotte Islands, and from there to the western limit of the Canada–U.S. boundary.

In view of these two coastal concerns, Canada unilaterally established a twelve-mile fisheries zone in 1964, and enacted legislation declaring boundaries on the west and east coasts that were in immediate conflict with boundaries claimed by the U.S. Rounds of bilateral talks were scheduled with all affected countries, but the upshot of these talks was that virtually no contending nations were prevented from entering Canadian waters despite the legislation. In 1966, Canada signed a reciprocal agreement with the U.S. following the American declaration of a nine-mile fisheries zone. Fry (1975:136–151) speculates on the futility of the 1964 legislation in which the new Pearson government was so committed to multilateral action that it simply could not uphold the interests of its own nationals. Johnson (1977:67) argues that the real problem might have been the inadequate preparation of the Pearson government.

By 1970, fifty-seven nations had claimed the twelve-mile territorial sea limit and some had gone further. International opinion was mounting on two seemingly contradictory fronts: (1) to create an international authority over ocean resources; and (2) to extend to two hundred miles the economic authority of coastal states. Canadian negotiators in the Pearson era had remained strongly committed to limited economic jurisdictions at the general level, but to greater jurisdiction for fisheries specifically.

U.N. Deliberations on Seabed Resources, 1968–1970

The United Nations 1968 conference on the seabed re-ignited issues raised at the UNCLOS conferences and heated up the debates on international versus national authorities. Here the divergent interests of the United States and Third World countries were clearly articulated, with the former claiming rights to ocean, mineral, or other resources and the latter arguing that the riches of the ocean should be the common property of all. It was this dispute that initiated the United States' withdrawal from UNCLOS in 1981, and though the issue had serious repercussions for international fisheries agreements, its nub was elsewhere in the sea: the minerals on its bed.

At the early conferences, however, the United States took a more conciliatory line. In 1958 and 1960, its primary domestic concerns were with its shipping, defence, and oil interests, all of which seemed best guarded by an open seas policy combined with control over the continental shelf and margins.

As late as 1970, the U.S. voted in favour of a United Nations General Assembly resolution designating mineral seabed resources as "the common heritage of mankind" to be exploited "in accordance with an international regime to be created" (Hollick, 1974; Clarkson, 1982:213). A Third World bloc of seventy-seven countries argued that mining by a U.N.–sponsored agency could ensure that profits be shared worldwide as part of the new international economic order. In return for cooperation in this direction, the United States extracted numerous concessions, including the preservation of navigational and air-flight rights. These applied to the Strait of Gibraltar and other strategic areas within the twelve-mile territorial limits of other countries. As perceived by the Pentagon:

> The U.S. Joint Chiefs of Staff are eager to have these navigation rights. With them, the U.S. Navy can steam uncontested through strategic straits like Gibraltar and Hormuz. With them its submarines can remain under water as they pass through coastal seas that other nations are increasingly claiming as sovereign. "Creeping jurisdiction" — the unilateral extension of territorial waters—has become a serious problem in the last 20 years. The treaty would lay it to rest, enhancing U.S. national security. (Burnett, cited by Clarkson, 1982:213–14)

Up to this point, military interests were paramount as far as the deep ocean bed was concerned, partly because of cold war issues and the strength of the U.S. Defense Department, but as well because petroleum deposits and manganese nodules were assumed to be generally limited to the continental shelf and its margins. In 1967 and early 1968, deeper water operations and new technologies revealed the probability of manganese deposits at much farther reaches (Hollick, 1974:16–24). A policy that protected only the continental shelf and failed to preserve the high seas bed for private exploitation was not in the interests of the most powerful American mining corporations. The oil companies were still concerned with preserving national control over the shelf and margins, and these two powerful groups were in conflict over American policy. The Defense Department still had the greatest bargaining clout, and between these three groups American policy wavered and changed over the next few years.

American mining companies, and joint ventures with mining companies in other developed countries, had invested in creating new technology for seabed mining. They were not altruistically inclined to share and campaigned vociferously against a proposed treaty that gave the high seas authority over to an international agency under United Nations patronage. In addition, an international agency with genuine economic clout in this industrial sector could establish a precedent for a new international order, and all the rhetoric spilt over that notion could not allay American fears of such a dramatic blow to capitalism. Supporting the American companies were the major mining companies with home bases in Canada but with international interests: Noranda and Inco (Buzan and Middlemiss, 1977:8; Clarkson, 1983:216; *The Globe and Mail*, May 18, 1981).

Two other groups besides fishing communities had an interest in American and international policy: marine scientists and environmentalists. Both groups argued in favour of open access for research and international authority over deep sea resources, but open access for disinterested research was inevitably tied to the interested access for military research and applications. In the U.S. Draft Seabed Treaty of 1970, the scientific community did not gain guarantees for scientific freedom (Hollick, 1974:26).

Canadian Offshore Oil Interests

Offshore oil exploration was meanwhile occurring in many world regions, including the waters off British Columbia. Between 1945 and 1960, the B.C. government had issued permits for over 1.4 million offshore acres (Buzan and Middlemass, 1977:9). This exploration raised wholly new jurisdictional problems over seabed ownership within Canada, erupting in 1965 in bitter contests between B.C. and the federal government. A 1967 Supreme Court advisory opinion favoured the federal position, but the opinion was not binding and was vague in reference to specific coastal areas. Both governments attempted to define these through legislation in their respective interests (Ibid.). Similar disputes were simmering elsewhere, affecting Quebec, the Atlantic provinces, Newfoundland, and the Arctic. Whatever decisions might emerge from these, it was becoming clear that international negotiations over fisheries and boundaries would have implications for provinces that had not, to this point, been involved in the international arena.

Pressures toward Two-Hundred-Mile "Economic Zones"

While the seabed debates were in progress, there were other issues in conflict. Among these were the long-standing complaints by Canada about violations on its Atlantic coast that finally erupted in 1975 with the barring of Russian ships from Canadian ports. This — and other considerations discussed below — persuaded Canadian negotiators that an extension of territorial control over fishing rights was required.

The first of these other considerations was the threat posed to Canadian sovereignty by the passage of the S.S. *Manhattan* through the Northwest Passage in 1969. By this time, the Trudeau government had assumed power, and the nationalist movement in Canada had gained some momentum.

In addition to public outrage over perceived violation of Canadian territorial rights, public awareness and concern with environmental protection was increasing. The sinking of a Liberian oil tanker in Chedabucto Bay in Nova Scotia increased public awareness of the damage such ocean traffic could do. Indeed, the world at large became more concerned with this issue after the sinking of the Torrey Canyon in 1967. Arctic waters were regarded as particularly fragile, and the two matters—the *Manhattan*'s passage and Arctic environmental issues — supported the passage of the

Arctic Waters Pollution Prevention Act in 1970 (McRae, 1980; McGonigle and Zacher, 1977; Johnson, 1977). Related legislation extended Canada's territorial sea limit from three to twelve miles for all economic purposes, thus establishing the Northwest Passage (which is less than twenty-four miles in width) as Canadian, not international, territory. As well, Canada advanced a claim to a hundred-mile pollution safety zone. The United States strongly objected to Canada's new Arctic policy, but while discussions ensued, the Canadian government proceeded to treat the Arctic as its property. In particular it began issuing permits for oil and gas exploration, and these became components of its subsequent jurisdictional claims. However, Canada's legal claims were not backed up by military strength, since the United States effectively controlled much of Canada's armed force via bilateral agreements, NATO, NORAD, and other military conventions.

The accompanying legislation established new "fishery closing lines" for exclusive Canadian fishery jurisdictions, including lines for Queen Charlotte Sound and Dixon Entrance in B.C. These lines were subsequently discussed with other countries, and with the exception of the United States, they were accepted pending various phase-out agreements (Johnson, 1977:68–70). The United States ignored the lines in Dixon Entrance, and its trawl fleets continued to enter that and Queen Charlotte Sound. Bilateral negotiations were agreed to while a surge of international opinion favouring extensions beyond the twelve-mile limit occurred, along with mounting international concern for water and other pollution, the dangers of oil tankers, and seabed mineral rights.

At the 1971 Seabed Committee meetings, Canada argued in favour of coastal states having preferential but not exclusive rights over a two-hundred-mile area. Brazil had already proclaimed its jurisdiction over two hundred miles, and other Latin American states supported that position, though many Third World countries were strongly opposed to it. The proposal did not gain acceptance.

The Fisheries Council of Canada resolved in 1972 that "the coastal state owns the fishery resources on and over the Continental Shelf and Slope" and that fishing by nationals of other countries should require the permission of the coastal state (Bulletin 1972:11, cited in Johnson, 1977:75). The continental shelf, however, varies in width from a few miles to over one thousand miles, averaging forty miles around major land masses; its ownership would therefore hold great advantage for some nations and harm others. The proposal was defeated when presented at the Seabed Committee meetings. Another proposal, to develop international law on a fish-specific basis, also failed.

In these few years of debate, the inshore fisheries off the Atlantic coast were declining, and the Canadian east coast fishery was pressing for greater national jurisdiction to stop the overfishing in high seas. Canadian negotiators became more insistent on the rights of coastal states to manage the fisheries. As well, there was pressure from the west coast fisheries to extend jurisdiction beyond even two hundred miles because of the peculiar migratory patterns of salmon (Johnson, 1977:78–82).

Further negotiations produced a compromise supported by moderate Latin American states and African coastal states: a two-hundred-mile economic zone with coastal states having exclusive rights to resources but with freedom of navigation assured. This position was on the agenda for the third UNCLOS meeting in 1974, and the probability that a two-hundred-mile economic zone would eventually be agreed upon was generally assumed, but there were many particulars still in dispute and little likelihood of an early convention. The coastal states developed considerable agreement among themselves and proceeded in the following years to engage in bilateral negotiations. The United States had, by 1974, shifted toward the two-hundred-mile economic zone concept though with some continuing reservations (Johnson, 1977:84–88).

In 1975, bilateral negotiations between the United States and Canada resulted in a treaty that recognized territorial control to the two-hundred-mile limit. On the basis of this vital agreement, other countries followed suit. By 1976, most countries fishing off the east coast had agreed to Canada's two-hundred-mile jurisdiction. Canada's two-hundred-mile zone legislation went into effect in January 1977. The passage of U.S. legislation establishing the two-hundred-mile zone, effective in the spring of 1977, was important to Canada, because otherwise Canada had no capacity to enforce the zone boundaries. The net effect of the two-hundred-mile zone agreements ensured that each coastal nation had the right to establish quotas or other limitations on access to fish. Canada had argued for exclusive exploitation rights of anadromous species but did not succeed: high seas capture is still uncontrolled.

All of these events are frequently cited, with a generally favourable press for Canada written into the record, though at the time the "coastal states group," of which Canada was a leading member, was sometimes characterized as greedy. The lifting of that label resulted from many compromises agreed to in the process of negotiation and the extensive research and leadership offered toward conservationist measures.[4]

However, to cite these events with this bias is to ignore the larger context and the other agendas of participating nations. The U.S.S.R. and the U.S. were exploring the possibilities for international agreements throughout the 1960s. As interpreted by McRae (1980:162): "Their interest was largely motivated by a desire to secure freedom of movement for their submarines and warships." The entire process from then to the final agreements was always subject to military and industrial concerns unconnected to the resource, while fisheries biologists, negotiators concerned about overfishing and habitat pollution, and the fishing industry struggled to preserve it. The question of which country should pay for cleaning up oil spills in particular held an edge beyond pollution concerns. Such concerns, much more pressing than the problems of fish ownership, undermined the idealistic positions of those who argued that the seabed should be humankind's common property.

Offshore Rights for Foreign Vessels during the 1970s

Foreign vessels may fish in Canadian waters subject to bilateral agreements. These rights have been negotiated between Canada and numerous other countries, including, for the Pacific waters, Japan, Poland, the U.S.S.R., and Korea. Vessels of these countries must conform to Canandian law, quotas, and licensing requirements.

Under the International North Pacific Fisheries Convention negotiations of 1978, Japanese high seas salmon-fishing rights were further delimited to 175 degrees east longitude in the north Pacific. The Canadian minister of fisheries, Romeo LeBlanc, noted: "This means there will be virtually no salmon of B.C. origin available to the Japanese fishery" (Langdon, 1983:67). Japan has protested these and earlier restrictions, arguing that the 1978 limitations reduced its high seas catch by about 40 per cent. Japanese sources estimate that up to eight thousand Japanese fishers were pushed out of the industry in that period (*Japan Times Weekly*, 1978, cited by Langdon, 1983:68). In consequence of the limitations, the Japanese government invested heavily in research on new species in its territorial waters and in fish farming. As well, Japanese merchants replaced Japanese fishers seeking sources of food for the Japanese market, and other fish to export. To this end, they began investing in plants elsewhere, including British Columbia.

BILATERAL NEGOTIATIONS, 1970–1985

The long history of jurisdictional disputes between Canada and the United States over management and marketing of salmon appeared to be settled between 1970 and 1973 with agreements permitting nationals of both countries to fish between three and twelve miles off either coastline. International proposals to extend fishing zones to two hundred miles in 1977 were anticipated in a 1976 agreement that allowed Canadian salmon trollers greater access to the northwestern United States coastal fishery, until bilateral negotiations could be established for the extended jurisdictions. The bilateral negotiations, however, were complicated in the mid-1970s by the Boldt decision and its aftermath, and by continuing friction over the Fraser River catch.

In 1974, Justice Boldt of the U.S. Federal District Court ruled that nineteenth-century American treaty rights obliged Washington State to cede half the salmon fishery to Indian inshore fishers. In the resulting dramatic shifts, not only were non-Indian American fishers squeezed by lower fish quotas, but so too were Canadian fishers in U.S. waters. On request by the U.S. government in 1977, the Canadian government ordered its fishers to withdraw from American waters (Greene and Keating, 1980: 734–735; Schmidhauser, 1976:144–71). The impact of this decision went beyond the American fisheries: it clearly had implications for how Canada dealt with its land and fisheries claims and affected the expectations and demands of natives in B.C. The Canadian government subsequently closed the shrimp grounds off Vancouver Island, ostensibly for conservation reasons, thereby prohibiting American as well as Canadian shrimp fishing. This move was generally regarded as retaliatory in nature, and there ensued a period widely designated as "the fish wars." These wars involved a number of separate but not always separated issues.

Dixon Entrance

Canada has long held that the waters of Dixon Entrance are Canadian, but the U.S. contests this and has on occasion arrested Canadian fishers in the area. The boundary in dispute was established by the 1902 Alaska Boundary Tribunal (Logan, 1974:60). Although close to Hecate Strait where oil drilling occurred in the 1960s, the area has not so far emerged as a potentially rich seabed region; thus the issues are more closely related to fisheries than other resources.

Swiftsure Bank

Debate over Swiftsure Bank is more heated than over Dixon Entrance. This is a major feeding ground for salmon spawned in American waters but lies entirely within Canadian waters. An interim agreement in 1978 was put forward that would have permitted Canadian salmon trollers some access to American waters, but on condition that Swiftsure Bank could be closed by American authorities if they deemed it necessary for purposes of conservation. The agreement was challenged by Washington State trollers (already severely restricted because of the Boldt decision), and it appeared unlikely that it would gain U.S. government approval.

Two requests by the American government in May 1978 to close Swiftsure Bank were rejected by the Canadian government on the grounds that these were based more on political than conservation considerations. The UFAWU and the Pacific Trollers Association (PTA) demanded that American fishers be banned from Canadian waters (*Vancouver Sun*, May 6, 1978; *The Globe and Mail*, May 9, 1978). The provincial government, as well, called on the federal government to take retaliatory measures against U.S. fishers. Despite these pressures, the Canadian government chose to close Swiftsure a month later, incurring a volume of wrath from fishers' organizations. It attempted to recoup its losses then by requiring reciprocity from the U.S., in which attempt it was unsuccessful. Simultaneously, a parallel dispute was raging over the Georges Bank area of the east coast, and neither dispute was resolved by the end of May 1978. In June, the Canadian government announced its decision to suspend the 1978 interim agreement. The respective waters of the U.S. and Canada were then closed to non-nationals.

Fraser River Salmon Fishery

Under the International Pacific Salmon Fisheries Convention, Canada and the United States were to co-manage the harvest of Fraser River pink and sockeye salmon in adjacent Canadian waters. These negotiations have been re-entered annually, with escalating conflict over jurisdictions and harvesting quotas. Canadian fishers (and sometimes Canadian governments) claim that the cost of maintaining the Fraser River fishery is dispropor- tionately borne by Canada, and includes both the direct costs of enhance- ment and the indirect costs of foregone opportunities for hydro-electric installations on the Fraser. The river lies entirely within Canadian terri- tory, but U.S. fishers obtain half the harvest.

Trading Relations

Fish harvesting rights are persistently confused with trade relations. When the Canadian government announced its decision to suspend the interim agreement of 1978, largely because it anticipated that the U.S. Congress would not approve it (Greene and Keating, 1980:739), the U.S. raised new objections to Canadian fish exports. During the 1970s (as documented in Chapter 6), the Canadian government provided extensive subsidies to Canadian fishers through such programs as the Fishing Vessels Assistance Program and Department of Regional Economic Expansion grants. These aids were believed by American fishers to permit Canadian fish to enter the United States at low cost. They petitioned the American secretary of the treasury to impose countervailing duties, and investigations of 1977 and 1978 heightened the tension between the two countries. The most bitter squabbles were over the Groundfish Temporary Assistance Program (GTAP), which New England fishers referred to in a demand for a total boycott of Canadian fresh fish. American negotiators linked the issue and any resulting solution to the maritime boundary disputes. The Canadian government finally agreed to accept modifications of its stance over the boundary issue, but the American treasury turned them down. Canada was eventually forced to end GTAP in October 1978 in order to prevent the imposition of countervailing duties (Greene and Keating, 1980:739–42).

In the course of these negotiations, provincial governments became more actively involved in fisheries policy because countervailing duties, federal subsidies, and the impacts of these on the processing sector (under provincial jurisdiction) necessarily brought them into the debates. However, it would probably be an error to interpret the increasing provincial participation as peculiar to fisheries, since this same period was fraught with federal-provincial disputes over respective jurisdictions and powers. Of particular importance were the struggles over offshore oil rights on the Atlantic and in the Arctic, proposed pipelines and tanker routes through the north and on the west coast, and oil royalties and taxes in Alberta. The outcomes of these disputes, including but not restricted to court decisions on jurisdictions, affect all resource industries.

As noted by Greene and Keating (1980), the processes involved in the development of the two-hundred-mile limit and the bilateral negotiations have greatly increased the political salience of fisheries relations. Government departments, concerned with maritime issues in general and fisheries in particular, gained greater public attention than at any previous period, and fisheries issues necessarily became more important on government political-priority lists. This is evident in various government policy papers arguing for "more direct intervention" (Chapter 1). The same themes are evident in American policy papers. As well, the issues became focal points for organization of interest groups, and all of these groups — processors' associations, unions, and the many associations of fishers — became more involved, more politically active, and more articulate about their needs.

THE REAGAN ERA

In 1981, President Reagan abruptly and apparently without warning fired the American representative to the UNCLOS and withdrew from pending bilateral treaties between Canada and the United States. The context for the new president's actions included numerous trade disputes with its neighbour, the National Energy Program and American oil companies' strong opposition to it, loud public protests over acid rain and general environmental pollution, and other issues beyond the fisheries. In the international arena, they included the successful negotiation by the United States of military free passage through certain international straits and numerous other conflicts over American claims (Clarkson, 1982). In the fisheries, the context included new fish stock breeding programs such as the Salmonid Enhancement Program. Fish spawned through public expenditures in the country of origin migrate as do wild fish, and each country argues that its expenditures should be rewarded with a guaranteed catch of the additional stocks.

But the central issue for the United States was not fish or any of these other matters. It was, quite simply, that the mining lobbies had succeeded in their opposition to the UNCLOS negotiations. The mining industry had gained licences in 1982 and was scheduled to begin seabed exploitation before the end of the decade (Clarkson, 1982:218). International pressure, including Canadian voices, persuaded the United States to reconsider these rash actions, though it took nine months before it re-established its membership in UNCLOS, and it did so conditional on new provisions regarding seabed mining. In 1982, the United States, Turkey, Venezuela, and Israel voted against the global treaty on the Law of the Sea.

With respect to bilateral issues and the Pacific fisheries, negotiations continued to stall. In April 1983, Department of Fisheries and Oceans Pacific region director-general Wayne Shinners said that, because the changes desired by the United States were unacceptable, fishery management plans outside the IPSFC treaty framework would be implemented for west coast fishermen. These would allow Canadian vessels to catch the fish before they entered convention waters, thus pre-empting much of the American catch. Among the results would be an increase in allowable vessels from about 1,000 trollers to 1,500 (*Province*, April 28, 1983:B1).

Further fruitless negotiations throughout 1983 and 1984 focussed on interceptions of fish spawned in the waters of one nation but migrants in the waters of the other. Negotiations were frustrated by the numerous levels of government through which U.S. legislation passes. Of the whole bitter history, Peter Larkin, professor at University of British Columbia's Institute of Animal Resource Ecology, said:

> We're arguing about how to cut the pie while we're foregoing an opportunity to make the pie twice as big. . . . For the last 20 years the Canadian government has said there will be no salmon enhancement on the Fraser River because the Americans would get half the fish. And now I hear the Americans are concerned about spending millions of dollars to enhance the Columbia River when Canadians would get a lot of the fish. (*Vancouver Sun*, Dec. 8, 1984)

In December 1984, the various governments finally reached some agreement, and a treaty was ratified in January 1985. This provided a higher catch quota to Americans of Fraser River spawners and restricted Canadian trollers on Vancouver Island and in northern B.C. It required that management plans of both countries be submitted to a newly formed Pacific Salmon Commission and that both governments engage in conservation and enhancement activities. A further treaty signed in March 1985 provided for harvesting limitations in both countries on chinooks, affecting particularly the trollers and sports fishery in the Gulf of Georgia.

This is surely not the conclusion of events. While the international community attempted to create a global law and much energy was devoted to the problems of anadromous and other fish, some nations — notably Norway and Japan — have simply found other means of fishing. Their fresh/frozen pen-reared salmon are now on world markets and could well displace the wild salmon around which so much of this history has revolved. Global tensions have re-created the cold war, and Japan has quietly emerged as one of the world's great powers, while everyone was arguing about other matters.

NOTES

1. Alicja Muszynski did the research on sessional papers used in the first section of this chapter.
2. Logan (1974:43) claims that this was the first international treaty designed for conservation purposes.
3. Manganese nodules are lumps of minerals tied to a manganese base, which occur in concentrated form in the seabed. The minerals include cobalt, nickel, copper, manganese, iron, silicon, and aluminum. According to Hollick (1974:22, with reference to David C. Brookes, "Deep Sea Manganese Nodules: From Scientific Phenomenon to World Resource," *Natural Resources Journal*, vol. 8, pp. 406–407, and other scientific sources): "While managanese nodules are scattered widely over the ocean floor, the nodules with the greatest proportion of commercially attractive cobalt, nickel, and copper are generally found in the deepest parts of the oceans."
4. See Copes (1979–80) for a particularly flattering view of Canada's negotiating stance.

PART 2

LABOUR AND ORGANIZATION

8

Labouring at Sea:
Harvesting Uncommon Property

NEIL GUPPY

This part of the book highlights labour issues. The work of fishing is organized to create a division of labour between fishers and fish processors. That division, and its attendant complement and contrast, is one major theme. A second theme is found in the exploration of divisions within the entire workforce across ethnic, ancestoral, and gender lines. A final theme focusses on the struggles of workers to organize around their common interests despite divisions that impede cooperative action.

FISHERS

Behind the occupational label, fisher, one might presume there existed a homogeneous collection of skills, techniques, work relations, and occupational careers. Such is not the case. While having much in common, significant cleavages between fishers persist. This chapter focusses upon the work experiences of fishers, seeking to uncover factors that make a systematic difference to the variegated manner in which their working lives have unfolded.

By way of comparison, the next chapter takes up a similar task, although attention centres on workers in the fish-processing side of the industry. Comparing fishers and processing workers reveals the consequences for people's working lives of being involved in one or the other sector of the industry. Such a comparison also highlights the significant kinship connections that exist on both sides of the industry, connections that are central to an appreciation of its development.

Much of the information reported in this and the subsequent chapter comes from formal interviews conducted with fishers and shoreworkers. Details concerning the breadth of coverage and the representative accuracy of the people we talked with, the types of questions they were asked, and the general nature of these interviews are documented in Appendix A.

Industry Recruitment and
Family Connections

Some Canadians attribute their position in the occupational world (or lack thereof) to a random process where blind luck, good or bad, is thought to play the determining role. Fate, many believe, governs a person's occupational destiny; it is "in the cards." While for individuals, fortune or misfortune may appear random, in the occupational world luck is not so blind. Career politicians and corporate executives rarely come from working-class origins, and likewise fishers are seldom the sons or daughters of those who roam the corridors of political or economic power.

What are the family origins of people in the fishing industry? What are their backgrounds? From where do they come? In the corporate and political worlds we know of the importance attached to connections; is this true, too, in fishing, and if so, what types of connections are important; what networks are useful when embarking on a fishing career?

The centrality of fishing in the lives of the parents of current fishers is at once remarkable and unsurprising. It is unsurprising in that fishing is often considered a traditional occupation, like farming, where sons follow fathers. Nevertheless, for current participants in the industry, the importance of fishing in the lives of ancestors is remarkable because of its consistency across industry participants. Fully three-quarters of our respondents reported that fishing was a "central pursuit" for their parents or grandparents. For Indians it was the rule almost without exception. Of those in the industry whom we interviewed, over two-thirds of fishers said they had at least one family member currently active in the industry. In few other lines of work in Canada could such a pattern be anticipated.

The occupational inheritance so characteristic of the fishing industry can also be illustrated by comparing it with the general Canadian pattern of father and son occupational identity. Even using crude skill groupings (e.g., professional, technical, semiskilled) we find only about one-quarter of Canadian sons follow in the orbit of their father's general industrial work grouping.[1] Among upper management and professional occupations, where inheritance is greatest, we find less than half of all sons follow their fathers into managerial or professional jobs. Only in farming is there a pattern resembling fishing — in Canada, approximately 85 per cent of current farmers come from families where farming was the central occupation (Pineo, 1983).

Among Indian fishers, 96 per cent of those interviewed reported their father's main occupation was fishing, while 89 per cent of their mothers also worked in the industry. For non-Indians the rate of occupational inheritance is lower (around 50 per cent for sons and fathers and sons and mothers), although if we limit the analysis to people born in B.C., a stronger link between parents and sons exists.

The importance of family networks in fishing, atypical of recruitment patterns in most other occupations, is characteristic of small business operations. Family connections are important in the industry not only because it is from their fathers that young fishers can learn skills associated with navigation, gear, fishing locations, but also because it is from the family that the amount of capital necessary to become a fisher can be found. While people can still enter the industry as small-scale fishers, the increasing prominence in the fleet of expensive seiners and freezer-trollers makes entry at this latter level difficult unless one has either family connections or a small fortune.

These findings parallel a rather well-established sociological generalization, discussed by Sorokin (1964:419, quoting the Italian sociologist Chessa): "Hereditary transmission of occupation is stronger in those occupations which demand a greater technical experience and specialization or a more or less large amount of money for their performance." Occupational inheritance is thus facilitated by the transmission of knowledge (specialized skill training in particular), network ties (e.g., insider information about training opportunities and job openings), and financial clout (access to capital and investment advice).

Nepotism is important in this occupational niche, as in many small businesses. The connection of industry participants is obscured by those who rely on imagery of rugged individualism as an ideological gloss for the industry. While individualism has its place, families are fundamental.

Current changes in the fishing industry, should they continue, may escalate the importance of family connections. If, as has been the recent pattern, large fish-processing companies continue to sell their corporate fleets, fewer opportunities will exist to learn the trade on company boats or to enter the industry via company rental fleets. The costs associated with vessels and licences have so increased in the last decade that inheritance of the family boat is becoming essential.

Although this focus on intergenerational occupational immobility has emphasized the status quo, the role of the family has altered in significant ways. Historically, on the B.C. coast, processing has been organized by corporate capital in industrial settings (unlike the family-centred production process of the Atlantic coast), where the recruitment of families was important for companies to ensure both a male fishing labour force and a female canning crew. In recent decades, this family recruitment has become less prominent as processing has grown both centralized and concentrated and as labour force recruitment has become easier.

Traditionally both female and male members of the family have played an equally important, if segmented, role in the industry. It is in this context that Thompson et al. (1984:167) point out that while "fishing is commonly thought of as a man's trade," it has been "an occupation peculiarly dependent on the work of women." The mythology of rugged individualism surrounding the industry hides the central place of women and family at the industry's very core (Maril, 1983:89–92). Whether one interprets past practice as processors recruiting good fishers by offering cannery jobs to their families, or processors insuring good canning crews by throwing in fishing jobs for the men, the centrality of the family, and thus of women, is crucial.

Aside from family recruitment patterns and the inheritance of capital and knowledge in small businesses, there is yet another salient reason for continuity in fishing families. Sons follow fathers because their alternatives are so few.[2] To focus on fathers as role models tells only a small part of the story. The lack of other role models, a consequence of few other job opportunities, puts generational continuity in the context of absences, as opposed to premises of socialization and role modelling. In remote fishing communities such as Namu or Winter Harbour, labour force options are severely constrained, often amounting to either working in the fishing industry or leaving the community in search of alternatives.

Education

In many cases entry into fishing occurs very early, well before completion of high school. While urban youth may delay career decisions until high school graduation, in the fishing industry the majority of current participants began fishing on a serious basis prior to their seventeenth birthday. For Indians the age of entry is even earlier; the majority of current Indian fishers began their active careers prior to becoming teenagers. For non-Indians the average entry age into fishing is 18.9 years, but for non-Indians born in B.C. the average drops to 16.9 years of age.

Romanticizing the lure of the sea in explaining this early career initiation would be easy. Other factors are, however, more plausible. First, as the lack of visible significant role models (especially in rural areas) indicates, the relevance of formal education appears remote. The need for Latin or calculus is not obvious when contemplating a job in the fishing industry. Second, and more to the point, early entry to the fishery provides a family labour force to work on boats, benefiting the fishing enterprise while adding little or nothing to the cost. As with farming and some other small businesses, family labour represents a cheap workforce.

One major consequence stemming from this early entry into the industry is the foregone possibilities of pursuing education. Table 8.1 compares the formal educational qualifications of current industry participants with five other groups: (1) their fathers; (2) a 1970 sample of fishers; (3) a representative B.C. sample of farmers; (4) a random sample of B.C. male blue-collar workers; and (5) a similar sample of workers in primary occupations (information for the latter three groups is from the 1981 Canadian census).

Over time the educational qualifications of all British Columbians have increased, a change that has also occurred in the B.C. fishing sector. Fishers currently in the industry averaged 10.7 years of schooling, and whereas 59 per cent of their fathers had no more than elementary educations, only 27.1 per cent of current participants had not begun grade nine. Furthermore, younger participants in the industry have more education

Table 8.1
Education Levels of Selected B.C. Workers

School levels	Interview sample		1970 sample*		Canadian census†	
	Fathers	Current participants	1970 Fishers	1981 Farmers	1981 Blue-collar workers	1981 Primary Occupations
1–8	59.0%	27.1%	38.2%	23.8%	15.6%	22.7%
9–13	28.8%	42.0%	41.4%	38.3%	37.5%	46.9%
Community College	4.1%	23.2%	12.9%	23.8%	36.6%	22.2%
University	8.2%	7.8%	7.5%	14.1%	10.3%	8.1%
Sample Size	146	181	2,279	298	6,502	594

* Based on figures in Wilson (1971:25).
† Based on a random sample of B.C. men from the 1981 Canadian census. Farmers are excluded from the primary occupations category for comparison purposes.

than their older counterparts (the correlation between age and level of schooling is −0.44 in our sample, indicating that older fishers have less education).

Educational upgrading is evident in comparisons between the results of our survey and a DFO study in 1970 (Wilson, 1971). While in 1970 only one in every five fishers had some postsecondary educational experience, almost one in three current participants have gone beyond high school. Similarly, now 27.1 per cent of fishers have no more than elementary schooling, while in 1970 38.2 per cent of fishers had not gone beyond grade eight.

While the formal levels of education among fishers have risen, their average qualifications still fall below male B.C. workers in either the blue-collar labour force or workers in primary occupations. Census results from 1981 show that only 15.6 per cent of B.C.'s blue-collar workers had not gone beyond elementary school (versus 27.1 per cent in fishing). Compared to B.C. farmers, fishers have a similar level of educational attainment, although slightly more farmers report some university experience. Considering other primary occupations, 77.2 per cent of workers in this sector had gone beyond elementary school, compared to 73 per cent in fishing.

If fishers were competing with other primary sector or blue-collar workers for permanent jobs, then educational disparities between these groups could serve as a disadvantage for fishers. This, of course, does not regularly occur since fishers typically seek other work opportunities only over the winter season when there are no herring and salmon fisheries. However, should a significant policy change on the part of government, or a long-term decline in fishing occur, then fishers might find themselves disadvantaged if forced to seek a second career. Since an individual's level of education continues to serve as a significant market signal for those seeking employment, any government policy restricting employment in fishing would need to be coupled with manpower retraining programs if fishers were to gain the skills, abilities, and credentials needed to compete.

Education, Income, and UIC

Within the fishing occupation, education has little discernible impact on success, and this is undoubtedly one reason why people planning a fishing career see no economic advantage to prolonged schooling. The relationship between "net" fishing income (or earnings from fishing) and education reveals no significant pattern. For example, when our respondents were asked to report their estimated pretax earnings from fishing for the 1981 season, the following averages for three levels of education were given: $23,000 for those with elementary educations; $27,000 for those with high school experience; and $26,000 for those who have some postsecondary schooling. Using a variety of indicators of economic returns, the relationship of these indicators with education is highly variable, showing no consistent pattern. This finding corresponds with the results reported in the 1970 survey of the Department of Fisheries (Wilson, 1971).

Education becomes important to the income of fishers in nonfishing earnings. As Table 8.2 shows, the higher an individual's level of education, the smaller contribution earnings from fishing makes to total personal income. For men with elementary education, fishing contributes 96 per cent of their earned income, while for fishers with some postsecondary educational experience the contribution from work in fishing to their personal income is 74 per cent. Once more this result parallels the 1970 Fisheries survey, where nonfishing income was found to be directly related to level of educational attainment.

Several factors are responsible for this pattern. First, the opportunities open to fishers for seasonal employment are greater, in urban areas especially, for those with more schooling. Second, since income security from fishing is more readily attained by vessel owners, and since older fishers are more likely to be owners, older fishers (who tend to have less education) have less need for offseason employment to supplement their income.

Related to this is a historical change that appears as a shift in the availability of seasonal work for fishers. While the pattern is difficult to quantify, it is apparent in charting the work histories of fishers that in the 1950s (and earlier) fishing was often an eight-month activity (preparations

Table 8.2
Contribution of Fisheries Earnings to Total Personal
and Household Income by Level of Education

Level of education	Fisheries earnings as % of personal income	Fisheries earnings as % of household income
Elementary	96	80
High school	86	68
Postsecondary	74	63

included) coupled with a winter occupation for a three- or four-month period. While winter jobs varied, they almost invariably had some connection with fishing — diesel mechanic, electrician, shipbuilder, carpenter, plumber. Recently, however, such seasonal jobs have become more difficult to acquire. The change has occurred not because fishers are any less skilled, but because the structure of other industries and union seniority rules present barriers to entry, and because there are fewer temporary jobs in the economy overall.

What this has meant is an increasing dependence by fishers on unemployment insurance as a form of income supplement. The extent of this reliance is apparent when comparing UIC benefits reported by fishers to similar reports by farmers, other workers in primary occupations, and blue-collar workers. This comparison is shown in Table 8.3 where estimates are provided on the percentage of occupational members claiming UIC benefits, as well as the average annual value of such benefits. These estimates show that, as a group, fishers rely far more on UIC than the other workers. Nearly one-half of our sample of fishers reported receiving some UIC income, while one in four primary-sector workers and only about one in ten farmers reported UIC income. In addition, the average annual amount of such benefits is higher for fishers ($2,800) than for other groups (averaging about $1,500).[3]

Three separate pieces of evidence point to an increasing reliance on UIC benefits by fishers. First, fishers were excluded in the original UIC provisions, so it was not until 1957 that they were even eligible for benefits. As noted previously, our work-history information suggests increasing use of UIC payments in more recent years as support over winter months. Second, this pattern is consistent with comparisons of our results to the 1970 Department of Fisheries survey where 27 per cent (as compared to our 48 per cent) of fishers reported drawing on UIC. Third, between 1980 and 1984 the number of fishers receiving UIC benefits rose from 3,176 to 6,140 (an increase of 93 per cent — based on figures from the annual UIC Statistical Report).

Table 8.3
Comparing UIC Benefits for Fishers and Other Workers

Category	Per cent reporting UIC claims	Average annual value of benefits
All B.C. male workers*	9	$1,700
B.C. male blue-collar*	14	$1,600
B.C. male primary workers*	26	$1,800
B.C. farmers*	11	$1,300
B.C. fishers†	48	$2,800

* Based on a random sample of B.C. men from the 1981 census. Farmers are excluded from the primary worker category for purposes of comparison.
† Based on interview sample of fishers. Given the different samples, the precise size of the values for fishers should be treated cautiously (but see note 3).

One reason for this change stems from alterations in UIC regulations. Initially, fishers had to make UIC contributions in fifteen separate weeks, whereas now the period of contribution is much shorter (eight weeks). In the late 1950s and 1960s, it was, however, somewhat easier to work fifteen weeks, while now closures have restricted opportunities to catch fish and therefore accumulate insured weeks. Recent levels of high unemployment have meant that fishers have little alternative but to rely on the state insurance program.

Incomes

Pearse (1982:vii) began his commission report on the Pacific fisheries by stating that the "economic circumstances of the commercial fisheries are exceptionally bleak." He continued by noting that "many commercial fishermen . . . are near bankruptcy" (Ibid.:3). The commission was created precisely because of this economic chaos — not only fishers were in trouble; so too were banks and processing firms. Also it was a rather sudden chaos, occurring after a period in the 1970s where confidence in the industry was buoyed by, among other things, licence limitation, salmonoid enhancement, the two-hundred-mile limit, and the roe herring bonanza. Such are the vagaries of the industry that at one moment buyers offer fishers satchels of cash, while at the next, collection agents and sheriffs play a leading role.

Table 8.4
Fishing Incomes and Earnings: the Accountants' Reports

Gear-Type		Top 25%	Middle 50%	Bottom 25%	Overall
Troll	Gross $	$ 56,847	$ 28,629	$13,536	$ 31,910
	Net $	$ 20,008	$ 6,973	$ 3,116	$ 9,267
	Sample N	31	62	31	124
Gillnet	Gross $	$ 41,453	$ 20,330	$ 7,459	$ 22,437
	Net $	$ 17,473	$ 7,519	$ 858	$ 8,360
	Sample N	12	23	12	47
Combinations	Gross $	$ 84,780	$ 46,613	$20,973	$ 49,860
	Net $	$ 23,410	$ 18,355	$ 1,819	$ 15,377
	Sample N	35	65	35	135
Seine	Gross $*	$218,433	$110,522	$52,197	$121,965
	Net $	$ 91,977	$ 35,106	− $4,341	$ 39,126
	Sample N	15	35	15	65

* Boat and net share only, crew shares are excluded. Only seiners fishing both herring and salmon are included (only a small number of salmon-only seiners are reported in the accountant data).
Source: Cruickshank (1982).

Given the sudden shifts in the fortunes of the industry, it should be of little surprise to find that only two years before Pearse's report, another University of British Columbia academic, Peter Larkin (1980:14), was remarking that "salmon fishermen are better off now than they have ever been." These changing tides, to borrow Pearse's metaphor, have been a historical constant in a turbulent industry.

The severe oscillations in the economic well-being of the industry are reflected not only in this historical pattern, but also at any one time period, where the extremes in the earnings of industry participants are large. These disparities in the earnings of current fishers underscore much of the tension so evident between industry participants. Table 8.4 presents 1981 earnings data from public accountants who serve as financial advisers for fishers.

The figures are divided, by gear-type, into three groups: (1) highliners — fishers whose earnings fall in the top 25 per cent of their gear-type; (2) average earners — the middle 50 per cent of earners in each gear-type; and (3) lowliners — those fishers earning in the bottom quarter of their gear class. The magnitude of the extremes in earnings can be quickly seen by comparing the ratio of earnings for highliners relative to lowliners. For trollers, where the greatest equality of earnings appears, the ratio is 6.42:1, indicating that highliners receive six times as much as lowliners. For fishers with combination boats, the disparity is just over twelvefold, while for gillnetters it is just over twentyfold, and for seiners the ratio is technically undefined (given the negative earnings of lowliners), although this latter sector is clearly the most unequal. These figures reflect extreme earnings disparities *within* gear sectors.

This inequality can be expressed a second way. For trollers the top 25 per cent of earners received 54 per cent of the net fishing income. Conversely, the bottom quarter of earners received only 8.4 per cent of the total troll-fleet earnings. For fishers in the gillnet, combination, and seine categories, similar inequalities occur—the top 25 per cent of earners receive a disproportionate share of the net fishing income. The extent of such inequality can be put in some general perspective by recalling that for total family income in Canada, the richest 20 per cent of families receive about 42 per cent of the income distributed to Canadians (Marchak, 1981:23). Earnings disparities within gear sectors in fishing exceed the general levels of income inequality found in Canadian society as a whole.

Earnings inequalities exist not only within gear sectors, but disparities in income also exist *between* fishers in different gear categories. Restricting attention to highliners, even among the most economically successful fishers in any one year, the average earnings of seine boat owners exceed by a factor of 5.3 the mean earnings of their highliner counterparts among gillnetters. Even among the middle group of 50 per cent, seine skippers out-earn their gillnet colleagues by a factor of 4.7. Such a pattern does not, however, carry over into the bottom 25 per cent where we find that seine skippers, although catching almost as much as highliner trollers, end up with a deficit once expenses are considered. Nevertheless, the accountants' data reveal sharp disjunctures in earnings when comparing across gear-types.

The extent of inequality *within* sectors of the fleet as well as disparities *between* sectors of the fleet are important parameters for understanding the tensions surrounding the industry. Between-sector controversies flair over issues such as strikes, catch allocation, duration, timing of fishing openings, and prices paid for fish. Within-sector inequalities are important for understanding solidarity among fishers of the same gear-type. While economic inequality is not the only variable of consequence in explaining tensions in the industry, it is of some significance, as shown below.

Before turning to that, however, it is necessary to comment on certain issues surrounding the accountants' data. First, the data from the public accountants, as presented in Table 8.4, ignore the long-term earnings of fishers. While participants and observers alike suggest that yearly success is highly correlated, so that most highliners remain in that category annually, we know of no published data that systematically establish that pattern. Our own calculations on unpublished DFO information suggest that for the salmon fishery the correlation between a vessel's catch value in 1979 and 1980 was 0.72 for the entire salmon fleet: 0.45 for seiners; 0.52 for gillnetters; 0.78 for trollers; and 0.56 for combination boats. However, even these figures only suggest that gross earnings are highly correlated annually, and levels of take-home pay may not be consistently related over time (if an individual's expenses remain stable over time then one might assume earnings, too, were highly correlated).

The Fleet Rationalization Committee, which commissioned the data reported in Table 8.4, commented on this relationship between gross and net incomes. They claimed that the net fishing income of each gear-type bore "virtually no relationship to the gross fishing income of the same group" (Cruickshank, 1982:29). Curiously though, their very own data show this to be incorrect, at least when aggregate values are used. Gross average income and net average income are highly correlated (r = 0.90) suggesting that skippers who catch a lot, earn more. Once again this is a pattern that holds across time since 1983 figures reveal exactly the same pattern.[4]

This suggests that both gross and net incomes remain relatively consistent over time. The significance of this is that not only are there large disparities in earnings among fishers, but a person's position in that pecking order of disparity has some durability. It is this durability of placement in the hierarchy of earners that is important to the development of allegiances among colleagues similarly located.

Table 8.4 can, however, also be questioned for relying on information gleaned from the records of public accountants. Accountants are a resource that some fishers cannot afford. Of the boat owners we interviewed, 71.5 per cent used the services of an accountant in the previous year (although for 40 per cent of these fishers it was a company accountant, not a public accountant). While this suggests the data from public accountants may underestimate the financial crisis of the industry, this is probably the most accurate data available.[5]

Public accountants aside, some fishers treat fishing as a livelihood, not as a business. One consequence of this is the inconsistency of detail fishers as a group have about their personal financial situation. While some fishers referred us to their accountants and others allowed us to copy their detailed tax returns, still others could provide us with only the wildest of guesses as to how much they might have earned after expenses (or even what their expenses were). For the latter group especially, fishing was something you did, and catching fish did not entail any form of actuarial accounting or detailed fiscal management. You paid the bills as and when you could, and lived on what was left. Record keeping, paper pushing, financial planning — those were affairs of business you escaped by becoming a fisher.

Objections about the use of single-year data from public accountants, or the fact that some fishers treat their work as a way of life rather than as a business, do nothing to undermine the existence of disparities within the fleet. No systematic evidence is available suggesting the accountants' data distort the general picture of economic disparities among fishers. In fact a prominent catch-phrase in the industry, used repeatedly by the Fleet Rationalization Committee, emphasizes the point about inequalities — "too many boats chasing too few fish" or "20 per cent of the boats catching 80 per cent of the fish." While the latter slogan exaggerates the reality, the underlying premise reflects an important tension in the industry.

Using unpublished catch-statistics information from the Department of Fisheries and Occans, we calculated the share of fish caught by the top 20 per cent of fishers. By ranking boats from the most productive to the least productive (where productivity is measured by either catch value or weight of catch), we established the total percentage of fish caught by the most productive 20 per cent. The figures are displayed in Table 8.5. Considering all the boats on the coast involved in the salmon fishery, we found that in the 1967 to 1970 period the most productive 20 per cent of the fleet caught, on average, 53.6 per cent of the fish (by value). There was no appreciable change in this figure over a decade since from 1978 to 1981 the average volume of fish caught (by value) by the top 20 per cent was 54.3 per cent. When these calculations are recomputed within each gear-type, it is within the seine fleet that the most equitable distribution occurs (parallelling the pattern of inequalities found in the accountants' earnings data).

Once more, however, comparisons *within* gear-types obscure profound differences which exist *between* gear-types. While it is important to remember that among the gear categories seiners generally have the most equitable distribution of gross earnings and catch, the seiners are increasingly dominating the fishery. For example, Pearse (1982:100) reports average 1980 gross earnings for seiners of $115,280, as compared to average troll earnings of $22,500, and gillnet earnings of $18,665 (see also Table 8.4). Over the period of the 1970s, although the actual number of seiners remained relatively constant, the catching power and subsequent share of the salmon catch for this gear-type gradually increased. The consequences of this changing balance in the fishery are increasingly apparent in the political activities of fishers.

Inequalities and Political Tensions

These inequalities translate into significant cleavages within the fishing sector. Seine boats, the big-boat fleet of the Pacific coast, are often million dollar investments, and the vast majority are owned by incorporated businesses, in stark contrast to the remainder of the fleet. Furthermore, it is generally with the seine fleet that processing companies are involved in equity positions. Schwindt (1982:48) suggests that in 1981, B.C. Packers

Table 8.5
Salmon Landings by Highliners: Percentage Shares to the Top 20%

By value	Time Period*	
	1967–1970	1978–1981
All gears	53.6%	54.3%
Gillnet	42.9%	40.0%
Seine	36.2%	34.5%
Troll		
Ice	52.3%	43.0%
Freezer	N/A	34.5%
Combinations	42.6%	35.4%
By weight		
All gears	59.4%	61.6%
Gillnet	43.6%	39.4%
Seine	36.1%	34.6%
Troll		
Ice	50.2%	42.7%
Freezer	N/A	35.0%
Combinations	43.7%	36.1%

* The percentages are averages for the four-year time period.
Source: Calculated from unpublished DFO Catch Statistics data.

alone had full or partial ownership in about one hundred seine vessels, or approximately 20 per cent of the seine fleet.

As the power of the seine fleet has increased, gillnet fishers have agitated for a more equitable distribution of the catch. For example, at a November 1983 forum billed "A Fair Share of the Catch," the gillnet fishers in attendance unanimously proposed not only that their share of the salmon catch be increased to 33.3 per cent (from a reported 17 per cent), but also that their representation on industry advisory committees be increased and that DFO management policies be altered so that gillnetters could obtain "their fair share of the catch."

Concerns about catch sizes and DFO regulations have also been voiced by trollers. While historically the troll fishery has been open throughout the summer, in recent years DFO officials have closed this fishery, both in specific areas and for certain periods during the summer months. Recently, trollers have not only defied the orders of government regulatory officials, but have also gone to court to force the DFO to alter its practices.

These political actions are mirrored by the differing perceptions fishers have of how the fishery should best be managed. Referring respectively to gillnetters, seiners, and trollers, the authors of the Fleet Rationalization report note that "the 'rag pickers,' 'vacuum cleaners,' or 'swivel necks' each regard their sector as having made disproportionate contributions to the cause of conservation through efficiency restrictions" (Cruickshank, 1982:3). In practice these very real concerns about advantages or "disproportionate contributions" accruing to certain groups have translated into political tensions. Some fishers have sought to circumvent these differential advantages by entering multiple gear categories in an attempt to hedge their bets and offset any shifts in DFO management strategy (a tactic making management more complex).

This action of seeking multiple gear protection is only one consequence of changing power relations between the gears. When asked about catch allocation, for example, seine skippers were less likely to agree to such actions than were either gillnet or troll fishers. When we asked respondents if DFO policies favoured certain fisheries, 63 per cent agreed that some form of differential treatment existed. Many net fishers believe that DFO management practices favour trollers, while trollers themselves tend to see seiners as the advantaged group (see Table 8.6).

The degree to which these tensions, either in the form of ideology or as political mobilization, are fought out between groups of fishers or between fishers and other groups (e.g., processors, the state), varies historically. For instance, our interviews with industry participants and our observations of the industry confirm the degree of distrust between fishers and state officials. The "faceless bureaucrats of 1090 West Pender" (the address of DFO's Pacific office) are frequently cursed by frustrated fishers. At times, however, some internal differences between fishers can be set aside, as evidenced in political lobbying efforts (e.g., the Fishermen's Survival Coalition) or fishers policy workshops (e.g., the Western Fisheries Federation). However, these moments of harmony seldom span the entire collection of Pacific fishers (neither the coalition nor the federation represented all fishers) and they rarely have any longevity.

Table 8.6
Perceptions of Department of Fisheries and Oceans Policies and Regulations

	Gear-type of respondent		
% reporting DFO policies favour certain fisheries	Gillnet	Seine	Troll
	61.5%	72.2%	52.8%
Gear-type			
Gillnet	0	23.1%	5.3%
Seine	37.5%	0	57.9%
Troll	56.3%	61.5%	0
Sample size	26	18	36

Independence among Fishers

Tensions extend beyond conflicts between the state and fishers. Industrial disputes also erupt between fishers and fish processing companies. These conflicts occur in the context of attempts by processors to control raw-fish markets, thereby constraining the options of fishers.

When asked whether they felt large processors had "too much control" in the industry, most fishers (75.4 per cent) agreed. Seine skippers, however, seemed less convinced than their counterparts in other gear categories (60 per cent agreement among seiners as opposed to almost 80 per cent agreement among others). Similarly when asked a question worded in the opposite direction, "Were large processors beneficial to the industry?" most disagreed, but again seiners were less emphatic (the joint financing of scine boats is important here).

An antipathy to large corporations stems at least in part from the individualism and independence often associated with fishers. A sense of adventure and rugged independence is frequently used to characterize fishers. We asked fishers to respond to a series of ideological statements focussed on independence: "A person's fate is decided at birth" (only 5 per cent agreed); "Life is a matter of luck" (only 20 per cent agreed); and "Success depends upon ability and ambition" (over 75 per cent agreed). Not surprisingly there is little sympathy for fatalism and a major emphasis on individual ambition and independence. However, these social psychological dispositions must be lived out within a context where governments and corporations increasingly intervene.

The antagonism between fishers and companies underscores the changing nature of paternalism in the industry. It is in the interests of fish-processing companies to insure that good relations are maintained with fishers who deliver to their plants. Delivery guarantees are insured by a variety of techniques (e.g., price, elimination of competition), and tight control of fishers has always been one tactic. As Marchak (1984:33) says, "The processors circumscribe the markets and so ensure their capture not so much of fish as of fishermen. Fishermen are much less mobile."

Control strategies are established through a variety of mechanisms including company provisions of preseason finance; year-round accounting, servicing, and repairing; inseason fish packing and ice supplies; and long-term loans or credit-rating guarantees for gear and/or vessel expenses (Schwindt, 1982:34). For example, almost half (48 per cent) of all vessel owners reported relying on a company to either guarantee or sponsor their loan with a bank or credit union.

In an industry where the private ownership of the resource has proven difficult, the competition for fish among processors *and* fishers is important. Economists' use of common-property language often obscures their analysis of company-harvester linkages. Pearse, however, suggests that "significantly . . . vessel owners appear to have become less dependent on processing companies for financial support, and the companies have preferred to withdraw from financial commitments to fishermen, so that the control of the fleet by processors has almost certainly declined."[6] So, too, has the need for direct control declined as raw-fish outlets dwindle and as banks enter as financial players. The decline has not gone unnoticed by fishers.

That the tradition of paternalism has changed is seen not only in the absolute level of reported antagonism between processors and fishers. Perhaps of greater concern here are the differences between young and old fishers and between Indian and non-Indian fishers in their perceptions of the benefits of large processors. Older fishers and non-Indian fishers, who have gained the most historically from paternalistic practices, view large processing companies much more positively than their younger or Indian counterparts. With ever greater concentration and centralization of processing plants, the need for paternalism weakens and processing companies have responded by tightening up their fish-harvesting operations, relying on independent fishers to bear the risks.

But to what degree are fishers able to act independently of processors? In certain areas the degree of latitude fishers have in decision making is large (e.g., equipment investments) although, as in any business, constraints exist. However, a key test of fishers' independence comes in ascertaining the degree of freedom they feel they have in selling to the highest bidder. In a truly free-enterprise, competitive-market arrangement, price should be the major determinant of sales. When we asked unaffiliated fishers the question, "What determines to whom you sell your catch?" only 38 per cent responded that price was the major factor. Company services (e.g., ice, bait), debt relations, joint-ownership arrangements, and "tradition" were the main factors noted as determinants of selling decisions. Comparing across the gear-types, trollers and combination-boat skippers more frequently reported price as a primary inducement (44 per cent), but even among these fishers, the majority claimed factors other than price were most important.

While price might not be a prime determinant for all fishers, it is perhaps even more surprising to find that for many it is not even the second most important reason. For the 62 per cent of fishers who reported that price was not the prime influence, a further 81 per cent said it also was not the second most important reason.

Three additional bits of evidence bearing on the degree of independence that fishers enjoy require brief mention. First, many fishers commented that price was often not a factor in deciding to whom fish were to be sold because there was very little variation among companies in the price offered for fish. While it would be wrong to make the blanket assertion that no price competition exists, most fishers who commented on this issue acknowledged that B.C. Packers normally dictates the price range, and it is a very small range with a few cents separating major buyers at most.

Second, another indicator of independence is the proportion of fishers who switch the company they sell to from season to season. Comparing the 1980 and 1981 seasons, only 13 per cent of fishers switched their sales to another company, and, among those selling to a major producer in 1980, only 7 per cent switched (similar findings occur when comparing between 1979 and 1980 or 1979 and 1981).[7] Furthermore, within any one season, fishers report only rare, and generally small sales to secondary buyers. The normal pattern is to sell to one and only one company throughout a season. This suggests at least a necessary, although not sufficient, condition for the lack-of-independence argument — not only does very little inseason deviation in selling occur, but also very little change occurs across time in the company to whom a fisher sells.

Third, from a small-company perspective, the large fish processors are adept at controlling fishers so that small outfits or cash buyers have difficulty competing for the purchase of fish. While our evidence here is largely anecdotal, marginal comments made by one fisher illustrate the point. This fisher, a troller, depended upon ice for refrigerating his catch while at sea, and, although a cash buyer was on the Tofino docks when he arrived in port, he couldn't obtain ice from a transient cash buyer, so he delivered, at a lower price, to B.C. Packers where a *regular* supply of ice was available (see also Schwindt, 1982:42).[8]

From the perspective of small processing outlets, high prices help in acquiring fish, but price alone is often not a sufficient criterion. It is in this sense that fishers are at one and the same time independent risk takers and captured fish sellers. They are indeed independent commodity *producers* but not independent commodity *sellers*.

Finally this issue of independence must be connected to the economists' conception of "common property." Pearse (1982:76) provides a succinct description of this idea of "rule of capture": "Fish in the sea are not assigned through property rights or licences to any particular users; each user competes directly with all the others for a share of the catch, and has no right to any particular quantity until he has landed it." One effect of this, Pearse continues, is that competing fishers, each trying to maximize his share, put such intense pressure on fish stocks that the fishery could be decimated.

The pricing structure of the industry is not, however, important to this process of intense competitive fishing. The rule of capture and an oligopsonistic market interact to intensify competition. In the absence of a competitive marketing system, fishers must depend on high-volume catches and rely on quantity — and to a much lesser degree quality — to realize what is lost through noncompetitive pricing (see also Maril, 1983: 156). In short, independence has repercussions not only for fishers, but also for fish stocks.

Fishers and Crew Members

An added twist to the issue of independence is that vessel owners are not entirely alone in bearing the risks of fishing. While crew members do not share all the risks (e.g., they have no capital investment), they have no guarantee of a fixed return on their labours when they go fishing. However, before turning to this issue of the financial risks associated with crewing, it is important to clarify the relations between skippers and deckhands.

Of current participants in fishing, fully three-quarters began their careers as crew members. Furthermore, of those who are now skippers, 65 per cent started as deckhands. As noted above, these initial opportunities are almost without exception provided by family (62 per cent) or close friends (24 per cent) — only 14 per cent of current industry participants who began as crew members started with someone other than family or friends.

Once again, however, an important trend may be emerging. Among young fishers, only 16 per cent reported beginning in the industry as skippers, in contrast to 37 per cent of older fishers. While this apparent trend may be more a function of who survives in the industry (i.e., those who begin as skippers may use this early advantage to outlast counterparts starting as deckhands), it may also signal an increasing importance to connections, especially family connections, necessary for survival in the fleet. Once more the theme of rugged individualism may be obscuring the actual decline of isolated individuals launching successful careers in fishing.

While for many industry participants crewing provides a route to eventual skipper status, there are many others who work as fishing crew members for very short periods of time. This transient nature of deckhands can most readily be illustrated by our problems in tracking down a representative sample of crew members for interviews. The UFAWU and the Co-operative Fishermen's Guild both kindly provided us with random samples of their membership who worked as crew, but in contrast to our success in locating fishers (over 80 per cent), we were able to find only half the deckhands whose names and addresses we had. More often than not the person had moved, and it was only because of family connections that we eventually were able to interview just over half of the names on our original list.

This transience is reflected also in the finding that most crew members had worked for their 1982 skipper for less than two years. In those circumstances where some longevity existed, it tended to come in two forms. Some skippers rely on kin to act as crew, so wives, sons, and daughters serve as deckhands. Especially among male relations, some pattern of crewing over the years tended to develop. The other branch of longevity among crews, although less frequent, came with salmon skippers acting as deckhands in the herring fishery.

The work involved in fishing varies from the intense to the monotonous. When seine openings occur, the crew and skipper must work diligently and cooperatively for the duration of the period. Likewise in trolling, when the fish start biting, action can be hectic. Conversely, long periods of idleness and inactivity must be endured.

Beyond coping with these extremes in physical and emotional intensity, workers must also adapt to uncertainties surrounding the duration of the voyage, the amount of money they might earn, and the precise timing of intense work periods. These work patterns and values are inimical to the normal process of purely capitalist relations of wage labour.

Both the mode of work and the familial relations among crew make fishing quite unlike other capitalist forms of work. The impersonal work relations of the shop floor are the reverse of the personal relationships and obligations felt in fishing. One of the skippers we interviewed spoke of these relations in the following terms:

> You have to worry about other guys, not just yourself; it was just too much for me. I worried all the time (saying to myself), "You've got to make money, their livelihoods depend on it," and this sort of thing. I had a nervous breakdown, six days in the hospital. . . . Every time we were out fishing — and anchored at night — I was up every five minutes looking out the window checking. Are we O.K.? Are we O.K.? The only time I would sleep would be when we were tied up in port on a weekend.

This skipper's worry over the pay for his crew highlights another interesting sociological aspect of fishing. Under capitalist relations of production, a labour force is hired by offering workers a wage. Labour is bought by capital and sold by workers. The actual price is negotiable and a worker is theoretically free to decline to work for wages deemed to be too low.

The most significant difference to this general pattern in the hiring of crew members in fishing is that skippers offer no guarantee of a fixed wage. An agreement to crew involves an arrangement to share, in some form (it varies across fisheries and gears), the money paid to the skipper for the fish caught. One might surmise that a form of wage labour closely associated with this share arrangement is a piece-rate system. Just as assembly-line workers are paid for volume produced, workers in fishing are paid on the basis of the volume of fish delivered. A fundamental distinction exists, however. On an assembly line, workers can reasonably expect employers will provide them with pieces to assemble. In fishing no such expectation is possible. The boat might sink, the net could split, the freezer could fail, the skipper might pick a poor spot; the list of compli-cations is long. In an assembly-line job, a worker punches the clock and is then assured of income in return for labour; in fishing hard work usually does pay, but it is a gamble.

Both skippers and crew are attracted by this lottery aspect, and as with most lotteries, the gamble can be rewarding. The exact size of the share, however, differs widely depending upon both the type of fishery and the gear-type. Salmon trollers, for instance, often pay crew a fixed percentage of the catch value, usually between 5 per cent and 15 per cent depending upon the deckhand's skill and experience (but also contingent on family status where spouses or offspring often receive little, if any-thing). When seining for roe herring, a more rigid structure exists whereby all trip participants (crew and skipper) divide 64 per cent of the union negotiated price/ton among themselves (after certain deductions). The remaining 36 per cent goes to the skipper/owner of the boat.

Especially among the net fleet, the share system is fairly rigid, and provided that something is caught, crews know almost exactly what they can expect to receive. In the roe herring fishery, the union negotiates a fixed price per ton of fish, and it is this price which is used to determine the crew share. In the salmon fishery the crew share is based on the posted price, which is the price processors advertise.

Slippage occurs in year-end payments to skippers from processors. In the roe herring fishery, where crew and skipper share catch proceeds based on the negotiated price, skippers are often paid more than this negotiated price as an inducement for delivering their catch (it is not common practice to share any of this year-end payment with the crew). Likewise in the salmon fishery, bonus payments, which are above and beyond the posted price used to arrive at the crew shares, are given to skippers (Schwindt, 1982:67–68, 105).

The size of the year-end payments has been increasing over time (see Table 8.7). Since these forms of payment are provided on an individual basis in company-skipper agreements, no public reporting occurs as with posted or negotiated prices. Nevertheless Pearse and Wilen (1979:766) estimate that between 1967 and 1977 bonus payments in salmon increased sixfold. On a percentage basis, between these same ten years bonus pay-ments rose from 5 per cent to 11 per cent of the total value of the catch. While we know of no firm figures for current bonus levels, no one disputes the fact they have risen in both salmon and roe herring over the past decade.

Several issues connected with these payments merit attention. The existence of bonus payments would seem to offer tangible evidence that some independence between fishers and processors must exist. If skip-pers were nothing more than "disguised wage labourers," then there would presumably be no need for such payments. Here especially it is

Table 8.7
The Trend in Bonus Payments in the Salmon Fishery

| Year | Revenues (in millions) | | | |
	Landed value ($)	Bonuses ($)	Total value ($)	Column 2 as % of Column 3*
1967	36.00	1.88	37.88	5.0
1968	44.89	2.89	47.78	6.0
1969	27.81	2.14	29.95	7.1
1970	45.08	4.05	49.13	8.2
1971	44.48	4.00	48.48	8.3
1972	50.34	4.53	54.87	8.3
1973	99.99	25.90	125.99	20.6
1974	73.99	6.20	80.20	7.7
1975	46.91	4.90	51.80	9.5
1976	91.94	11.70	103.60	11.3
1977	108.72	13.40	122.10	11.0

* This column represents the bonus payments as a percentage of the total payment for the catch (i.e., [(bonus payment ÷ (land value + bonus payment)) × 100]).
Source: Pearse and Wilen, 1979:766.

important to note the 1973 aggregate bonus payment that represented just over 20 per cent of the total value of the catch. The 1973 salmon season was exceptionally good. Fishers were making very good money, and processors worked hard to ensure they obtained substantial catch volumes by competing through delivery bonuses.

Increasing bonus payments must also be seen in the light of increased fleet capitalization (McMullan, Chapter 6). It is this link between vessel/ licence values and year-end payments that Pearse and Wilen (1979:768) emphasize:

[These are] lump-sum payments which are not normally shared with the crew. . . . These rising bonus payments are almost certainly linked to the increasing capitalization of the fleet and reflect the inflexibility of the traditional share system in the face of changing proportions of labor and capital costs. In effect, bonuses are a means of increasing payments to the owners of capital, by circumventing rigid sharing of revenues from landings as the fleet becomes more capital-intensive.

By increasing payments to "owners of capital" processors are effectively diluting the potential earnings of crew members who no longer share in the full value of the catch. One response to this comes in Pearse's and Wilen's claim that the share arrangement was too inflexible to accommodate the "changing proportions of labor and capital costs." The irony is that crew members are effectively prohibited from fully realizing additional earnings because vessel owners need increased payments to offset the cost of capitalization. Yet, as both Pearse and Wilen and Hayward (1981b) show, capitalization, and the resulting fleet overcapacity, was at least in part a function of a failed government program. Labour is effectively penalized for business decisions over which it had no input and that, with hindsight, were ill-advised. While the financial woes of bankers, vessel owners, and processors have been well aired, the situation of crew members has received far less attention.

Casualties and Safety in the Fishery

One other area in which both crew and skipper share the risks of fishing is in accidents and mishaps. Discomforts and dangers are an intimate aspect of fishing. Sleeping space is cramped, often hot, usually stuffy; walking space is minimal; eating is precarious; reading induces illness; food is haphazardly prepared and rarely lavish; people living in close quarters for long periods may be short-tempered: overall, fishing is not a particularly comfortable activity. Setting and retrieving nets or lines is physically tiring. Much of the task is tricky because it requires a worker to do several things simultaneously, keeping various conditions and alternatives in mind. To work troll lines, for example, requires quick eyesight and reasonable coordination of hand and eye, the development of a rhythm for pulling in the line and removing hooks, an agility in landing a fish after estimating its size and noting its species (since fish of the wrong size and species must be thrown back after the hook is removed), and then much care in throwing out the hooks as the line is rolled back out. A careless worker could lose a hand or an eye, and could, as well, endanger others. In addition to the normal round of tasks, there are the extraordinary tasks and dangers involved when the ship encounters unexpectedly heavy seas without time to find safe harbour. These add a very significant danger to fishing that is lacking in most other forms of work (Sinclair, Hale, and Karjala, 1984).

Indeed, no other industrial sector in Canada currently has as high a fatality rate as fishing. As Table 8.8 shows, in comparison to the other most hazardous industries in Canada, fishing proves to be the most dangerous. The number of deaths from causes directly related to commercial fishing is almost sixteen times the national average for occupationally related fatalities. Furthermore, the rate in fishing is substantially above the rates in both forestry and mining, the next most life-threatening jobs in Canadian industry.

Not only is fishing a dangerous occupational pursuit, but fishing on the Pacific coast is more dangerous than on the Atlantic coast. As can be seen in Table 8.9, in each of the three years between 1979 and 1981, at least one-half of all fishing-related deaths occurred on the west coast. So while

Table 8.8
Canadian Fatality Rates in Selected Industries

	Fatality rates per 100,000 workers				
Year	Fishing	Forestry	Mining	Construction	Total
1979	125	151	93	38	12
1980	147	111	99	40	12
1981	147	92	71	37	10
1982	158	119	84	31	10

Source: Labour Canada, 1984:57.

only about one-quarter of the fishing labour force works in the Pacific region, more than 50 per cent of all fishing fatalities occur in the west.

The explanation for this markedly higher rate is not easy to adduce. The fatality rates are parallelled by the higher incidence of vessel losses in the west. Deschenes (1984) reports that between 1975 and 1982 just over 43 per cent of the commercial fishing vessels lost in Canada were on the west coast. This frequency of loss is twice as high as the proportion of boats located on the west coast over the same time period (20 per cent). It does not appear therefore that there is anything anomalous about the data used to calculate fatality rates. Nor would it appear that the explanation lies in less sturdy or sophisticated boats on the west coast, since the value of the western fleet represents almost half the monetary value of the Canadian fleet.

Three possible explanations come to mind. First, fishing on the west coast is more competitive and therefore the risks fishers take are higher. Second, Pacific storms, although no more severe than on the Atlantic coast, may strike with more suddenness and less predictability. Third, the recent surge in the Pacific roe herring fishery has increased the number of inexperienced fishers and these individuals, unwary of the dangers, fall victim to the hazards of the industry.

Table 8.9
Percentage of Canadian Fishing Fatalities in B.C.

Year	% of fishing deaths in B.C.*	% of fishers in B.C.†
1979	60	26
1980	55	24
1981	50	25

Source: * Labour Canada, 1984.
 † DFO, Annual Statistical Review.

Looking to the Future

One of the more curious aspects of the fishing industry is the confusing role of state policy. At the same time that one branch or another of government is encouraging fleet upgrading by offering fishers loans and subsidies for boat building and vessel improvements, another arm of the state is restraining the ability of fishers to catch fish by reducing opening times, regulating technological changes, and requiring restricted licensing. Increasingly, this contradiction between the promotion of overcapacity and the sharp reduction in fishing opportunities has meant the very antithesis of what fishers want. It has meant a greater reliance on state support in the form of UIC, welfare, and job retraining. In a very real sense, then, the idea of independent commodity production is bound up not only with relations between fishers and processors, but also with relations between fishers and the state.

Despite these and other frustrations, however, fishers remain committed to the industry. Given the turmoil within the fishery, it would not be astonishing to learn that many people anticipated other jobs in the future. However, given the stark reality of limited job opportunities elsewhere, it is not surprising to learn that few see viable options outside fishing. For example, only 11 per cent of our respondents thought they had a reasonably good chance of finding a better job elsewhere.

In the 1970 DFO survey we referred to earlier, fishers were asked if they anticipated continuing in the industry. Slightly more than 80 per cent of those asked indicated they thought they would continue fishing as a major occupation. In our survey, 79 per cent of fishers felt they would be fishing in five years, and for those who anticipated leaving, the majority thought they would retire because of age.

As shown in Table 8.10, we also asked whether fishers "considered leaving the fishing industry." Once again most participants in the industry showed a high level of commitment to staying, although as expected, crew members showed less intention of remaining. As would be expected, more people said they thought about leaving than said they would actu-

Table 8.10
Future Commitments to the Fishing Industry by Age and Position

| | Young | | Old | |
	Crew	Skipper	Crew	Skipper
Not fishing in 5 Years	28%	18%	17%	14%
Consider leaving	35%	23%	38%	20%
Sample N	31	53	16	61

ally leave within five years. Significantly, it was younger participants, as either crew members or skippers, who thought they would not be fishing in five years. The insecurity and dissatisfaction underlying this feeling may create greater tensions if the pattern should continue or worsen.

The oscillating fortunes of industry participants continue, and today's pessimism quickly becomes tomorrow's optimism. The uncertainties of the industry, from income to accident, are in this sense both the attraction and the worry.

CONCLUSION

In addition to providing descriptive detail on some of the salient characteristics of participants in the fishing side of the industry, this chapter has highlighted three issues — the myth of rugged individualism, the economic inequalities between fishers, and the tensions between independence and dependence faced by fishers.

The masculine image of adventure, survival, and rugged independence obscures much of the core aspects of fishing. Family connections in securing money for a boat or in inheriting a family boat play a crucial role in the lives of numerous skippers. Many skippers also relied on opportunities to crew with family or close friends to learn the fundamentals of the industry (and many now likewise rely on family members as crew). Equally important here are the important family linkages between shoreworkers and fishers that continue to be crucial to UFAWU success (Muszynski, Chapter 12).

While family connections are central to a certain harmony in the industry, the significant cleavages in the industry relate in important ways to the inequalities that characterize the fishery. Most significant here is the changing pattern of economic fortunes that befall certain sectors of the fleet (notably gillnetters). Strike actions and court battles play a basic role in the industry (Marchak, Chapter 10). In this chapter we have tried to outline economic disparities that underly some of the political divisions in the industry.

Finally, on the issue of fishers as "petty capital" or "disguised labour," we have explored the contradictory forces that at one and the same time put fishermen in the position of being propertied labour (e.g., the lack of a dynamic, competitive marketing system for their fish sales) and aggressive entrepreneurs (e.g., investing heavily to outproduce or at least match the harvesting competition). The nature of the resource and the nature of the oligopsonistic marketing arrangements will continue to make for an uneasy tension in the fishery. However, the "pleasure of larger capital" may slowly shift if the future is thought to lie in aquaculture and not in fishers.

NOTES

1. The pattern for daughters is significantly different given the dramatic increases in women's labour force participation in the last few decades.

2. While there are women working as fishers, their small numbers precluded us from systematically including them in our sample. Rather than speaking of "offspring following parents," the generalization is restricted to men.

3. These appear to be quite robust estimates for fishing. In 1981, 3,429 fishers received an average of $3,200 in unemployment insurance benefits ($400 above our estimate of $2,800 based on self-reports). Taxation statistics show that 6,600 fishers filed tax returns that same year. The ratio of UIC recipients to tax filers suggests that slightly more than 50 per cent of B.C. fishers collected UIC in 1981.

4. Using unpublished DFO data from 1983, the estimated correlation between the average gross and net incomes in ten gear categories is again 0.90, suggesting higher expenses do not, on average, completely erode the higher gross earnings of seiners and freezer-trollers.

5. The data from public accountants were collected because little quality information existed on fishers' incomes. This lack of comparative data makes it difficult to estimate the reliability of the accountant information. One crude method of establishing its reliability is to compare the average after-tax income as reported by accountants (approximately $7,800) with an equivalent figure from 1981 taxation records ($7,550). Given the crudeness of this method, the averages are remarkably close.

6. Since company-harvester links are maintained by more than finances, it does not necessarily follow that a "withdrawal from financial commitments" translates into an almost certain decline in processor control (to paraphrase Pearse). Other control mechanisms could have been strengthened to insure that reduced financial ties did not effect processor control. Schwindt (1982:35) is another economist suggesting this "non-price competition is now changing." He offers as evidence the increasing number of cash buyers. Curiously, however, in a different passage he maintains that a "most important" reason inducing a fisher to sell to a specific buyer is "the ability [of buyers] to accept all salmon offered for sale through the maintenance of substantial processing capacity." Price competition seems to be more gear specific than either Pearse or Schwindt implies, being most vigorous in the sectors of the troll fleet where *frozen* fish requiring no extra processing can be handled by cash buyers.

7. This was also a volatile time in fish processing when the future of Canadian Fish, a large processor, was in some doubt (see also Schwindt, 1982:35).

8. It is for this reason especially that freezer-trollers have become more powerful on the coast, since they can deliver a good quality, exportable product and they require nothing in services (e.g., ice) from cash buyers.

9

Labouring on Shore: Transforming Uncommon Property into Marketable Products

NEIL GUPPY

In research reports and other publications focussed on fishing, shoreworkers often constitute an invisible sector of the industry. As Garrod (1984:19) has suggested for the early years, government reports "described the life conditions of salmon, herring, clams, etc., in minute detail . . . [but] totally ignored the workers who processed the catch." Contemporary reports are no different in that workers in fish-processing plants remain peripheral, their work apparently assumed to be tangential to the main core of the industry, and their labour unworthy of analysis. The vigour of debate about "common property" or "resource management," which characterizes commentary on fishing, is in stark contrast to the silence on shorework. As a manufacturing enterprise, this assembly-line work presumably lacks something of the drama many see in high seas fishing.

Yet as this and the following chapters will show in some detail, shoreworkers have played an unambiguous and central role in the development of the industry. This role has come about not only from the value-added component of their onshore production, transforming a raw product into a marketable commodity, but also from their alliances, kinship and otherwise, with fishers in struggles with fish-processing corporations. In this latter regard, Marchak and Muszynski show in later chapters how those alliances were important in the development of unions in harvesting and in processing.

The following discussion relies heavily on structured interviews we conducted with a random sample of shoreworkers. The sample, representative of the industry, was drawn from lists supplied by the unions to which processing workers belong. More detailed information on the samples, our interviewing procedures, and the types of questions we used are supplied in Appendix A.

SHOREWORKERS: NUMBERS AND PRODUCTION

The seasonality, perishability, and unpredictability of the resource have a pronounced impact on how a shorework labour force is organized and maintained. Salmon is the main resource for the canning and freezing lines, and it is harvested only in the summer months. Salmon seasons vary in length and intensity as a function of the fish's life cycles, previous escapement levels, government harvesting restrictions, and so forth. This uncertainty aside, processing labour must be available almost instantly and workers must be prepared for strenuous and lengthy shifts when runs are heavy because fish are so perishable.

As quickly as the labour demand escalates, it deflates with equal suddenness. Employers require large quantities of labour for sudden, discrete time periods. If regional demands for labour were high, no surplus work force existed, and employers anticipated a need for labour in the future, there would be an obvious advantage in carrying the labour force over the slack period. But, in fact, processors hire many workers for extremely long hours during the high season and lay off most of them during the low seasons. A shorework labour force can shrink or rise by several hundred members within the space of a few hours. How is this possible?

The answer has to be that an enormous reserve supply exists in regions where processing plants are situated. Such a reserve exists in all regions of the province, especially for women, because job openings are far fewer than job seekers. In earlier eras, when many small canneries existed in remote areas, an imported labour force was necessary, and canners relied on Indian and immigrant labour for processing. The pattern of centralization of canneries has meant that now processing workers are typically drawn from areas where large labour surpluses exist. As long as surplus labour is available, plants can continue to employ workers on hourly wage rates, permitting the unions to insist on seniority in recalls but not employing workers on an annual salary or any other condition of long-term contractual obligation.

The seasonal nature of fishing and the reserve nature of the bulk of the workforce make an accurate estimate of the actual size of the shorework labour force extremely difficult. Especially at the height of activity, when catch volumes can exceed freezer or other refrigeration capacities, the labour force grows rapidly, work shifts are lengthened, and overtime hours maximized. This volatility can be a daily occurrence.

The only annual attempt to determine the size of the fish-processing labour force is undertaken by Statistics Canada. Under the auspices of the Census of Manufacturers, Statistics Canada collects information on the "Fish Products Industry." Table 9.1 provides the most recent information released on the B.C. fish-processing sector. In 1983, the Census of Manufacture information shows there were *on average* 2,415 production workers in the fish-products sector. While in many manufacturing sectors an annual average employment level is a sensible figure, in fishing such a number simply obscures the highly seasonal nature of employment.

Table 9.1
Establishments, Labour Force, and Wages in B.C. Fish Processing*

Year	Firms[1]	Output of top 3 firms	Average annual work-force[2]	Person hrs. paid[3] ('000s)	Total wages[4] ($'000s)	Average hourly wage rate
1975	47	54.8%	2,084	4,320	27,740	$ 6.42
1976	50	61.4%	2,794	5,824	39,407	$ 6.77
1977	49	62.6%	2,845	6,016	45,064	$ 7.49
1978	53	53.8%	3,363	6,978	55,730	$ 7.98
1979	61	50.9%	3,010	6,212	54,970	$ 8.85
1980	59	54.5%	4,138	6,278	55,736	$ 8.88
1981	52	52.7%	3,306	6,783	74,080	$10.92
1982	50	49.1%	2,244	4,550	51,739	$11.37
1983	49	n/a	2,415	4,793	57,208	$11.94

[1]The Census of Manufacture uses a minimum "shipment size" as a cutoff point, and thus many very small fish processors are not included.
[2]This figure represents "the average number employed during the year."
[3]Aggregate number of hours paid to all production workers.
[4]Total wages (overtime included) paid to hourly workers.

* Source: Data in column 2 and columns 4 to 7 are from the Statistics Canada annual, "Fish Products Industry" (Cat. #32-216). Information in column 3 is from Proverbs (1978; 1980; 1982). Both sources provide extensive and important notes on how the data are collected.

The salmon and roe herring fisheries are the major sources of raw product requiring substantial numbers of workers. Major herring roe work lasts for four to six weeks in the plants, while the main salmon canning and fresh and frozen salmon production period extends from mid-June to mid-September. There are about four months in which a large labour supply is required; thus the actual number of people working in the plants is more accurately reflected as about 500 to 1,000 relatively permanent production workers, and about 2,500 to 3,500 seasonal workers. Thus only about 25 per cent of the labour force can count on any extended period of employment throughout the year.

The elasticity of this labour force is graphically depicted in Figure 9.1. The top panel of the figure shows the monthly fluctuation in the hourly paid labour force. In 1984, a relatively placid year by industry standards, the hourly paid labour force fluctuated from a low of 1,000 in February to a seasonal high of 3,500 in July. For 1983, a more hectic year in the industry, the number of production workers varied from a low of 900 in February to a high of 4,600 in August. The expansion and contraction of this shorework labour force is complemented by a constant salaried administrative and supervisory staff of approximately 550 people.

Figure 9.1
Seasonal Labour Force Changes in Fish Processing

Hourly Paid Labour Force Size by Month: 1984

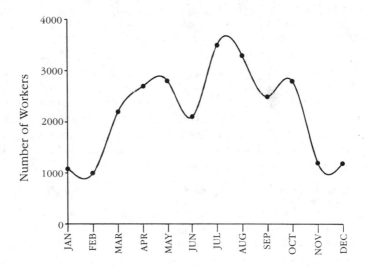

Overtime Hours per Worker by Month: 1984

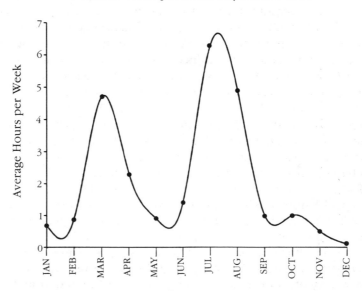

Source: Statistics Canada, #72-002.

Not only is the size of the labour force employed in any given month highly variable, so too are the hours of work for these employees. The average number of hours worked per week in 1984 varied from a low of 19.7 in November to a high of 29.1 in July. A typical work week for salaried fish-processing staff averages thirty-eight hours, and the same is true for some fish-production workers. The reasons for this *average* of less than thirty hours per week for hourly paid staff stem from the rapid alteration day to day in the number of workers required, the ever-changing number of hours worked per shift, and the significant variation between plants located in different regions of the coast. Each of these factors contributes to the volatile nature of fish production work. The bottom panel of Figure 9.1 captures this volatility in another dimension by charting the typical number of overtime hours worked in any month. The sharp increases in overtime worked in March and then again in July and August of 1984 are a consequence of the high demand for labour in herring roe season and then in the summer for salmon processing.

Further perspective on these labour dynamics can be gained by contrasting fish-processing work with both food processing and the aggregate of all manufacturing jobs (the former two are included as minor components in the latter). Table 9.2 provides details of this comparison along three separate dimensions: annual variations in (1) the size of the hourly paid labour force; (2) the average number of hours worked per week; and (3) the typical number of overtime hours worked per week. Each separate indicator clearly reveals that work in fish processing is characterized by far more variation than is seen in either food processing or other manufacturing activities. The straightforward comparison of minimum and maximum values on each dimension provides some measure of this variation, but given the differences of magnitude between the three sectors, a better measure is the coefficient of variation ("CofV" in the table), which increases in size as variability rises. For each of the six possible comparisons, this measure indicates far greater variability in fish processing.

For most people in fish processing, the work period is not only seasonal, but also erratic within seasons, as both Figure 9.1 and Table 9.2 illustrate. It is nevertheless regularly erratic seasonal employment, the turbulent nature of the hours of work being a recurring feature of shorework. To describe the work as merely seasonal misses the volatility occurring within seasons, while to emphasize inseason volatility minimizes the stability that surrounds the patterning of work over an extended period. For example, of the people we interviewed, and especially for women, many have worked in fish processing for several consecutive years, although for most the period of employment lasts only a few months, and people are never exactly sure when in a year the season will start or how long it will last.

Table 9.2

Indicators of Variation in Fish-Processing Work*

	Size of hourly paid labour force			Average # of hours worked (per week)			Average # of overtime hours (per week)		
	Min.[1] (000s)	Max.[2] (000s)	CofV[3]	Min.	Max.	CofV	Min.	Max.	CofV
Fish processing	1.0 (Feb.)[4]	3.5 (Jul.)	0.43	19.7 (Nov.)	29.1 (Jul.)	0.14	0.1 (Dec.)	6.3 (Jul.)	1.0
Food processing	9.5 (Feb.)	14.1 (Aug.)	0.13	30.3 (May)	33.2 (Apr.)	0.03	0.7 (Jan.)	2.2 (Jul.)	0.41
All manu-facturing	76.1 (Feb.)	98.9 (Aug.)	0.08	34.6 (Dec.)	36.0 (Jan.)	0.01	1.0 (Mar.)	1.5 (Oct.)	0.13

[1]This refers to the annual minimum for a particular category.
[2]This refers to the annual maximum for a particular category.
[3]This refers to the coefficient of variation (the standard deviation of a distribution divided by the mean of the distribution). Higher values represent greater amounts of relative variation.
[4]The months in parentheses are the months in which the minimums and maximums occurred.
* Source: These data are from the Statistics Canada monthly, "Employment, Earnings, and Hours" (Cat. #72-002).

Estimating the relative size of the male and female labour force is more problematic than determining the total number of workers. Using the annual *average* number of workers, men compose about 75 per cent of the administrative staff and about 45 per cent of the production staff. However, this *average* is meaningless for the relative number of women and men employed in wage labour, since most women can count on at most six months work (and are therefore undercounted by Statistics Canada). A more reasonable estimate would see the relative share of women in fish production work at about 70 per cent of the labour force.[1]

One method processing companies have used in recent years to attract seasonal workers, especially women, has been a reasonably high wage. The urbanization of processing has meant competition for workers with other businesses. While a reserve labour force exists that can be tapped on short notice, high wages add to the attractiveness of shorework and have consequently added to the continuity of the labour force.

Fish-processing work is relatively well paid as the figures in Tables 9.1 and 9.3 indicate. In fact, B.C. fish-processing workers are often described as the most highly paid shoreworkers in the world, although such international comparisons are always fraught with dangers (e.g., hours worked, inflation rates, standard of living). Table 9.3 uses as a comparison workers in allied industries (food processing and manufacturing) and fish-plant workers on the Atlantic coast. Using take-home pay (including overtime), B.C. fish-processing workers receive an hourly wage comparable to other manufacturing workers in B.C. and significantly above shoreworkers in Atlantic Canada.

Fish-processing and food-processing workers earn about the same in each separate province. However, fish-processing workers in B.C. earn about 90 per cent of the average manufacturing wage in the province whereas their counterparts in Atlantic Canada earn only between 60 per cent and 80 per cent of the average manufacturing wage. While there are numerous factors that may account for this difference, among them the strength of unionization and the percentage of the labour force composed of women (see below), it is clear that B.C. shoreworkers are relatively well off in comparison to their Atlantic counterparts, but no better off than their B.C. peers in other food-processing work.

These are again averages, and, given the sizable employment fluctuations in the industry, $12.00 an hour wages do not translate into annual wages of $25,000 a year as one might expect if the workforce had fifty-two weeks of forty hours per week. As we will see below, annual wages for most shoreworkers fall well short of this mark. However, once again, if annual wages are calculated from data similar to that in Table 9.1, Atlantic region workers earn only about 60 to 70 per cent of the annual salaries of B.C. shoreworkers (these calculations are based on data from Statistics Canada's "Fish Products Industry.")

Table 9.3
Comparison of Pacific and Atlantic Average Hourly Wages
in Three Industrial Sectors for Selected Months (1984)*

	B.C. Feb. $	B.C. Aug. $	N.S. Feb. $	N.S. Aug. $	N.B. Feb. $	N.B. Aug. $	Nfld. Feb. $	Nfld. Aug. $
Fish processing	12.46	14.24	7.56	7.24	7.49	6.30	8.32	7.55
Food processing	12.60	12.42	7.73	7.43	7.74	7.37	8.31	7.64
All Manufacturing	14.03	14.62	9.95	9.86	10.32	9.89	9.79	9.93

* Source: Statistics Canada monthly, "Employment, Earnings, and Hours" (#72-002).

Although B.C. fish-processing workers appear to receive relatively good hourly wages, these wages are not equal for all processing workers. It is important to recognize that these average wage rates obscure variations, explored below, between annual and seasonal workers, and especially between men and women.

PRODUCTION COSTS AND VALUE ADDED IN FISH PROCESSING

Before exploring divisions within the B.C. shoreworker labour force, some discussion of the cost to processing companies for labour is essential (even if it can only be tentative given the quality of the data). This is important because the relative cost of processing labour is significant for the organization of shorework (e.g., job tenure, technological innovation).

Table 9.4 presents information on the cost of shorework labour, energy, and raw materials in conjunction with the total value of west coast fish production from 1975 to 1983 (in constant dollars). Over the nine-year period, the average proportion of labour costs, relative to the total value of production, has been 14.9 per cent as compared to 62.0 per cent for the cost of raw material, the basic component of which is fish, but also includes salt and aluminum. Using information on landed value and total value of output from "Fisheries Production Statistics of British Columbia," the cost of fish accounts for the major portion of raw material expenses and approximately 50 per cent of the total value of production. Fish therefore are the most costly factor of production, with labour costing only about one-quarter the amount of the basic raw material. Furthermore, note that, over time, labour costs are more stable than are costs for fish.

For different species with different onshore processing techniques, these labour costs represent a greater or lesser percentage of all expenses. Salmon canning uses more labour than the production of fresh and frozen salmon where onshore labour costs are very low, especially for salmon delivered from freezer-trollers. The roe herring process also requires a substantial amount of labour since the roe is "picked," "pulled," or "popped" by hand.

If the share of production for each processing technique remained constant over time, then the labour costs would remain stable. However, herring roe production is very erratic and salmon canning is slowly losing its dominance. One consequence of this, explored below, is the increasing amount of the salmon catch being sold fresh or frozen and thus requiring relatively less onshore labour input.

In summary, fish processing is a manufacturing activity dominated almost exclusively by assembly-line work. Whether in terms of salmon canning, fresh and frozen fish, or roe herring, the final product is transformed into a marketable commodity by a set of discrete, sequential activities. The erratic demand for labour makes shorework unique among

Table 9.4
Input Costs and Value of Production in B.C. Fish Processing
(in Constant 1976 Dollars)*

Year	Labour costs $	Energy costs $	Raw material costs $	Total value $	Value added $
1975	29,691 (15.1%)[1]	1,778 (0.9%)	100,709 (51.3%)	196,434	69,076
1976	39,407 (16.1%)	2,152 (0.8%)	161,108 (65.9%)	244,572	106,372
1977	41,729 (13.8%)	2,485 (0.8%)	188,475 (62.4%)	302,170	144,007
1978	47,371 (14.3%)	2,608 (0.8%)	238,128 (67.0%)	355,603	134,524
1979	42,822 (12.9%)	2,672 (0.8%)	225,341 (68.0%)	331,277	112,613
1980	39,405 (16.9%)	2,569 (1.1%)	158,431 (68.0%)	232,954	70,725
1981	46,596 (16.9%)	2,840 (1.0%)	170,842 (61.8%)	276,521	103,887
1982	29,336 (13.5%)	2,869 (1.3%)	129,650 (59.9%)	216,570	93,101
1983	30,663 (15.3%)	2,425 (1.2%)	107,666 (53.6%)	200,799	77,830
Mean	38,553 (14.9%)	2,483 (1.0%)	164,479 (62.0%)	261,874	98,011
St D[2]	7,053	344	47,828	57,708	22,135
CofV[3]	0.18	0.14	0.29	0.22	0.23

[1]The figures in parentheses are the percentage of each cost factor as a function of the total value of manufacturing production.
[2]The standard deviation of the distribution.
[3]The coefficient of variation (see note 4, Table 9.2).
 * Source: Statistics Canada annual, "Fish Products Industry" (#32-216).

manufacturing activities. Not only do most fish-processing workers labour on a seasonal basis, but even within seasons there is an ebb and flow to labour demand. While the volatility noted above remains high, the historical trend has been toward a more stable demand for shorework labour. Especially as a consequence of refrigeration techniques, production schedules can be better planned and less rigidly constrained by fish-catch volumes and delivery times.

In common with other workers in food processing, the social divisions in the labour force play a crucial role. Immigrant labour and Indian workers have historically been the major groups from which labour has been recruited. Over the past decade, and at least partially as a function of both wages and processing centralization, this has slowly begun to change. A second feature of the social division of labour is the difference in the work of women and men. A sharp differentiation exists between the activities of women and men in fish processing.

Before turning to this discussion of the social division of labour, some description of the technical aspects of fish processing is required. While space permits only a cursory overview of what occurs in fish processing, this basic context is essential for a full appreciation of the fault lines characterizing the social division of labour.

THE TECHNICAL DIVISION OF LABOUR

While salmon and roe herring dominate discussions of west coast fishing, B.C. fish processors market over twenty different varieties of seafood products, from geoducks to abalone to black cod. For each, the type and intensity of onshore processing differ. However, since salmon and herring processing represent the major financial backbone of the industry, these types of processing will be focussed on here.

The Salmon-Canning Process

The factory setting of fish production relies heavily on what might be best described as a "disassembly" line where various pieces of machinery perform much of the work in cutting, packaging, and cooking salmon. In sharp contrast to work on fishing boats, the actual labour in fish-packing plants is routinized and repetitive.

As noted in Chapter 2, fish are transported from "the grounds" to the plants either by fishers themselves or in fish packers. The boats use a variety of refrigeration systems to help preserve freshness (e.g., refrigerated sea water, ice). Upon arrival at the plants, fish are usually unloaded via an automatic pumping mechanism, often by way of a vacuum or syphoning system, although cranes and buckets are still used at times, especially for troll-caught fish. The unloading activity itself allows workers a reasonable degree of discretion over the job task, although as with any job, when the workload is intense, as it often is when many deliveries occur simultaneously, the volume of fish to be unloaded swamps any immediate choices over work pace or work priorities. Nevertheless, the unloading system is controlled by the small work group involved, which is always a male group, and the opportunity to interact with others, fellow workers or vessel crew, is very high.

Once unloaded the fish move on conveyor belts past a grading and sorting area. Normal practice is to consign an entire load of fish to fresh/frozen or canned production, depending upon species and quality (e.g., freshness, size, body bruises, cuts). For canning, the main grading of fish is by species; different fish are used for different qualities of canned product. At this point the load is weighed and thus a value determined for payment of the catch. This is again a job in which men predominate, although some women are involved in assistant grading positions.

After grading and weighing, the fish are usually stored temporarily in iced holding bins. These bins (or totes) are moved about plants by fork-lift trucks. Fork-lift drivers, virtually all men, play an integral role in the dynamics of a plant, adding to the general din and hustle with their incessant honking and darting movements. More importantly, for the pace of work, the fork lifts link unloaded fish with the first of the can-processing procedures, the butchering machines. These drivers have a key responsibility in transporting fish around the plant, insuring a steady supply of fish to all lines and an adequate number of empty totes for the grading and weighing operation.

The actual processing operation begins when the fish are brought to the butchering machines. These machines mechanically remove the head, fins, and tail and eviscerate the salmon, a feat accomplished swiftly but loudly. The din is such that even a shouted conversation is difficult, so the small crew of three to four people silently work at their repetitive task of lining the salmon up for the butchers' guillotine.

Three separate destinations exist for various parts of the fish after the butchering operation. The body section proceeds to the washing area, where it is given final attention prior to canning. The neck section is sometimes sent to a special flaking machine, where a specialty canned product is made from the oil-rich neck area. In addition, livers, roe, and other body parts are sometimes extracted for special processing. The remaining portions of the fish (head, fins, etc.) are sent to a reduction area where fish oil and meal are produced. This sorting procedure is accomplished automatically.

Immediately prior to entering the canning and filling machines, the salmon are washed by a large crew of women who work in front of long washing tables. Again the decibel level from surrounding noise is such that many women wear ear protectors in an attempt to shut out the noise. The fish arrive on conveyor belts directly from the butchering machines and are cleaned in cold running water on the washing tables. The women, typically clad in rubber boots and waterproof, numbered aprons, can clean an average fish in less than sixty seconds (although it varies by species and weight). With thirty to fifty women at each table (and up to ten tables in some plants), over two thousand fish an hour can be washed. The washing itself consists of removing blood, dirt, facia, and membrane from the salmon.

At this point a highly mechanized process is involved as the salmon is mechanically cut and deposited into cans, which are automatically fed into the cutting and filling machines after arriving on a separate conveyor belt from a can loft. These lines run at relatively high speeds. For half-pound cans, a typical line can run at a rate of 250 cans per minute, requiring 125 pounds of fish per minute or 7,500 pounds of fish each hour. Given that many canneries have multiple lines, several of which are often running concurrently, this requires a high volume of fish and a large labour component.

The filled cans proceed automatically to a weighing machine where underfilled cans are detected and rerouted to a "patching" table where a female crew is responsible for adding, by hand, a small quantity of fish in order to correct the weight. This female crew of eight to ten members is also responsible for a quick visual inspection of each can. Cans may be pulled from the line and repacked if too much bone or skin is visible.

All cans then move along to a clinching machine, where the can is capped, and then to a closing machine, where the can is vacuum sealed. The final detail prior to cooking is washing the can. This entire process, from initial filling of the can to final washing, is virtually automatic save for the patching table where a hand operation is involved.

The canned salmon is then cooked in a steam oven, a retort, for more than two hours depending on the size of the can. After cooking, the can is passed through a "double dud detector" where both the weight and vacuum seal are checked. Finally the cans are labelled and boxed, ready for shipment to market.

Fresh and Frozen Salmon Processing

In contrast to the canning process, salmon that are designated for either the fresh or frozen market receive far less onshore processing. Especially among trollers, fish can be dressed (i.e., eviscerated) at sea so that for a fresh market almost nothing beyond unloading and packaging is required. In freezer-trollers, the fish can be both dressed and properly frozen while at sea (this involves dipping the fish in a brine solution when freezing so as to prevent drying out).

While at-sea processing does occur, there remains an important onshore processing of fresh and frozen salmon. Fish, some from ice trollers and some from the net fleet, are unloaded either by mechanical pumps or by the traditional bucket and crane method. Fish are inspected and graded, and unsuitable ones are sometimes sent for canning if the plant has a cannery attached. The fish are then transported to the butchering lines where, if necessary, the head is removed and the body dressed. This process is normally accomplished entirely by hand, although more recently some companies have been relying on a mechanical heading machine.

Most of the fish are left in dressed condition for marketing although some salmon are also cut into steaks for sale on either fresh or frozen markets. One crucial difference in the handling of salmon for freezing is the extra care that must be taken in preparing the fish. Washing must be done more carefully in freezing to insure that a clean fish is frozen, whereas in canning, the cooking process destroys any unwanted bacteria that might remain after washing. After washing, the fish are either packaged for immediate shipment to a fresh market or are frozen for export or for shipment to domestic markets.

Onshore processing of freezing fish from the net or troll fleet is a very labour-intensive operation. After the salmon is dressed, each fish is handled individually a number of times throughout the freezing process as the proper glazing is applied to protect the fish from dehydration.

Roe Herring Processing

Since the early 1970s, the marketing of roe herring has played a key role in the industry. Pacific herring weigh about 500 grams at full maturity. While the volume of the catch has varied significantly in the past decade from a high in 1977 of 97,000 metric tonnes to a low in 1980 of 25,000 metric tonnes, even 25,000 metric tonnes translates into a minimum of 50,000,000 fish. While only half of these will be female, and thus contain eggs, processing nevertheless requires the handling of an incredible number of fish.

The volume of fish was one stimulus to developing new mechanical pumping systems on the coast that could quickly and efficiently unload large quantities of fish. Initially the fish went directly into the production stream and were salted or brined to soften the carcass and firm the roe. More recently, the fish have first been frozen and then thawed when the time came for processing. The freezing process allowed companies to stretch the processing season out over a longer period of time, thereby stabilizing plant operations and allowing greater planning and scheduling of processing. The freezing process also acted to firm the roe and soften the herring bodies, making the extraction of the roe easier and enhancing the "recovery" process (i.e., the amount of intact roe recovered from each fish).

Since no prior screening occurs, both male and female herring are frozen. As the schedule permits, herring are removed from the freezers and allowed to thaw (some plants now have automated thawing machines to speed this process). Once again it is fork-lift trucks that are used to transport the herring about the plant. Upon thawing, the herring are usually washed or brined prior to being dumped onto a conveyor belt and moved to tables where women extract the roe from the females. The conveyance system is aided by cold water that moves the fish along and keeps them cold. The roe is removed from the belly of the fish with a quick flick of the wrist. The carcasses are transported to a reduction pile, and the roe are moved to a grading table for a preliminary grading.

A very intricate system of grading occurs, and only roe of the proper size and colour receive a top grade. Top quality roe must be unbroken, so care is required at all stages from capture to the actual extraction to ensure that the cluster of eggs each female carries is kept intact. After the initial grading, the roe are put through a series of brine solutions, the first a mild brine to leech out any blood and then further brines to help in cleaning and preserving. These final brines also help in sharpening the colour to a golden hue, which will help sell the roe at top dollar on the market.

After this brining process is completed, the roe are given a final grading. In some plants this final grading is done on a mechanized shaker where the roe are graded by size. The final product is put into plastic pails, along with a brine solution, and shipped to Japan where it undergoes still more grading and then repackaging before it is sold on local markets.

While these separate processing operations are discussed here as independent events, there is an interesting linkage between them. First, the canning and fresh and frozen production of salmon occur over the same period. Most of this production occurs simultaneously in the same plants, although some firms do only fresh and/or frozen processing. Where the processes occur together, fish, fork lifts, and labour may be shared depending upon the volume in each operation.

More important, however, is the link between herring roe and salmon production. Again the major processors engage in both operations, and since salmon canning was the traditional process, roe herring production has been adapted to the canneries' physical layout. While none of the actual canning machinery is used, the two operations share, among other things, unloading facilities, freezers, conveyor belts, and perhaps most significantly, workers. The establishment of big and prosperous roe herring operations has provided processing labour with longer periods of employment and higher incomes.

The salmon-canning process traditionally requires the most labour, not only because it lasts for a longer time span than herring roe production, but also because the raw product is transformed into a marketable commodity. With herring roe the season is shorter, and a final product is not produced in Canadian plants. Fresh and frozen salmon are prepared over the same time frame as canning, but the labour input from shore-workers is minimal in comparison. This latter difference is evident because more salmon are canned than frozen, and because with some frozen salmon, the dressing and freezing occur on board vessels rather than onshore.

The importance of recent changes in the industry is the gradual decline in the volume of roe herring being processed and the rising volume of salmon not destined for the canned market. The implications of these changes are that shorework processing has been falling over time. For individual shoreworkers, this has meant the loss of jobs, for communities it has meant fewer job opportunities for residents, and for the union, it has meant a shrinking dues-paying membership (see Muszynski, Chapter 3).

THE SOCIAL DIVISION OF LABOUR

The nature of the division of labour and the actual work people do depend on how capital organizes technology and labour toward specific kinds of production. As the discussion of the production process in salmon and roc herring processing has made clear, the work is organized around social divisions in job allocations. For example, women typically find themselves in jobs that are routine and monotonous, stationary and seasonal.

Beyond this sexual division of labour, there are also ethnic divisions among shoreworkers. Historically the temporary, seasonal shorework jobs were undertaken by indigenous Indian and immigrant Asian women, and while these divisions have altered more recently, different regions still depend on shoreworkers from different ethnic backgrounds. For example, in the smaller plants on Vancouver Island, especially in Ucluelet, Indian women are in the minority, whereas in some of the northern plants in Prince Rupert, Indian women constitute about half of the workforce. In contrast, monthly rated men (those with high seniority and relatively good job security) tend to be non-Indian Canadians or northern Europeans.

One explanation for these differential allocations of workers to jobs on the basis of gender and citizenship is that different social groups possess different levels of skills, education, experience, or seniority. That is, market attributes such as abilities and knowledge distinguish labour-market participants. The obvious social division of labour occurring in fish plants is merely a consequence of the capacities different groups of workers possess.

In contrast to this explanation, it could be that nonmarket statuses determine the distribution of individuals in the labour process. Workers are assigned certain tasks not because they lack the skills and abilities necessary for other work, but simply because they are from a particular social category. Gender and citizenship are the central categories among fish-processing workers that correlate highly with job assignments.

This latter argument can be extended further. It is not merely a passive, unreflective decision based on societal stereotypes that is at work here, but rather a conscious strategy to employ certain workers *only* for certain jobs. That is, factors external to the market, gender and citizenship, offer distinct advantages to employees in the "creation and maintenance of different labour processes" (Thomas, 1982:88).

People are not randomly assigned the positions they occupy in the division of labour. If that were so, some women would be driving fork lifts while still others would be in upper-management positions. Fish-plant workers are allocated their jobs on the basis of long established hiring practices. What is at issue here is what factors underly these practices and why they are maintained.

The Sexual Division of Labour

Table 9.5 summarizes the basic division of labour by sex. In addition to noting jobs in which women are and are not typically found, this table presents information on four basic dimensions of job tasks, each a subcomponent of the general level of freedom at work that people enjoy. These dimensions incorporate the degree of physical movement about the plant (spatial discretion), the extent the work is paced by a machine (machine paced), the amount of repetitive, routine movement in the task requirements (relative monotony), and the degree of verbal interaction the work task permits (worker interaction). While it is important to recall that most factory labour is boring and spatially confining, and fish-plant work is no exception, there are degrees of latitude on each dimension.

What the table illustrates is, first, that women and men are assigned to different jobs. Second, and more significantly, these jobs vary conspicuously in terms of the relative pleasantness and freedom they entail. While some women do have jobs equal to those of men, most women do not. In contrast, the majority of "good" jobs are occupied by men, although

Table 9.5
Fish-Plant Jobs and Worker Discretion: A Summary*

Job	Women involved	Spatial discretion	Machine paced	Relative monotony	Worker interaction
Unloading	No	Yes	No	Low	High
Fork-lift driving	No	Yes	No	Low	High
Freezing	Few	Yes	No	Low	High
Grading	Few	Yes	Yes	Low	High
Butchering	Few	No	No	High	Low
Washing	Yes	No	No	High	Low
Canning line	Yes	No	Yes	High	Low
Egg pulling	Yes	No	No	High	Low

* This is a generalized summary of a "typical" plant. Exceptions do occur.

again some males are in undesirable positions, along with the majority of women.

While the jobs differ, the backgrounds of the occupants differ on more than just gender. Table 9.6 compares the characteristics of male and female fish-plant workers. Women tend to be younger than the men in our sample, although women also tend to have greater longevity in the communities in which they now reside. This greater geographical mobility of men is related to the markedly different living situations of the two groups. Over 80 per cent of our female respondents were married or living in common-law relationships, while only just over 50 per cent of the men reported a marital or common-law partner.

These ties to community and family are further reinforced among women by dependent children living at home and spouses who work. The vast majority of women had employed spouses (79 per cent), whereas, for men, slightly more than 50 per cent had a working spouse. Family size also differed between the two groups. Married or common law men were in families with an average of 2.1 children, while their female fish-plant colleagues had an average of 2.4 children. Children of school age were an especially important factor for women who were divorced or separated; eight of fourteen women in this category had one or more children.

With respect to skills or abilities that might differentiate the two groups, notice that while men have more years of schooling on average (11.3 years versus 10.3 years), slightly more women are high school graduates.[2]

Table 9.6
Background Characteristics of Female and Male Fish-Plant Workers

	Men	Women
Age (yrs.)	51.2	45.8
% Indian	23.5%	27.7%
% Asian	11.8%	15.7%
Length of residence[1]	15.1	16.3
% Married or common law	51.1%	82.5%
% Working spouse[2]	54.5%	78.8%
Number of children	2.1	2.4
Years of schooling	11.3	10.3
% High school education	31.4%	35.8%
Self-reported formal skills[3]	2.8	2.7
Sample size	52	88

[1]Number of years of continuous residence in community where now living.
[2]For respondents married or living with common-law partners.
[3]Number of formal job skills respondents reported.

A second indicator of skill levels comes from a self-report question in which we asked respondents about their perceptions of their own skills. The average is nearly identical with men having only a slightly higher number of self-perceived marketable skills (2.8 versus 2.7). On these two measures, skills and education, it would appear that men and women have, at least on paper, reasonably similar skill levels.

Educational level and skill possession do not, however, translate into equal rewards for men and women. First, with respect to average annual earnings from shorework (1981) women receive $9,700 as compared to men who earn an average of $17,400. A small part of this difference occurs because women receive slightly less per hour than men (mainly because of seniority levels), but the bulk of the almost $8,000 discrepancy occurs because women work for far fewer months of the year. While for most women the annual wage is earned in four months, for most men it is earned over eight or more months. Once again this relates to the typical jobs of men and women. There are only large numbers of fish to wash in certain periods of the year, but there are always machines to maintain and repairs to be undertaken.

A second method of understanding this unevenness in employment comes in comparing the typical work histories of men and women. The typical pattern of short-term employment is repeated annually for most women. The monotony of women's work at the washing tables is repeated in the repetitiveness of their employment cycle.

How then is this sexual division of labour to be explained? The most obvious explanation is with physical strength. Women simply do not possess the brute strength required to handle certain of the tasks men undertake. While driving fork lifts and turning on pumps do not demand great physical exertion, there are undoubtedly times when shifting a tote by hand is necessary or when a pump needs realignment and physical strength is required. That does not, however, necessitate a complete complement of male fork lift drivers and unloading crews.

It is also important to recognize that women are not sheltered from physical exertion by being assigned soft jobs. Standing at a table for hours at a time washing fish in cold water is not an easy task. Nevertheless, the reality is that more men are injured in fish processing than are women. For the three years 1981 to 1983 the percentage of wage-loss accidents reported by men was 60 per cent, 66 per cent, and 64 per cent respectively.[3] For men, typical injuries came from being struck by moving material or from strains and muscle pulls caused by overexertion. For women, typical injuries were cuts, lacerations, or tenosynuvitis developed from repetitive forearm movement straining tendons across the wrist.

A second explanation for the differential job allocation and the consequent imbalance in material rewards is that the skills of men and women differ as substantially as the job structure. However, there is little direct evidence to suggest this is the case. Education levels, often used as credentials in decisions about job allocation, do not differ markedly between men and women. Self-reports of jobs skills also suggest little variation between the sexes.

Perhaps the best illustration of the arbitrary nature of some job allocation comes in the case of filleting. On the B.C. coast, virtually all filleters are women; on the Atlantic coast filleting has typically been a man's job. When people on either coast are asked why, workers and managers alike respond that it is either women's work or men's work (depending on the coast). When pressed, justifications can be offered: "The fish are too heavy," or "The task is too delicate," whichever rationalization is needed. However, to paraphrase Thomas (1982:107), there is "nothing feminine about the job of washing fish or filleting, though many people would assert that women are better suited to do the work."

An alternative strategy in explaining the sex typing of jobs, of filleting in particular, is to argue historical precedent. If it has always been that way, to alter the pattern would be to disrupt long-established customs and traditions. This, however, merely begs the question of why it was ever established and maintained as a custom.

Reformulating the question points toward a large part of the answer. Given the seasonal nature of the raw material being processed, how can fish-plant managers find sufficient labour and maintain any type of workforce stability? While some of this seasonal labour is composed of young males, the vast majority of seasonal workers are female. How then can fish-plant managers recruit a seasonal labour force to cope with peak processing volumes in March and April and then again in July and August?

The strategy is one familiar to most Canadian women. Create "women's work," a limited range of jobs for which there is a large pool of applicants. As one plant manager explained:

> Most of the men have been involved with maintenance work and keeping the machinery together, and because the machinery has been important during the season, we've had to carry the men, where the women have been more or less available for the jobs when the work is there.

In other words, men are "carried," but for women a large pool of labour is at hand when the need occurs, and this reserve workforce can be called in as required.

The conditions that exist to create this are several. First, management actively recruits women for the seasonal jobs available. The family and community ties, noted above, enforce a greater geographical stability on women, thereby increasing their availability on a regular, seasonal basis. Furthermore, the general pattern of job segmentation in the entire labour force restricts the options women do have when they seek employment.

Connelly and MacDonald (1983:67) provide a similar profile for east coast women in fish processing: "Women have always constituted an available cheap labour reserve and . . . they have consistently responded to any and every wage-labour opportunity open to them." Here it is important to underscore the point that while hourly wages are relatively high, the annual earnings for female shoreworkers are relatively low. Women thereby constitute a cheap labour pool in that the employer has no long-term obligation.

It is not, however, just management practice that has created this situation. Jobs in female enclaves become defined, especially by men, as women's work. As Kanter (1977) argues, "The occupation comes to reflect the status of the occupants, not the skills or aptitudes requisite for the work they perform." Transient, seasonal workers do not have the same "perks" and privileges as regular, full-time employees. Their jobs are not seen as being as crucial to the long-run success of fish-processing operations. Women themselves reinforce these boundaries by recruiting other women from among their personal networks. In short, a female enclave of seasonal employees does not just happen, it is the active result of practices by management *and* by male and female shoreworkers.

Finally it is paramount here to emphasize a crucial aspect of skill that has so far been neglected. While most women occupy jobs for which others could be trained in a short period of time, not all jobs women occupy require such little knowledge or skill. The best example of this comes in trying to explain why so much overtime is paid. Why doesn't management simply call in yet another group of women and run a third shift in the cannery as opposed to using two shifts with overtime?

It is not a function of any mechanical problem in starting or stopping the line. The problem is that a sufficient number of some key personnel are just not available for a third shift. The canning line and herring roe "popping" always depend on a mix of experience levels, from seasoned veterans to recently trained newcomers. When the mix becomes too diluted with inexperienced workers the result is a poor product — more broken roe. Rather than run the extra shifts, which would only be for very short periods, it is cheaper to pay regular crews overtime.

The Ethnic Division of Labour

The traditional fish-processing labour force has been composed of indigenous Indians and Asian immigrants. Indian and Japanese women often worked in canneries while their husbands fished. Contractors were also used to recruit both Indian and Chinese labour forces for cannery work.

As Table 9.6 shows, this traditional cannery labour force has begun to erode as more and more people of European ancestry enter the shorework labour force. Based on our sample of the entire coast, now fewer than 50 per cent of shoreworkers have Asian or Indian backgrounds. Regional influences have had a major impact on this change in labour as the centralization of fish-plant work to more urban environments has meant a more diverse pool of potential workers. Prince Rupert and Vancouver both provide far different labour settings than processing sites such as Namu or Bella Bella, where one or at most two ethnic groups dominate the population. Coupled with this, rising levels of unemployment over the last decade have meant an increasing pool of people from all ethnic groups searching for work.

There is also a great deal of variation in the ethnic composition of the labour force between plants in the same locale. Female shoreworkers in the Prince Rupert Co-operative plant are more likely to be of European origin than are B.C. Packers shoreworkers, who are more likely to be Indian women. These patterns reflect both traditional hiring practices and contemporary hiring networks.

While one explanation for the shift away from such a heavy reliance on Indian and Asian women is the centralization of plants, another is the rise in hourly wages. In fact personnel officials in the plants were most likely to cite wages as the major factor in accounting for the gradual increase in "white" women (as many in the industry refer to women of Anglo-Canadian or European ancestry).

While there has been a diversification in the ethnic composition of the workforce, jobs in the plants are still stratified on the basis of ethnic background. Not only are non-Indian, non-Asian workers more likely to hold supervisory positions (over 70 per cent of such jobs), they are also likely to direct more people when they do occupy supervisory jobs. Even on those occasions when Indian, Japanese, or Chinese women do move into supervisory roles, the justification is so they "can look after their own," as several respondents and personnel officers reported.

The division of work on the basis of ethnicity is also reflected in the average earnings of fish-plant workers. Table 9.7 presents information from our sample on earnings from fish processing and personal income from all sources for men and women from three separate ethnic categories. Incomes clearly differ by ethnic group, with Anglo-Canadians and Europeans receiving larger shares. More specifically, it is Anglo-Canadian and European men who earn the top annual salaries largely because they are employed in the fish plants throughout the year. Seasonal employment and seniority explain the remaining variation among Asian and Indian men and among all women. The larger earnings for Asian women is mainly a consequence of seniority in that Asian women in our sample reported slightly longer periods of employment than the other two groups of women.

The third row of the table reports average unemployment insurance earnings, and large differences are evident here. In terms of average payments, Indian people receive greater amounts of unemployment insurance than does either of the other two groups of shoreworkers. Some people, however, do not receive any UIC payments, but even when this fact is considered, the differences in average earnings for the three groups (either

Table 9.7
Average Earnings by Gender and Ethnic Background

	Men			Women		
	Anglo-Canadian European[1]	Asian	Indian	Anglo-Canadian European[1]	Asian	Indian
Fish-plant earnings	$20,000	$11,400	$13,300	$ 8,400	$13,900	$ 9,000
Gross personal earnings	$22,300	$15,800	$15,300	$10,600	$15,900	$12,005
UIC earnings[2]	$ 700	$ 200	$ 1,500	$ 1,000	$ 1,000	$ 2,000

[1]Respondents from Anglo-Canadian and European ancestry.
[2]Unemployment Insurance Commission earnings.

male or female) remain. In addition, regional variation does occur. Vancouver fish-plant workers, for example, receive less UIC money, but even controls for both gender and location do not eliminate the ethnic differences in UIC earnings.

Although the ethnic composition of fish-plant labour has changed since the early years of the industry, an ethnically divided labour force still exists. While differences in pay levels occur between women and men, equally large discrepancies occur between workers from different ethnic groups. This ethnic division of labour is also reflected in the work people do in the plants, especially in terms of supervisory responsibilities.

FAMILY CONNECTIONS

In Chapter 8 family connections were shown to be significant for people engaged in fishing. Fishing was a "central" family activity for nearly all current fishers, and for many, knowledge and property passed from generation to generation.

In fish processing the historical record reveals a similar importance to family networks. Especially among the Japanese and Indians, women and children worked in canneries while their husbands or fathers were out fishing. The use of family labour was crucial for canners in securing a labour force, and it, in part, explains some of the historical pattern of women's cannery labour. As one person told us, "We have a large Japanese labour force here because their husbands were fishermen. Their men were out fishing in the summertime. They had nothing else to do but to work too."

The significance of this family network seems to be eroding among shoreworkers. As a consequence of the continuing closure of rural processing plants and the relative growth of urban-based processing, shoreworkers are increasingly coming from nonfishing families. This pattern is seen in Table 9.8 where the main labour-force occupations of the mothers, fathers, and spouses (if married or living with common-law partners) of shoreworkers and fishers are shown.

The most revealing comparison comes in contrasting the percentage of mothers and fathers who did not work in the fishing industry. In about 65 per cent of the cases, the parents of shoreworkers were not involved in fishing, as compared to about 44 per cent for fishers. Also, the spouses of both shoreworkers and fishers are most likely not involved in fishing occupations. In fact, for the husbands of female shoreworkers, fewer than 25 per cent are now involved in the fishing industry and fewer than one in five are actually fishers.

Table 9.8
Paid Jobs of Respondents' Mothers, Fathers, and Spouses

	Shoreworkers			Fishers		
	Mothers	Fathers	Spouses	Mothers	Fathers	Spouses
Fishing	7.8%	24.6%	14.3%	14.1%	52.5%	12.1%
Processing	31.3%	5.1%	17.9%	42.3%	3.0%	24.2%
Nonfishing	60.9%	70.3%	67.9%	43.6%	44.4%	63.7%
Sample	64	118	84	78	162	124

There have always been certain strains in the union between shore-workers and fishermen. While it has been the unity of these two groups that has led to significant changes in the industry, the potential does exist for a schism to grow, a schism that previously was inhibited by family networks that spanned the industry.

CONCLUSIONS

This chapter has described the organizational context and practice of work in the fish-processing sector of the industry. Four aspects of the industry have been lighlighted—the volatile nature of work, the technical aspects of production, the social divisions of labour, and the family connections among industry participants.

In contrast to other forms of food processing and manufacturing in general, fish processing is volatile work. While fish-plant labour is often characterized as seasonal work, the volatility of the industry cuts more deeply than seasons. The volume of raw product arriving on fish-plant docks can oscillate wildly. The incoming fish must all be handled and processed, whether the immediate destination is the freezer, the can, or the roe bucket. Shift work prevails, overtime mounts, the days all run together, and then suddenly it all stops.

The actual work process itself resembles factory work, except that here it is food that is being processed, and this requires certain standards of sanitation and cleanliness that do not characterize other forms of manufacturing. Botulism, a disease that develops in spoiled food (either canned or smoked) and that can cause death, is a major concern of fish-processing people, workers, and managers alike.

Fish-processing work also involves a social division of labour. The volatile, short-term jobs are dominated by women, as are the jobs that involve less freedom and worker discretion. An ethnic division of labour also occurs, and significant differences in incomes and work responsibility correlate with ethnic membership.

Finally, while family networks have previously characterized the industry, an emerging trend is that more and more shoreworkers are coming from nonfishing backgrounds. What impact this will have on the industry is difficult to judge at this point, although it may require more union effort to maintain solidarity across the harvesting and processing sectors.

NOTES

1. The UFAWU membership of 5,796 in 1982 consisted of 1,961 women (33.8 per cent; see Corporations and Labour Union Returns Act (CALURA), 1983). Prince Rupert Amalgamated Shoreworkers and Clerks Union (PRASCU) had approximately 800 members of whom 400 were women. These two figures suggest a 70 per cent estimate is reasonable, given the data in Table 9.1
2. The apparent inconsistency between years of schooling and high school graduation comes from women having more variable educational backgrounds.
3. In terms of aggregate hours worked by all employees, women as a group accumulated more hours than did men. Therefore while men work slightly less than 50 per cent of the total hours of all fish workers, they account for over 60 per cent of accidents.

10

Organization of Divided Fishers

PATRICIA MARCHAK

Fishing is a way of life. But it is no longer a subsistence activity. As fish became commodities, fishing became a commercial business. "I would argue that it is time we listened to people instead of theories and arguments which when followed to their conclusion create EXXON and New York. It is time we gave real value to the lives and aspirations of people who live and work in the hinterland" was the way fishing was understood by the spokesman for the Pacific Trollers Association in his brief to the Royal Commission on Pacific Fisheries Policy of 1981–82 (April 5, 1981). The Pacific Coast Salmon Seiners brief echoed that: "We are not serfs or greedy exploiters. We have invested our life in fishing and while a bureaucrat has his pension we only have our earnings which many put into a fishing vessel. For many of us, our pension is our vessel" (June 4, 1981:15).

But the sentiments so strongly expressed and fully believed are at odds with an industry organized entirely according to normal rules of accumulation. Those rules pit fishers against one another, and the lives and aspirations of people who live and work in the hinterland (or, for that matter, in the urban centres) are subordinated to the pursuit of profit. Nowhere is this more evident than in the history of the west coast fisheries.

Within weeks of the publication of the Pearse Commission, hundreds of vessel owners, crew members, shoreworkers, and small processing firms were organized in opposition to implementation of its recommendations. But they were not organized together under a single umbrella. Nor were they unanimous in the precise recommendations they opposed. Indeed, the Pearse Commission was occasion for highlighting divisions that had persisted throughout the fishers' twentieth-century history.

As outlined in Chapter 1, some B.C. fishers are union members, thereby declaring their identity as labour. But others are co-owners of processing facilities as well as vessels, thereby declaring themselves capital, though in the largest co-op, crews are also co-op members through an affiliated guild. Others besides co-op members own vessels and are members of vessel owners' associations, again as capital, yet these same individuals may, on occasion, accept the union's leadership and obey its strikes over price bargaining with processors. Crews on these vessels may also belong to associations rather than the union and may identify themselves as either capital or labour, varying by circumstance. Shoreworkers are union members, but they are not all members of the major union, UFAWU; some are members of a separate union within the Prince Rupert Fishermen's Cooperative Association (PRFCA).

That only touches the surface of the divisions in the fleet. A persistent division throughout history has been along ethnic lines; originally between Japanese and all others, then between native Indians and all others. This division does not overlap with class, it bisects class cleavages. Some native Indians are members of the co-op, some are vessel owners, some are crew members. Most are members of the Native Brotherhood of B.C. (NBBC), some are members of other native associations, which on occasion are antagonistic to the NBBC, and some are members of the UFAWU. Indians may be commercial fishers either all or most of the time, but there are other native fishers whose interest is the food-fishery, and whose aboriginal claim is ancestral rights to capture fish at will for subsistence.

There is a gender division, with most women assigned to shorework, and their concerns thereby become major concerns of a union in which most fishers are men.

Another division occurs along regional lines, sectionalizing the fleet into northern, central coast, west coast Island, and south coast components (and on occasion, further subdividing the regional groups).

And yet further divisions occur by gear-type and (though this has diminished over time as the separate fisheries have been integrated) by fishery.

All these divisions take place between commercial fishers. A further, and increasing line of tension, arises between the commercial fishers as a whole and the sports fishers, who include both individuals who fish for recreation and tourist resort owners who sell recreational fishing as a commodity.

These multiple lines of cleavage between fishers frustrate any attempt to classify them neatly into any one category. A union, necessarily seeing the major struggle of history in class terms, is confronted with a practical as well as theoretical problem when trying to persuade fishers (as well as governments and processors) that they are labour under another name. They may be indirectly subordinated to capital via marketing controls, but indirect subordination is not the definition of labour (see Chapter 1), and many fishers clearly reject the category.

Though the divisions are not simply along class lines, and it is inadequate to reduce ethnic and other tensions to a labour-capital conflict, processors hold an advantage because of the multiple divisions. The creation of the labour force by canneries, with different conditions attached to gender, race, gear-types, and regions, defined and intensified these other divisions. Further action by both capital and the state sustained them.

This chapter is about the internal divisions among fishers. It provides brief histories of the major organizations, by way of showing how the diversity developed and how current dilemmas have been shaped by past events. Muszynski and Pinkerton in following chapters provide further historical details: the former from the perspective of the union, and the latter with reference to the native fishers.

FISHERS AS INDEPENDENT COMMODITY PRODUCERS

The first question concerns the definition of fishers, raised initially in Chapter 1. While the union sees market controls as equivalent to the conditions experienced by wage workers, many vessel owners and crew members strongly disagree, even when they accept the argument that they are exploited in their market relations with processors.

They base their counter arguments on the nature of their work process, their greater independence than other workers from direct supervision, their methods of payment in shares rather than in wages, and for those who own vessels, their property rights and their control of the surplus value from their own labour. As well, for that portion of the fleet that has market options, the freedom to sell commodities elsewhere than to major B.C. commercial processors is evidence of their independent status.

There is a dramatic contrast between the labour processes of shore-workers and fishermen. A vessel owner who fishes with one or more crew members has the authority and responsibility to make decisions regarding the pace, duration, and other aspects of timing of the undertaking; the intensity of effort; the distance travelled; the maximum quantity to be caught; the duties of various crew members; and all other conditions of operation while the vessel is on the landing grounds. A skipper can determine whether the fish are of adequate quality and proper size and type to keep, and choose the method of preservation subject to the technical constraints of the vessel. As well, he or she controls a vessel's technology, subject to the financial capacity to obtain technical equipment.

Crew members are subject to the decisions of skippers with respect to the labour process, but, in practice, a large proportion of these decisions are made cooperatively, with the skipper having the final responsibility. They are rarely directly supervised and once they have moved from the "novice" to the "experienced" category, they are normally treated as coworkers. Whether as skippers or crew, fishers respond not to clocks but to natural conditions of sunlight and weather. Much of their supervision is built into interdependent relations and the threat of danger if they are inattentive. However, these generalizations must be tempered by consideration of differences between the crews of small vessels (trollers and gillnets) and the crews of large vessels (longliners, seiners, and trawlers). While there is no equivalent on the Pacific coast to the factory-ship trawler, the larger vessels do employ more crew members. A six-person crew is more subject to supervision than a single-member crew, and skippers of large boats may act more formally toward labour than the skipper of a small vessel.

Crews in the Pacific salmon fishery are normally employed on a share basis. But there are seasons when a combination of low catches and low prices means that both captain and crew are paid primarily in fish rather than cash. The impersonal rules of capitalism encourage the payment of labour at the lowest possible price within the context of labour and other market conditions, but they do not encourage payment in kind or such highly variable payment that workers cannot rationally estimate their incomes over time. Since capital can obtain fish (and very few other products) by relying on the noncapitalist rules of kinship and friendship networks that sustain fishers, they have a positive incentive to sustain the independent system. While this system may, then, best advantage the processors, the relationship between crew and skipper is closer, more personal, and more cooperative as a consequence. These relationships sustain the view that fishing is a way of life, even when it is also a commercial undertaking. (During a troller voyage I was privileged to share, the crew member's steady work intrigued me. I asked if he did not want a break. He looked genuinely puzzled and said, "This is my life. I'll keep working as long as I'm here. There is nothing else." This was all the more remarkable because it was clear that the catch on that trip would not return much on his labour: his payment was largely in the form of fish, taken home and distributed to family and friends.)

Throughout the past century, the different gear-types have had different market situations. The net gears in general have much less flexibility than trollers. Trollers, carrying a high-quality fish, can market through brokers and buyers in the United States as well as Canada, although their alternative options have fluctuated with international agreements and the interest of wholesalers. In general, the net gears have been more constrained and more dependent on traditional processors. With the development of advanced freezers and other methods of preserving fish on board vessels, and with improvements in gear that have greatly reduced the quality differences between gear- and troll-caught fish, these traditional differences are now declining. Groundfish trawlers remain largely dependent on existing processors. Halibut longliners have sold through an exchange since the 1920s. This effectively establishes competition between buyers, and eliminates the need for collective resistance to processors. These real market differences are central to different perceptions of class situation. The greater the independence, the greater is the likelihood that vessel owners — and often their crews as well — will designate themselves as small capital or entrepreneurs. As dependence increases, the similarity to labour in other industries increases, and vessel owners and crews are more likely to turn to collective bargaining as the means of dealing with capital in the processing sector.

The differences continue to be important, but another trend has also emerged throughout postwar history: the integration of the fleet. Increasing numbers of vessels have combination gears, so that when the salmon fishery is closed they can continue to fish for groundfish and/or herring. Combination gear vessels are heavily capitalized, increasing the pressure on owners to fish throughout the year, and to resist attempts — such as recommended by Pearse — to license vessels separately for different fisheries. They also resist union limitations on their fishing behaviour because they cannot afford strikes with so much capital tied up in vessels. Thus the trend reduces the capacity of the union to create a bargaining association for the entire fishery.

Fishers as Co-Op Owners

The problem of categorizing fishers disappears when we consider those fishers who own processing firms. For them, the market subservience is overcome. Moreover, they are employers of labour, and in that respect, unambiguously capital.

Fishers who own vessels employing more than two persons are not eligible to join the major union: for them there is no question about their possible status. In addition to these, however, many who are eligible have opted to join cooperatives. At the present time (1984) approximately fourteen hundred are members of the Prince Rupert Fishermens' Co-operative Association (PRFCA), and about sixteen hundred belong to smaller cooperatives. A founding member of the PRFCA expressed his understanding of the reasoning this way:

> The fishermen who founded our organization were seldom starry-eyed theorists intent on persuading society to change its economic habits, but were in the main producers of needed commodities who had become increasingly fed up with the inefficiency and exploitation they saw in the distribution system which affected them as a group. Self-help gradually dawned on their imagination as a logical way to solve their dilemma. (Leslie H. C. Phillips, in Foreword to PRFCA Annual Report, 1979)

A similar understanding came out of a meeting of the Cooperative Future Directions Congress, June 1982, where members of the PRFCA, along with others, adopted the following statement:

> Canadian co-operators share a vision of people working together to achieve their potential, to enhance their social and economic well-being, and to produce and consume what they need through democratic institutions that root social and economic power in local and community organizations. We recognize the independence of people, and of organizations; and the need for effective, responsive linkages. We pursue our vision through our co-operatives — organizations based on equity, equality, and mutual self-help. (PRFCA newsletter *Co-Pilot*, 1983)

These statements suggest a very different ideological perspective on the status of the fisher and the "appropriate" methods for dealing with what members of both the union and the cooperative perceive as market exploitation.

ETHNIC AND GEAR DIFFERENCES IN EARLY UNIONS

Contemporary divisions in the fleet have long histories, beginning with competition between ethnic groups contingent on the different working conditions provided them by early canneries.

By the 1890s there were fishers of European, American, Japanese, and native Indian descent around the Fraser River. The Japanese and Indians were more frequently tied to the canners by cannery-held licences; others held independent licences. Racism directed especially toward persons of Asiatic descent was widespread, permeating the fishing as well as other industries. The Fraser River Fishermen's Protective Union was organized during the depression of 1893, claiming a membership of 1,600 but excluding the Japanese, and engaged in a strike for price increases. Canneries attempted to replace European fishers with Japanese and Indians. Indians joined the Europeans, and according to the union secretary, "refrained from fishing to a man" during the strike. Indeed, they showed more solidarity than the whites (Jamieson and Gladstone, 1950; Jamieson, 1968:137).

The depression was succeeded by the gold rush and an exodus of Europeans from the Fraser fishery to the Yukon. Canners replaced them with Japanese. By 1901 the Japanese, according to government calculations, held 1,958 out of a total of 4,722 fishing licences, and of the 1,090 licences issued to canneries, many were used to employ Japanese fishers. The Royal Commission on Chinese and Japanese Immigration (1902) estimated that some four thousand Japanese were employed in the fisheries, mainly in the Fraser River region (Jamieson, 1968:138).

Further organizing by non-Japanese fishers led to the formation of the B.C. Fishermen's Union in 1899. In self-defence, the Japanese organized a union, the Fishermen's Benevolent Society. The conflicts between these groups, together with strikes over prices, generated violence in strikes of 1900 and 1901 (Ralston, 1965). Several years later, in 1913, a strike was called by the Japanese — who now numbered 1,088 compared to 832 whites and 430 Indians holding licences — to protest a price cut by the Fraser River Canners' Association. Jamieson notes that the roles of Japanese and others were now reversed: it was the Japanese who organized the strike, the whites and Indians who were unable to bring their members into line, and the Japanese who were reported to have used violence and intimidation against strikebreakers (1968:142–43).

Continuing tension between the Japanese and other fishers was acknowledged in favour of non-Japanese in the 1920s by a reduction in the number of licences for Japanese by the federal government, an early indication of how important a role the government played in the organization of the industry. Tensions remained through the next two decades until, in 1942, Japanese fishers, together with other residents of Japanese descent in British Columbia, were forcibly evicted from the coast. Their vessels were confiscated.

One needs to be cautioned against seeing this history as "merely" a conflict between interest groups in the fisheries, or as a camouflaged protection of capital. The expulsion of the Japanese and the many forms of anti-Oriental legislation over the first half of the twentieth century were official acts of governments, and were supported by workers in many industries and unions throughout British Columbia. Moreover, the eviction of Japanese fishers was not in the interests of processing capital; it rested rather on perceptions and deeply ingrained prejudices held by the non-Japanese working class. Yet at the same time, these working-class conflicts were fuelled by differential conditions of employment and wages accorded different ethnic groups in the fisheries as in other industries, so that we cannot isolate ethnic divisions from property relations.

Other Organizations Prior to 1945

The first gear-specific union was the Pacific Halibut Fishermen's Union, organized in Vancouver and Prince Rupert in 1909 and affiliated with the International Seamen's Union, American Federation of Labour. It was reorganized as the Deep Sea Fishermen's Union in 1912, with headquarters at Prince Rupert. With the establishment of the halibut exchange after World War I, there was no further need for union negotiations over minimum prices; the union's main role thereafter was to monitor share agreements between owners and crews. Yet the Deep Sea Fishermen's Federation Union, which grew out of the early union of halibut fishers, survived to play quite other roles in later history as a rival and opponent to the industrial union that formed in 1945, and as a major component in the Prince Rupert Fishermen's Cooperative Association (PRFCA).

The Fishing Vessel Owners' Association was established in 1935 to represent owners in any fishery requiring a crew of three or more in addition to the skipper. No owners with interests in processing plants were eligible. Since gillnets and trollers rarely have crews of three or more, most members were owners of longline vessels and seiners, with a minority engaged in trawling.

Ancestors to the UFAWU included the B.C. Seiners Association, later named the United Fishermen's Federal Union, and the B.C. Fishermen's Protective Union. Unlike the second of these parents, the Fishermen's

Industrial Union, formed in 1931, allowed Japanese fishers to join and adopted the industrial unionism form of the Workers' Unity League as organized by communists throughout Canada in the 1930s. This union became the Fishermen and Cannery Workers' Industrial Union in 1933 and began signing up shoreworkers (Muszynski, Chapter 12). When the Communist Party disbanded the league in 1935, the Canadian Trades and Labour Congress recognized the B.C. Fishermen's Protective Union and the United Fishermen's Federal Union rather than the larger but now orphaned Fishermen's Industrial Union. The wartime demand for fish enabled these groups to mount a single, stronger union embracing fishers, tendermen, and shoreworkers: the United Fishermen and Allied Workers' Union.

The cooperative movement emerged with greater tranquility. A number of fresh-fish agents became established in the northwestern United States in the late 1920s and 1930s. They purchased cod, halibut, and troll-caught salmon. They chilled, smoked, or salted the salmon in preparation for eastern American markets. To take advantage of this alternative market, trollers on the west coast of Vancouver Island and the Queen Charlotte Islands first chartered and then purchased fish packers through cooperative arrangements. By 1934 there were four cooperatives registered under the Provincial Co-operatives Act, one of these being the PRFCA established in 1931 to represent troll-salmon fishermen.

The PRFCA increased its membership by joining with the halibut fishers, later adding members in the trawl and seine salmon sectors. Together with the Deep Sea Fishermen's Union in 1940, it ventured into processing with a liver-oil reduction plant at Prince Rupert. This expansion was facilitated by financial aid and organizing skills advanced via the federal government. Membership in the PRFCA was de facto restricted to fishermen with some financial independence since they had to wait up to eighteen months for settlements of annual accounts and to invest share capital in the cooperative. This effectively reduced participation of fishermen with small vessels and most particularly, many native Indians whose capacity to obtain financing at that time was restricted because they had no private property collateral if they lived on reserves.

THE UNION AT ITS PEAK

Membership in the UFAWU was about four thousand in the early 1950s, and this number rose to about seven thousand by the early 1960s. Between the union and the Native Brotherhood, representation by the end of the 1950s covered all or nearly all the crew members on salmon seine, herring seine, and halibut longline vessels, 70 to 75 per cent of salmon gillnetters, but less than 20 per cent of the salmon trollers (Federal-Provincial Committee on Wage and Price Disputes in the Fishing Industry, 1964:79). The representation demonstrates the difference in conditions for fishers. The more numerous crew members in the seine and longline fleets were closer to the traditional conditions of supervised labour than the independent trollers and their occasional crew members. While crews on gillnets were rarely in that situation, gillnet owners were much more vulnerable than trollers because they had to market their product to commercial processors.

With the formation of this union and its initial contracts, a new phase of Canadian as well as British Columbia history was opened. As noted in Chapter 1, all of the 9,571 Canadian fishers under collective agreements in 1945 resided in B.C. Of these, 9.7 per cent were covered by union shop conditions, 11.2 per cent under preferential agreements, and the remainder under entirely voluntary agreements. The majority of fishers covered by agreements were in the salmon sector (76.5 per cent including tendermen). Just under 10 per cent were in the halibut fishery, and smaller proportions were in herring, pilchard, and tuna fishing (*Labour Gazette*, 1947:1,426).

In addition to coping with legal challenges to its right to bargain on behalf of fishers, the union was confronted with challenges from the Canadian Trade Union Congress in 1949 on the grounds that its leadership was communist. Other B.C. unions (notably the International Woodworkers of America) went through similar battles during the 1940s. By 1948 the B.C. Federation of Labour was firmly under the control of noncommunist (and perhaps more accurately, anticommunist) leadership; the cold war took its toll throughout the labour movement. For the UFAWU, expulsion meant isolation during the critical years of its growth, yet it engaged, even so, in militant wage bargaining and strikes throughout its history.

Between 1945 and 1963, the UFAWU was involved in approximately twenty-five strikes (Jamieson and Gladstone, 1950, updated by Federal-Provincial Committee Report, 1964). After 1950 disputes occurred virtually every year, affecting the entire coast—in 1952, a four-day strike of 6,000 workers; in 1954, an eight-day strike of 5,620 workers; in 1957, a twenty-six day strike by 5,500 workers; and in 1959, a two-week strike by 5,000 workers were the larger confrontations (Jamieson, 1968:373). Two strikes of shoreworkers as well as net fishers in 1959 and 1963 shut down the entire industry. Others closed particular fisheries for varying periods. The 1963 strike tied up fishers from mid-July to early August, a total of twenty-two days during the salmon season.

The successful organization of both fishers and processing workers set four spirals in motion. The first was the impact on processing firms. Since the wholesale price of canned salmon was determined in a world market and could not be increased dramatically by B.C. producers, escalations in prices for fish and labour provided canners' motivation to seek ''more efficient'' methods of operation. The second was the creation of tensions between the two sectors, since, if the price of the raw product increased, canners would naturally seek to reduce the total cost of labour in processing to balance it; if the cost of labour increased, they would seek to maintain low prices for the raw product.

Simultaneously the organization of a union that intended to unite longline and trawl, as well as net and troll fisheries, began another spiral of conflicts between different groups in the fisheries whose interests were not compatible.

Finally, these various tensions combined with strikes provided much of the impetus for expansion of producers' cooperatives to process and market the catch of independent vessel owners.

CONFLICTS BETWEEN ORGANIZATIONS

Conflicts and simmering antagonisms between fishers affiliated with different organizations became more commonplace. The herring fishery provided an instance of this in the later 1950s. Herring had replaced pilchard in the reduction industry after 1940. With the introduction of sonar equipment, herring seemed destined to follow pilchard into extinction. The equipment itself had different effects on natives and non-natives because natives were unable to command the loans or possess the capital to add sonar equipment and thus effectively compete. But the major open conflict occurred between the UFAWU and independent seiners when the union struck an agreement in 1959 with the Fisheries Association (representing the larger companies) to limit participation to company vessels with union crews, ostensibly to reduce overfishing. Independent seiners could not effectively challenge the agreement since only unionized reduction plants purchased herring. Independents sought an outlet for herring in the PRFCA, which had constructed a reduction plant in 1955. The Federal-Provincial Committee (1964:117) suggests that the number seeking this option was greater than the PRFCA could handle. Its plant, and three new vessels constructed for the herring fishery with its aid, expanded fishing in direct contravention of the union-processors agreement. Depletion of the stocks finally occurred, and the fishery was closed between 1967 and 1972.

A similar incident occurred in the halibut fishery. Following World War II, the construction of large combination-gear vessels in Canada (apparently less frequently built in the United States) increased pressure on the stocks. About 125 longliners, many with combination gear, took 85 to 90 per cent of the total catch, the remainder being caught by small, converted gillnet and troll boats numbering about six hundred by 1963. In the waters northwest of Cape Spencer particularly, the annual catch increased from between four and five million pounds prior to 1952 to between sixteen and twenty-one million pounds in the early 1960s. Approximately 60 per cent of the total catch was exported, mainly to the United States, as fresh and frozen product (Federal-Provincial Committee, 1964:26–27, 56–57).

However, annual catches after 1969 rapidly decreased. The five-year average from 1961 to 1965 was 33.5 million pounds; from 1976 to 1980, 8.6 million pounds (brief to the Pearse Commission, Pacific Coast Fishing Vessel Owners Guild, prepared by Eric Wickham, Table 1, p. 6).

The UFAWU and American unions in the 1950s advocated curtailment programs together with minimum prices. American fishers did not agree on the minimum price proposals, but both they and Canadian fishers agreed in 1956 to rules for "lay-overs" which were designed to extend the halibut fishery over a longer period and permit fishers to rest between trips. Enforcement of these rules in Canada resided with the UFAWU, which reserved the right to clear vessels even where these were not under union jurisdiction. This involved an agreement between the major processors and the UFAWU, explained as a cooperative arrangement for preservation of the resource, but it was not universally perceived in that light. Vessel owners and longline crews sailing from Prince Rupert perceived another motivation — control of halibut fishing by the union. This and related issues simmered through the 1950s and erupted in the 1960s in jurisdictional battles between the UFAWU, the Prince Rupert Vessel Owners Association (PRVOA), the PRFCA, and the Deep Sea Fishermen's Union (DSFU). The DSFU emerged in 1956 as the new form of the earlier longliners' unions.

Reciprocal agreements were signed in 1959 between the UFAWU and the DSFU recognizing respective rights of representation when crew sailed on longline vessels under the other's jurisdiction. In the salmon and herring fisheries, however, jurisdictional disputes continued. Halibut crews normally working on PRFCA vessels were required to sign with the UFAWU when participating in these fisheries.

Industrywide strikes in 1959 and 1963 affected the PRFCA and vessel owners as well as other processing firms, and brought to the fore the cleavages between the PRFCA and UFAWU. These struggles were stated in terms of shoreworker wages, the PRFCA paying higher wages than those negotiated by the UFAWU with other processors. In consequence of the PRFCA settling with its employees within a week of the 1963 strike, co-op fishers were able to go out on the grounds two weeks before others settled with their employers.

The 1967 conflict began when the UFAWU tried to sign up all trawl crews and initiated strike action for formal share agreements. The Fishing Vessel Owners Association resisted. Their crews in Prince Rupert were also members of the PRFCA. Six PRFCA vessels refused to curtail their fishing to enable crew members to attend a meeting at which negotiations were to be undertaken, and their fish were declared "hot" by shoreworkers when they returned. The PRFCA locked out shoreworkers, who refused to unload the vessels, and succeeded in keeping its other operations working. The DSFU extended its representation to trawler crews and then to all crews on PRFCA–affiliated vessels. The UFAWU pointed out that the PRFCA was dominated by vessel owners and that it was acting entirely on behalf of these owners who controlled the board of directors and the DSFU. The UFAWU faced an injunction in the midst of this battle, refused to obey it, and both the president and secretary were jailed on contempt of court proceedings. As well, the UFAWU was obliged to pay damages to the vessel owners.

At this point the long-standing opposition of the Canadian Labour Congress became an additional problem for the UFAWU. Within a year of the 1967 strike, the co-op shoreworkers had voted for decertification from the UFAWU, and the PRFCA succeeded in ousting those who had obeyed the picket lines of the UFAWU during the strike. The DSFU created an umbrella for them in an in-plant union for the co-op, and they remained under that umbrella until 1974 with the blessing of the CLC (Muszynski, Chapter 12, provides the UFAWU version of these events).

The 1967 strike had established the battle lines between the PRFCA and the UFAWU, and throughout the next decade these lines hardened. Although the union won some battles, the co-op, together with new organizations representing fishers during the bonanza period of the 1970s, gained strength at the expense of the union. The union's strength was further eroded in the 1970s as it dealt with the problem of displaced shoreworkers in the salmon and roe herring sectors. The problem for shoreworkers was the obverse side of the bonanza for fishers. Fishers lost nothing from in-plant investment by the Japanese and high-demand markets for unprocessed fish. Thus, when the union defended its shoreworkers, it ran the risk of losing the support of fishers who did not want to strike on behalf of shoreworkers. For those who could do so, the attractive alternative now was the PRFCA, which took advantage of the situation to add facilities for the processing of salmon and herring.

Parallel developments reduced the bargaining unity between the union and native fishers. The cash prices paid on the grounds for herring roe in 1978 were exhorbitant and greatly exceeded the guaranteed minimums negotiated for that year. The Native Brotherhood of British Columbia/ UFAWU attempted to negotiate higher minimums in 1979 and 1980, failed, and in 1980 entered a prolonged strike for herring roe fishers that effectively lost them the season. Simultaneously, the Japanese market collapsed. In 1981, in consequence, the NBBC/UFAWU negotiators were dealing with a fleet that was overcapitalized and fraught with bankruptcies, and a processing sector that had suddenly declined to seven companies, several of which were in grave financial straits. These conditions affected the relationships between the NBBC and UFAWU, which steadily worsened throughout the next two years and erupted, in 1982, in a public dispute over shoreworker wage settlements versus price settlements for fishers.

This history only makes sense when we refer to the underlying cleavages between different gear-type fishers and the nature of property rights. The vessel owners, and the PRFCA in general by the mid-1960s, were more than petty capital when acting as employers in the processing sector. As long as they could persuade crews on the larger vessels and shoreworkers in their plant to support them, they could avoid the UFAWU and its strikes. The ideology of cooperatives was buttressed by acceptable wages and shares and by internal organization of committees to deal "in house" with labour disputes. The ideology of unionism could succeed only if these workers could be persuaded that they were industrial labour rather than members of cooperative groups. Up to the mid-1970s, the PRFCA ideology won out, and vessel owners both in their vessels and in the plant avoided confrontation with labour.

THE COOPERATIVE AT ITS PEAK

The PRFCA's membership among vessel owners in the salmon-net fishery and trawling sector increased steadily through the 1960s and 1970s, reaching about 1,400 by the end of the 1970s. The co-op began to process net-caught salmon and crabs in addition to herring and groundfish, investing in new plants and machinery and increasing its employment of processing workers.

The growth and investments in new technology created a different labour climate in the plants than had been possible with the smaller operations. In 1974 the shore-plant workers separated from the DSFU to form the Prince Rupert Amalgamated Shoreworkers and Clerks Union. In 1978 the new union and the co-op failed to come to a satisfactory agreement over the introduction of new filleting machinery, and the co-op locked out its employees. Production for that fiscal year dropped from an average of 39.2 million pounds over the previous five years to 28.5 million pounds (PRFCA Annual Report, 1979:3).

The loss of production workers depleted the DSFU of a substantial part of its membership. Although continuing to be directly affiliated with the Canadian Labour Congress, it changed its name to the Prince Rupert Fishermen's Guild, and, in this form, represented all crewmen on multi-crewed vessels in the cooperative fleet, including seiners, trawlers, longliners, trollers and gillnetters. However, since all crew members must also be members of the PRFCA, and vessel owners may also join (though most do not), the guild is not an effective bargaining unit on behalf of labour, and its membership does not hold a high proportion of potential members.

The revival of the herring fishery in 1972 resulted in heavy capitalization of combined-gear vessels and intensification of competition for the resource. The 1979 PRFCA annual report notes that

> almost chaotic conditions in the 1979 herring roe fishery resulted in unrealistic prices being paid by cash buyers in many instances. The long-standing

principle in our Association of returns to members being based on actual market prices, less expenses related to each species on an equitable basis, has not permitted fully competitive returns for roe herring, although returns on most other products are more than competitive. (p.3)

A 1983 association newsletter reminds members that they should bring all catch to the association (as required by the terms of membership). Apparently this requirement was not fully observed since the chief executive officer at that time noted that ". . . last year we lost approximately 10 per cent or 500 tons of herring to the cash buyers" (Co-Pilot, 1983). Such problems notwithstanding, the PRFCA had much optimism and wealth from its sales of herring and other species during the 1970s.

Buoyed by its success, it embarked on a program of expansion, including purchase of Canada's first freezer-trawler in 1978–1979. Since the vessel was restricted in potential catch to species not then exploited by the west coast trawl fleet, the PRFCA failed to profit on the purchase in a depressed market for the principal fish it could catch (hake). In 1980, the co-op purchased an interest in Dolphin Seafoods of Cleveland, Ohio, in order to obtain a market for oven-ready fish products. Both investments ran into continuing difficulties and losses as depressed markets for groundfish persisted.

At this stage, then, the PRFCA had become a corporate entity with the labour relations typical of corporate employers and with properties well beyond the normal sphere of a simple producers' cooperative. And with the new financial problems of the 1980s, it was a corporation in trouble. In 1982, the PRFCA reported a loss of $786,000 on the Dolphin investment. That company subsequently declared bankruptcy.

By 1980, as Pearse conducted his royal commission investigation, the two major organizations had passed their peak years. The cooperative, like other processing firms, could not expand its accumulation base with a declining fish stock and competition from non-Canadian fish-farm firms. The union could not expand its membership base when the employers for its members were themselves decreasing in number and encountering declining market demand.

NATIVE INDIANS, 1900–1980

Independent licences for northern fishers were restricted until 1923, and native fishers were further restricted in their capacities to obtain these because they lacked funds to purchase commercial vessels. As both became available for some members of northern native settlements, a basis existed for formation of a protective organization. The Native Brotherhood, fashioned after an earlier organization in Alaska, was established primarily for and by northern native fishers, though it always had other cultural and political functions as well. An earlier organization formed in Vancouver, the Allied Indian Tribes of British Columbia, had attempted since 1916 to create a pan-Indian association to pursue land claims. It had presented claims to the Canadian government, but these were rejected and natives were prohibited from fund raising for further activity on claims. The organization collapsed shortly before the NBBC was established in 1931.

Of some importance was the religious orientation of the northern Indians, particularly the Haida and Nishga, who established the NBBC. The organization was permeated with Protestant beliefs, symbols, and rituals (Tennant, 1982:28–29). Southern Indians, by contrast, were more affected by Catholicism, a difference due to the greater presence of Roman Catholic missionaries in the south, Protestant missionaries in the north. This, combined with geographical distance, reduced the ties between natives in different locations, and while the NBBC was the most successful and, until the 1950s, the only strong organization of B.C. Indians, it remained primarily a northern organization.

Indian fishers with licences totalled between two thousand and twenty-five hundred between 1938 and 1962. As a percentage of total fishers, however, their proportion declined from 22 to 14 per cent in that period, as total licences increased from 10,314 to 16,437 (Federal-Provincial Committee Report, 1964:64). In the Nass, Skeena, Fraser, and Rivers Inlet regions, they engaged mainly in gillnetting and purse seining; along the west coast of Vancouver Island and the Queen Charlotte Islands, in trolling and purse seining. However, a majority were still engaged in gillnetting in the 1950s. Gladstone (1953:24–25) numbered gillnet licences to Indians as 1,100 in 1950, compared to 922 for purse seining and 596 for salmon trolling. In 1950 as well, 263 halibut fishing licences were issued to Indians. Gladstone notes that while few Indians could obtain financing for longliners, an increasing number were purchasing larger seine boats, and some of the smaller gillnetter and trolling boats were being used as part of the halibut "mosquito fleet." Few Indian fishermen were engaged in the herring fishery because of the high investment requirement for sonic-equipped vessels.

The Native Brotherhood represented Indians and had bargaining powers, even though it included both vessel owners and crews or skippers of rental vessels. The legitimacy of its bargaining powers was challenged on the same grounds as was the union's, with which it worked closely throughout the 1950s and 1960s. Approximately 30 per cent of its member fishers were also members of UFAWU (Federal-Provincial Committee, 1964:84), though Gladstone (1953:27) notes that they joined unions "only as a last resort." The reasons for this, in his opinion, were continuing geographic isolation of many Indians and a lingering distrust of white unions, resulting from earlier discrimination. As well, he noted that if Indians had merged completely with the union, they would have been a minority group and those aspects of their lives directly linked to being Indians, in contrast to being fishers, would not have been represented.

While the NBBC was without doubt the most important Indian organization in British Columbia, its membership remained largely among Protestants in the north. As of 1964, it had a reported membership of two thousand, of whom about 65 per cent were fishers and the remainder shoreworkers or tendermen (Federal-Provincial Committee, 1964:84). In 1958, the Nuu-chah-nulth of the west coast of Vancouver Island formed the Allied Tribes of the West Coast. In 1964 Coast Salish bands formed the Southern Vancouver Island Tribal Federation. Neither of these was specifically concerned with fisheries, but since the fishing industry was a major employer and fishing continued to be a major occupation, the sometimes divergent interests of geographically dispersed and culturally diverse groups were beginning to be expressed and to affect Indian politics in the fisheries. Further changes encouraged this process in the 1960s. Federal government policies on land claims, following the Nishga court case, became more favourable. Funding became available for research on claims. As new organizations emerged and land claims gained greater acceptance in Canadian society as a whole, the Native Brotherhood expanded its representation of coastal fishers but simultaneously confronted new claims from rival groups.

Throughout the 1970s, the UFAWU stoutly defended native rights and represented its native fishers. But the attitude of Indian leaders emerging in both the NBBC and other organizations was less favourable than earlier leadership to unionization. Some native fishers who, with the financial backing of federal government loans, had become owners of substantial vessels did not entertain the self-image of wage labour; they began to make it clear that in their view they had more in common with capitalists than with unions. Others, noting that these financing schemes had actually benefited the few rather than the many and had increased the gulf between rich and poor within native communities, were most concerned with the development of self-sufficiency on tribal lands. To this end, land claims — including fisheries claims — had to be settled, and recommendations to the Pearse Commission were directed primarily to that end.

Land claims are, of course, claims against property rights; rights of a very different nature than the disputed rights of labour. As it became evident that the commission's recommendations were leaning toward special status for Indian fishers, the division between natives and non-natives intensified. In addition to land-claims issues, the federal government became increasingly engaged in financing or aiding in other ways the establishment of Indian fishers and plants. In 1975 the Shearwater plant next to the Indian village of Bella Bella on the central coast was scheduled for closure by its owners, Millbanke Industries. With government aid and investment by Marubeni, executive members of the NBBC and the Bella Bella band council established the Central Native Fishermen's Co-operative to take over this and a plant in Ucluelet on the west coast of Vancouver Island. The NBBC expressed anxiety over this development because it was unclear whether the new plant would operate with shoreworker agreements established by the NBBC and the UFAWU or become another cooperative similar to that at Prince Rupert. A similar situation developed near Port Simpson on the north coast with the establishment of the Pacific North Coast Native Co-op. In both cases, the organizing impetus arose with vessel owners whom some natives and the UFAWU executive perceived as an elite within the native community.

Conflict between Native Brotherhood and UFAWU

In price negotiations with the Fisheries Association, the Native Brotherhood and the UFAWU had cooperated for many years, with UFAWU taking the leadership role in negotiations. The relationship between the two was not formally established, but tensions were not aired publicly. A strike of both shoreworkers and fishers in 1982 divided the two groups: the NBBC was prepared earlier than the union and a higher proportion of its members favoured accepting price settlements for fishers. In workshops of the Native Brotherhood meetings in 1982, the leadership role and greater numbers of UFAWU negotiators were criticized ("The Negotiation Process," 1982). The union resolved to negotiate a formal agreement with the NBBC and expressed reluctance to negotiate jointly until such a document had been signed.

In the fall of 1982, the executive director of the NBBC was quoted in the Vancouver newspapers: "The Union sold the fishermen out to get a raise for shoreworkers" (*Vancouver Sun*, August 13:1). With no apologies or withdrawal forthcoming, the UFAWU wrote an open letter to fishing-industry workers, calling for open discussion of the divisive issues. The central issue perceived by both sides was the relationship between settlements for shoreworkers and fishers. The union's position was that "fishermen's prices are determined by their degree of militancy and unity, not by the size of the shoreworkers' settlement." The union argued that Alaskan salmon would have been processed in B.C. had it not been for the support of shoreworkers (Nov. 1, 1982). The union argued that

> labour in the fishing industry cannot be advanced . . . without taking into account the vital interests of Indian people; and Indian people in general cannot advance without taking into account the vital interests of working people in general and its organized section, the Trade Unions in particular. (UFAWU Convention, 1982)

The president of the Native Brotherhood expressed his feelings on the issues in a CBC interview:

> We have tried to work jointly with the UFAWU and I guess it has always been a very unstable relationship, but I feel that the UFAWU treat Indian people the

same as the Department treats us, like second-class citizens. What we say doesn't seem to matter. As a result of that we have a very uneasy working relationship. (November 26, 1982)

The Brotherhood suggested that it could market its own fish at better prices if it did not go through the processors at all. In another CBC interview (November 30, 1982), the executive director of the NBBC suggested that a trading company could be established for frozen salmon, agreeing in the course of the interview that this would reduce the amount of fish processed in the labour-intensive canning sector. Obviously this would by-pass unionized shoreworkers, including Indian women. Presumably they would be eventually brought into the new Indian-owned freezing operations.

An earlier split emerged over the federal government purchase of the gillnet fleet from B.C. Packers for the (newly established) Northern Native Fishermen's Corporation in April 1982. The union's assessment of this was that it was "a bonanza for the Weston subsidiary and a disaster for rental fishermen, Indian and non-Indian alike" (*The Fisherman*, April 23, 1982:5). Leading up to the sale were frustrating negotiations in which the union claimed that it had proposed that the government first buy the fleet and recover the cost through rentals, offering protection to Indian fishers in the process; subsequently that title to the fleet be transferred to the tribal councils but that the company be compelled to rent boats operated by non-Indians to those individuals until they retired; and that the tribal councils had refused to sit down with the union to discuss the proposals. The tribal councils responded that the union's opposition was tantamount to opposition to Indian land claims and a larger role in the fishing industry.

This history needs no further analysis: the claims and counter-claims are not subject to tests for "truth" because what they represent are underlying attitudes not toward Indians (the union's support for Indian rights *is* a matter of record and long precedes these disputes), but rather toward property, the division of property rights, and labour.

VESSEL OWNERS AND GEAR-TYPE DIFFERENCES

The original owners' associations overlap with the original union and cooperative association memberships: groups of fishers bound together for purposes of price negotiations. As the union became more powerful, larger-vessel owners became more distant from others, separated now into their own associations and the cooperative. But as the union and the cooperative began to lose their momentum, new associations formed to cope with price squeezes, seek out new markets for fresh fish, and, of particular interest, establish their separate interests as distinctive gear-type groups in the competitive fishery.

Pacific Trollers Association (PTA)

The Pacific Trollers Association (PTA) formed in 1956 and eventually gained nine hundred members (most residing on the west coast of Vancouver Island), 70 per cent of whom were dependent entirely on trolling. At that stage, the organization was not designed as a collective bargaining unit; its stated purpose was to represent trollers to government departments and international commissions. Some of its members were also members of the UFAWU, others of the PRFCA (Federal-Provincial Committee, 1964:85). However, its membership had frequent disagreements with the union, obeying strike calls with increasing reluctance until 1959. After that, and particularly in the 1963 strike, the trollers went their own way, linking up with halibut fishers who packed their catch to Seattle. The organization's strength increased with cooperative relations established with new regional organizations, the Northern Trollers Association (NTA) and the Gulf Trollers Association.

Pacific Gillnetters Association (PGA)

A similar organization was formed by about four hundred Fraser River gillnetters in 1952. This organization was briefly engaged in price negotiations with the Fisheries Association, and it formed in response to gillnetters' concerns about UFAWU activities. A condition of membership was that members could not belong to other organizations. After the original incidents which gave rise to the B.C. Gillnetters' Association, however, membership decreased. The organization was not revived until the mid-1970s.

The B.C. Gillnetters re-emerged as a larger organization in 1976, the Pacific Gillnetters Association. On challenge from the UFAWU, the Department of Fisheries and Oceans confirmed that it had approved the distribution of confidential mailing lists, including names of 3,800 herring and salmon gillnet fishers, to the PGA. Jack Nichol, president of the union, argued that

> since the 1978 UFAWU strike, it's been obvious the PGA was organized as nothing more than a strike-breaking front and an organization to divide fishermen. . . . Gillnetters are organized. They are in the UFAWU. When Fisheries starts taking sides like that, they in effect take the side of the companies, who argue that gillnetters are businessmen who don't need a union. (*The Fisherman*, vol. 43, no. 21:1–3)

At the heart of this schism is the fuzziness of class location and self-perception of it. The problem is built into the union in its limitation of membership to those vessel owners employing no more than two crew members: this acknowledges that some fishers are labour, some petty capital. The union cannot alter the classification, which, while arbitrary in

242 / Labour and Organization

specifics, is true to life in spirit. The intervention of the DFO in this case may have been the result of clumsiness rather than Machiavellian conspiracy, but it is a clumsiness that grows out of a structural cleavage between fishers themselves and makes them an ambiguous group, neither fully labour nor fully capital.

Other Vessel-Owner Associations

Shortly before the Pearse Commission hearings, some rejuvenated organizations and several new ones became active in preparing position papers and seeking representation where their members' interests were involved. Among these were the Pacific Coast Salmon Seiner's Association (PCSSA) and numerous groups, besides those noted above, organized around gear-types or regional locations or both. What appeared to be happening as the industry declined was a growth in the divisions between fishers as each group attempted to articulate its *particular* interests in anticipation of new rules for fleet reduction. In part this was a response to the success of the PTA and affiliated trollers' organizations, which left other gear-types relatively defenceless.

This development of diverse and special interest-group organizations is at variance with the other trend toward increasing integration of the fleet, as vessel owners added combination gear so that they could operate in different fisheries over the course of a year.

The two developments led to inevitable contradictions experienced at the individual level for many fishers and at the collective level where unified action was frustrated by diversity of interests. The UFAWU, attempting to unify fishers while at the same time trying to represent shoreworkers adequately, stood to lose members who straddled the petite bourgeoisie and labour classes and whose capital outlays on combination gears caused them to seek other kinds of protection in the face of threats no longer confined to price settlements.

Special-Interest Pleading

The definition of special interests by gear-type was prominently demonstrated throughout the Royal Commission hearings. It was becoming clear that Pearse would have to recommend ways of rationalizing the fleet, so each gear-type was obliged to fight for its survival.

The gillnetters were blunt: their recommendation to Pearse was to "reduce effectiveness of the troll and seine fleets" and as well to "grant a twelve-hour advance opening to gillnetters in all salmon net openings" (Pacific Gillnetters Association brief, June 3, 1981). They argued in favour of moving toward terminal fisheries. In their attack on the trolling fleet, they noted that gillnetters had been prevented by fisheries policies from becoming more effective, while both troll and seine fleets had been allowed to increase their effectiveness and their catch. "In summary," they argued, "the PGA strongly believes that the salmon troll and seine fleets should be curtailed for two reasons, firstly because the resource is being rapidly decimated due to their effectiveness and the great difficulty in controlling which salmon are caught and how many fish are caught, and secondly to restore a fair share of salmon catch to the gillnet fleet."

Trollers stood to gain most from a phase-out of the fleet on the basis of vessel size and capacity, but trollers were also heavily in debt and even within their ranks they recognized that if those who had gambled on new gear were to pay the price, it would include their own members. They argued their case on the basis of returns to investment and labour.

As the hearings proceeded, it was clear that there was no unity in this industry. Capture, of what Pearse, following historical precedents, described as a common property, had become such a competitive activity that participants could only succeed by limiting the access of others. For them as for Pearse himself, the question was: Which groups should be phased out?

REACTIONS TO THE PEARSE REPORT

Following are the major recommendations in the Pearse Commission report on fleet rationalization.
1. Commercial licences should be issued for each species separately.
2. Area zones should be established so that all fishers cannot move into all zones, although any fisher could purchase licences to move into all zones if he/she had the capital to do so and was prepared to pay for the fuel required for such coastal mobility.
3. Limited entry licences should be issued for salmon and roe herring fishing that specify the gear-type authorized.
4. Allowances should be made for companies to own licences, but a limitation placed on their proportions.
5. All such licences would have ten-year terms, renewable only by competitive bids.
6. Downsizing of the fleet should be made through reduction in the number of licences and regular review of the numbers, consistent with the resource supplies.
7. Initial preference for licences should go to existing licence holders, but licence holders would be free to transfer rights to third parties and presumably to sell these if a market for them emerged.
8. Prohibitions should be placed on the rights of financial firms to own productive property, meaning that properties on which they foreclose must be sold to bona fide fishers within a specified period of time.

These recommendations, if followed, would indeed reduce the size and capacity of the fleet and would, as well, increase the competition between independent commodity producers not only for capture of the resource but also for licences. They might reduce the incentive to over-capitalize vessels and would reduce the incentive to create vessels equipped to roam the entire coast. Thus they would ensure that vessels would be more closely tied to particular processing plants in particular regions. They would, in addition, increase the cost of fishing not only by the licences under a bidding system but also by royalties. The remaining fishers would necessarily be well capitalized and tied in closely with processors. Since licences would be scarce, the banks would recover their losses through resale of licences obtained in the event of foreclosures, thus obviating the problem they now have when they foreclose on vessels but cannot sell the licences that make the vessels valuable.

The debate following publication of the report has focussed on problems of justice. Should seiners be rewarded for overcapitalization on the grounds that they are "more efficient"? Should gillnetters be given advance time for openings because they have technical disadvantages? Should trollers be rewarded for their fuel efficiency? If someone must go, should it be the large vessels or the small ones? Should natives be treated any differently than other participants? All possible arguments have been mounted and defended. The burden fell on fishers: they were blamed for the tragedy of the commons; it was implied that therefore they must solve the problems, they must make the sacrifices.

By the time the final report was published, the diverse groups were seeking means of uniting in common cause; everyone was fearful of rationalization, increased costs of licensing, increased competition, and further dependence on the major remaining processor. But what could they be in common cause against? The Ministry of Fisheries and Oceans? The companies? The numerous polluters of habitat outside the fisheries? Pearse himself and his recommendations? It turned out that simply identifying the problem and specifying the enemy was a major undertaking throughout the months that followed.

Western Fishermen's Federation

The Pacific Trollers Association initiated formation of a federation with the hope of uniting all groups in the industry under a single umbrella. The primary objective as stated by President John Sanderson was to obtain a permanent consultative process that would make DFO accountable to fishers and give a small group of fishers, to be called the Pacific Region Fisheries Council, the power to form policy. The proposal put on the floor of the first meeting was that such a council would be appointed by the minister and an advisory group including the processors. Attending the inaugural meeting were representatives of the PTA, PGA, PCSSA, NBBC, FVOA, NTA, and Trawlermen. The UFAWU and Halibut Guild were invited but did not attend, though individual members of both subsequently took part in the federation's activities. The organizers sought federal funding for the federation and obtained some support. As well they obtained support from the University of Victoria, which provided the organization and facilities for conferences.

The federation sponsored two conferences during 1982 and 1983, the first dealing with the Pearse report (at which Pearse gave the opening address) and the second with the consultative process. The first of these was influential in obtaining a one-year moratorium on legislation to permit industry participants to study the Pearse recommendations.

The second conference adopted a position paper arguing in favour of a Pacific Fisheries Council with powers to advise the ministers and make policy recommendations, structured with two members each from the commercial fisheries, the UFAWU, processors, native fisheries, and sport fisheries, plus two other persons to be appointed by the minister to represent other fishing interests — environmental groups, aquaculture, salmon enhancement, habitat management, and the public at large (*Turning the Tide*, Final Resolution, Feb 4–6, 1983).

UFAWU

The format and objectives of the Western Fishermen's Federation were not coincident with union interests. Equal representation from all sectors of the industry and open alliance with processors were unacceptable. The union, therefore, did not support the federation and held its own conferences, likewise advertised as public meetings. The largest of these had representation from nine fishers' organizations, together with numerous environmental groups and small businesses connected to the fishing industry. Similar recommendations came out of this and the WFF conferences, with similar consensus on the need for better habitat protection, better management by DFO, the need for consultative processes on a continuing basis, greater democratization of decision making, and a moratorium to give time for study of the Pearse report. Diversity of opinion on specific recommendations regarding licensing, quotas, and the manner of fleet rationalization emerged, very similar to the diversity found among those who attended the WFF meetings.

Though a close scrutiny of the reports of the two sets of meetings discloses no giant differences between them in responses toward specific issues — no fishers agreed with competitive bidding for licences, quotas were generally disliked, and royalties were regarded as unfair — there was a considerable difference in the tone of the response. Arguing that the Pearse recommendations would most benefit the banks and processors, the UFAWU regarded the report as, in the words of one official, "skunk-cabbage." The union also warned that the benefits to Indians were not as great as Indians were inclined to believe.

Native Indian Responses

The Native Brotherhood also held meetings to consider the report. Pearse had recommended strengthening Indian fishing rights through giving top priority to band allocations, providing financial assistance, reserving a block of licences exclusively for Indians, and instituting an Indian Fishermen's Economic Development Program (a proposal originally put forward by the NBBC). The NBBC meetings were generally supportive of the recommendations, and supported the recommendations for a permanent advisory council as advanced by the Western Fishermen's Federation.

The Nishga Tribal Council linked the fisheries explicitly to land claims. They argued:

> The recommendations Pearse made with respect to Indian participation in the commercial fisheries were, while possibly better in structure, essentially similar to past assistance programs. Pearse's commercial fishing recommendations do not guarantee any Indian access to the fisheries. Buy-back would proceed without regard to any social objectives. The licensing system would not set aside any block of Indian licences — as Indian licences came up for renewal they could be bid on by anyone. . . . Until ownership of the fishery resource is resolved, the government has no right to collect revenues from them. (letter to Minister of Fisheries and Oceans, Jan. 3, 1983)

The Minister's Advisory Council (MAC)

Advisory councils had been appointed and had occasionally met in previous years, but no one felt confident that they were taken seriously by the minister or that they played any useful purpose. Early in 1983, the minister appointed a council with members representing the PRFCA, the Central Native Fishermen's Co-op, the Fisheries Association (processors), the Native Brotherhood, the sports fish advisory board (two members), and each of the following associations: Pacific Trollers, Northern Trollers, Pacific Gillnetters, Guild Trollers, Deep Sea Trawlers, Co-op Fishermen's Guild, Pacific Coast Fishing Vessel Owners Guild, Fishing Vessel Owners, and Prince Rupert Vessel Owners. This body met almost continuously for a month to determine which recommendations they could collectively agree to and to identify the areas of disagreement.

The council did, in fact, make considerable progress on recommendations concerning habitat and reached some agreement on other matters. But on the central issues of buy-back, royalties, licensing, quotas, and area licensing there were differences between them and no guarantee that the minister would act on their recommendations (MAC minutes). When they broke off discussions for the fishing season in 1983, it was unclear whether progress had been sufficient to provide guidelines for new legislation. Protests continued throughout 1983 as bankruptcies, foreclosures, and uncertainties mounted.

Like the fisheries they represented, the MAC Council was fraught with the divisions by gear-type, relative independence in markets for catch, vessel ownership, and the divisions between Indians and other groups. As well, it included the large processors' association but none of the small companies and cooperatives.

Yet a further division emerged between sports fishers and commercial fishers. This had not been a significant source of conflict prior to the 1980s, but it had the potential to eclipse other sources as commercial fishing and the tourist industry became rivals for capture of a limited resource. Indeed, over the course of negotiations, the increasing demands of organized sports fishers (including tour-guide companies) occasionally brought all commercial fishers together in a common cause: on this, at least, they were unambiguous in their self-perceptions. But sports fishers had a cause coincident with new provincial government policies. The tourist industry seemed a possible alternative to the crisis-striken forest industry. Sports fishing would provide a lure only if commercial fishing could be restricted. Commercial fishers, perhaps too immersed in the history of their conflicts, could not stay united to struggle for the industry; once the council had dealt with those issues it could negotiate, it was left with a considerable range of issues that could not be amicably settled. In particular, it could not agree on allocations by gear-types, and on this issue the council began to disintegrate.

In November 1983, the NBBC resolved that native Indians should assert ownership of 50 per cent of their fishery, and called for a joint management system between federal and provincial governments and the Indian people (53rd Annual Convention Resolutions).

In February 1984, all groups represented in MAC except the UFAWU met together with DFO officers and others with interests in the fishery to hammer out a set of proposals for reviving the industry (DFO, "The Year 2005"). These included legislation and financial aid for small-scale fish farms. During the *same* weekend, the UFAWU sponsored a demonstration in Ottawa to prevent implementation of legislation based on the Pearse report. Although both groups stated for public consumption that the two events had been accidentally scheduled for the same weekend, no participants believed it. By that time, though MAC negotiations stumbled on, it was clear that the union, the PTA and other vessel-owners associations, the Native Brotherhood, and the sports-fisheries associations were moving in contrary directions.

Post-Pearse: Unresolved Tensions

The federal government proposed aid of approximately $100 million toward fleet reduction and a restructuring of the industry in May 1984, but the proposal did not pass into legislation before the government went into an election. In September 1984, the entire process haltingly began yet again with a new minister, new research, new advisory committees, renewed struggles between gear-types and organizations. By the spring of 1985, the minister's advisory committee had clearly come unstuck. The PTA abandoned it from sheer frustration with the lack of action by the new minister and continuing conflict between trollers and net gears in the commercial fishery and trollers and sports fishers in the chinook fishery of the Gulf of Georgia (*Current*, PTA Newsletter, 2[1] 1985).

The NBBC followed suit, likewise frustrated with the formal deliberations that seemed to go nowhere, but as well because the government was moving toward further financing schemes to enable native fishers to increase their participation in the industry, apparently at the expense of the others. These other negotiations were not part of the MAC deliberations, and other participants have expressed their belief that they were not honestly dealt with.

One of our interviewees, who had been extremely active throughout 1982 and 1983, responded when asked to comment on the events of 1985: "Hell, I've no idea — I gave up, I'm going fishing." None of the problems had been solved. But as well, there was a temporary sense that maybe the problems did not have to be solved; the salmon runs of 1985 were much higher than anyone anticipated, and the scramble for the catch was "on" again.

11

Indians in the Fishing Industry

EVELYN PINKERTON

Indians have always played a central role in the British Columbia fishing industry and have pressed for an aboriginal claim. As such, they have constituted an important status group lobbying for its interests along ethnic lines. As well, Indians have allied with other vessel owners, crew, rental shippers, or shoreworkers in similar positions to themselves to press for common interests. This chapter focusses on the interplay between the way Indians have acted as a separate interest group and the way they have made common cause with other fishers or shoreworkers

Early forts, trading posts, and industrial sites were strategically located near Indian communities because these enterprises depended on Indians and their resources for development. As suppliers of fish, furs, and other commodities to the Hudson's Bay Company, Indians provided essential labour by which capital became established in British Columbia. When fish processing became fully industrialized via canning operations, the canning companies relied on Indian fishers and shoreworkers as a seasonally available workforce for plants in dispersed and geographically isolated sites along the coast.

Most B.C. Indian nations did not sign treaties with Canada, yet agreements made with Indians before canneries were built, subsequent negotiations with the Department of Fisheries, and the commission that set up Indian reserve lands all recognized Indian rights to capture and sell fish. But these rights are ambiguous in the legal and institutional framework of contemporary fisheries management, and so Indians lobby as a special interest or "status" group, claiming full recognition of their rights.

Indian commercial fishers often act as a status group for two additional reasons. Because they are legally wards of the state without the right to offer reserve land as collateral, Indian fishers have not had the same access to capital as many non-Indians, and have thus developed special relations of dependency with fish-processing companies. Programs to assist Indians in acquiring their own vessels have existed only in the last seventeen years and have helped mainly the most well established. In addition, the continuation of traditional cultural forms (often modified) has influenced the way many Indians act as commercial fishers, creating different patterns of capture, utilization of product and vessel, investment in vessels, and hiring of crew. These distinctive patterns, including the cooperative sharing of benefits within Indian communities, may have been important in bringing about a situation in which Indian fishers lost both licences and vessels. Tribal and pantribal organizations have formed to represent the special needs and interests of Indian fishers and Indian communities. These have created institutions to support aboriginal rights, cultural heritage and community life, and a desire to maintain a position in the fisheries, all simultaneously. In sum, aboriginal rights to fish, separate cultural traditions and communities, and a long history of dependence on the industry, especially in relations with processors, have caused Indians to perceive themselves as a special status group with unique interests and concerns in the fishery.

Despite the importance of achieving solidarity as Indians, such status identification has not always overridden divergent interests within Indian communities, between tribal units, and between Indian fishers and shoreworkers. Indian vessel owners, rental skippers, crew, and shoreworkers have different degrees and kinds of control over the means of production and, especially in recent years, have had different opportunities for capital accumulation and upward mobility. Like others, Indians in these positions may also have differing interests and different ideological commitments to change. The different political activities and affiliations of Indian fishers and shoreworkers reflect these cleavages. In recent years, Indian leaders have attempted to bridge these differences by creating new Indian cooperatives and corporations and by including some of these groups under one umbrella.

THE TRANSFORMATION OF THE ABORIGINAL SYSTEM BY INDUSTRIAL DEVELOPMENT

Although aboriginal social organization varied from group to group, property rights over fish were generally vested in the kin-based corporate group or extended family. For example, among central and nothern coastal tribes, the corporate group owned advantageous fishing sites such as river mouths and the weirs or fences constructed there. Among upriver and southern groups, smaller units owned fishing sites from which traps or nets were projected (Suttles, 1960; Maud, 1978; Cove, 1982). The more or less hierarchically differentiated head of the group had first rights of access and regulated other members' access to the site, ideally ensuring equity as well as conservation.[1] Thus, an entirely open-access "common property" situation never prevailed in the Indian fishery, nor does it today in the Indian "food" fishery, in which Indian groups continue to fish for subsistence according to traditional rights. Communal property (as defined in Chapter 1) might be a more appropriate description.

Preferential access rights were accompanied in some places by obligations to distribute fish at ceremonial feasts and at other times. Although a person's claim to such rights was inherited, the rights could not be exercised until validated by a feast at which fish and other property were distributed. An individual's personal status was largely a function of his or her redistributive activity within the community and to other communities. The feast and other ceremonies also served to reaffirm the group's relationship to the fish as supernatural beings who were believed to return to these fishing sites only if treated with proper respect. The way in which respect had to be shown and other aspects of the ceremonial complex appear to have precluded overfishing (Swezey and Heizer, 1977). The aboriginal system of property rights thus functioned to "manage" the resource, to regulate access, and regulate the distribution of surplus.

During the early years of contact with Europeans, the aboriginal system remained intact, and fish purchased from Indians were an important food source for Hudson's Bay posts. The aboriginal system of fish exchange among groups was simply extended to include sales to Europeans. When commercial fishing first began, the canneries tended to take the earlier salmon runs, leaving the later, less commercially valued fish to the Indians, who prized these for drying (Sessional Papers, 1882). Originally, Indians were willing to accept Department of Fisheries schedules for opening weirs on small streams during poor runs, because this did not entail great sacrifice, and it resembled aboriginal management (Sessional Papers, 1870). Indians were quite conscious of the importance of fishing closures (Rogers, 1979). However, in their early participation in commercial fishing, Indians on the Nass carried guns on their fishboats and could successfully resist being fined for failure to obtain fishing licences (Sessional Papers, 1890, 1892). The chief insisted that he had the right to manage the fishery and that this right had never been ceded to Europeans (Sessional Papers, 1889).

But as commercial canning operations expanded from nine canneries in 1880 to sixty-four in 1900, the canners increasingly sought the fish stocks on which the Indians depended, as well as the actual fishing sites (Fisher, 1977). Ironically, Indian fishing rights were officially recognized and confirmed during this period by both federal and provincial governments, largely because of disagreements over the size of Indian reserves. These were set out by joint Indian Reserve Commissions operating from 1876 to 1877, 1879 to 1880, 1880 to 1898 and by the joint Royal Commission operating from 1913 to 1916. The province persuaded federal officials that Indians in British Columbia did not require reserves as large as those set aside in other provinces, as long as Indian fisheries were protected. For their part, Indians agreed to the small reserves allocated them on the understanding that their rights in the fishery were guaranteed (Lane and Lane, 1978).

These rights were at least partially recognized by the first inspector of fisheries, Alex C. Anderson, also an Indian agent, appointed in 1876. Denying allegations that the Indians destroyed the salmon, he recommended that "any interference with the natives, therefore, under hastily formed or frivolous pretext, would be imprudent as well as unjust" (Sessional Papers, 1878). In Anderson's opinion, "the exercise of aboriginal fishing rights cannot be legally interfered with" (Sessional Papers, 1879). Under subsequent inspectors, however, the Fisheries Department appeared more sensitive to pressure from the canners, and insisted that Indians' aboriginal rights did not exempt them from regulation under the Fisheries Act. Two examples serve to illustrate the direction of policy in later years.

The Barricade Agreements

In 1904, canners on the Skeena River in northern B.C. began to lobby Ottawa to apply the Fisheries Act to Indians in order to prohibit their use of weirs and their sale of fish. The principal targets were the Babine Indians who took sockeye at a weir or "barricade" on the Babine River, part of the upper Skeena system. The canners suggested that the Indians were destroying the resource and that salmon runs would be eliminated if the use of weirs continued. A series of charges, imprisonments, and negotiations resulted finally in a 1906 agreement between the Babine tribes and the Department of Fisheries. The Department of Fisheries agreed to let the Indians fish and sell salmon from nontidal waters as a legitimate exception to the Fisheries Act. In return, the Indians agreed to destroy their weirs and use only nets supplied by the department. Later the department unilaterally rescinded the agreement to allow the sale of fish. A similar agreement was reached with the bands at Fort Fraser and Fort St. James on the upper Fraser River in 1911 (Lane, 1978).

Continued Efforts to Curtail
Indian River Fishing

Following these agreements, fishery inspectors' reports on the Indian fishery continued to demonstrate a sensitivity to pressure from the canners and even from sports fishers, but little appreciation of the Indians' legal rights or the cultural importance of fishing. For instance:

> There is, of course, another side to the question and the position of the Indians must be appreciated. Before the commercial fisheries assumed such large proportions there was no question as to the *propriety* of the Indians obtaining all the salmon they required in any manner they wished for the purpose of food for themselves and their dogs. In view of the intensive commercial fishing which has developed of recent years, and the greatly increased value of salmon, the operators feel that the catch of the Indians, on the spawning grounds particularly, should be curtailed *if not discontinued entirely*, but they realize that some adequate measures should be taken to the

end that Indians may not suffer. Several suggestions have been made to meet the situation, such as the *subsitution of canned pilchards* or salmon put up by other methods at the coast and shipped to the Indians. No concrete proposal, however, agreeable to the Indians has yet been forthcoming, but it is understood to be the intention of the salmon canners to suggest some form of co-operation with the fishery administration with a view to the substitution of some other suitable variety of food and the *non-interference with the salmon*. The numerous anglers who fish in Stuart Lake are also objecting to such large quantities of trout being taken by the Indians. (Department of Fisheries, Annual Department Report, 1929–30:105 [emphasis added])

Although the Department of Fisheries and Oceans has persistently questioned the rights of Indians to exemption from Fisheries Act regulations, neither the Indians' sense of proprietorship nor the legal standing of their claim has been effectively eroded. Instead, Indians as subsistence fishers have suffered de facto loss of fish to commercial fishing pressure. The Department of Fisheries and Oceans considers that the Fisheries Act supersedes any obligations incurred under treaties or commissions, while the Department of Indian Affairs upholds the rights of Indians to pass their own band by-laws concerning the management of fish on reserves. The legal opinion of the federal Department of Justice and recent court cases suggest that Indian claims to fish are still in good legal standing (Pearse, 1982:180; Harper, 1982; Mandel, 1983, personal communication). However, the power of the Department of Fisheries and Oceans to allocate fish to other users before they reach Indians underscores the practical importance of negotiating some kind of cooperative management status with the department, such as exists with some Washington tribes, with Micmac fishers in Restigouche, Quebec, and with native fishers in Ontario (Pibus, 1981; Harper, 1982; Restigouche Accords, 1982; Berkes and Pocock, 1983).

I have until now discussed the legal status of property rights in fish and the transformation of actual access to fish. It is also necessary to consider the transformation of the socio-economic system of aboriginal property rights. How has the practice of the aboriginal system persisted or been transformed in the twentieth century? There are two forms to be explored: first, the aboriginal system, usually called the "Indian food fishery" or the "Indian fishery," and second, traditional property rights in the Indian commercial high seas fishery. Both have persisted formally and informally as discussed below.

THE INDIAN FISHERY

One of the clearest examples of the maintenance of traditional property rights in a "food fishery" is provided by the Gitksan and Wet'suwet'en people of the upper Skeena and Bulkley Rivers in northern B.C. Here, families still own traditional fishing sites from which they project their nets, often suspended from poles, into the river, or gaff fish at advantageous locations. In order to maintain ownership in good standing of these sites, the family must still feast the inheritance of the site and still distribute what is taken from the site in specified ways. These people take less fish per capita now than their estimated aboriginal harvest, partially because Department of Fisheries and Oceans regulations make it difficult to distribute fish to off-reserve family members or to sell to nonfamily.

Where the taking of fish by Gitksan and Wet'suwet'en Indians has put them in conflict with Canadian law in recent years, there have been few successful prosecutions by the Department of Fisheries and Oceans. With the exception of the reduced quantity of fish taken and the energy and funds required to conduct legal defences and prepare a coordinated management system, the aboriginal system operates within the traditional framework.

The persistence of the aboriginal system creates unique opportunities for regional economic development. For several years the Gitksan-Wet'suwet'en Tribal Council has been conducting a professional "Fish Management Study" in order to generate a data base and management expertise superior to that possessed by the Department of Fisheries and Oceans for their area. They expect to co-manage, with the department, the fish stocks of the rivers in their territory and to take a greater porportion of these stocks in the "home territory" of the fish. They base their argument for the superiority of this form of management on the widely accepted view among fisheries biologists that stock-by-stock management of harvest is the most certain way to prevent the elimination of individual stocks. Only when fish are taken as they approach their own spawning grounds or wintering grounds is it possible to easily differentiate stocks and avoid the "mixed stock" fishery problems of high seas fishing, in which weaker stocks are overfished when taken with stronger stocks (Submission to Pearse Commission, Gitksan-Carrier Tribal Council, 1981; Clark, 1982; Morrell, 1985).

The Gitksan-Wet'suwet'en propose to use the traditional system as the basis for local economic development, in which the Tribal Council may provide marketing services and assist in establishing processing facilities in this area where unemployment among Indians varies seasonally from 60 percent to 90 percent (Gitksan-Carrier Tribal Council, unpublished census, 1979). Ideally, the aboriginal system of property rights is to be the framework for the development of a commercial operation in which rights to harvest and obligations to share benefits from the fishery are structured along traditional lines.

For many coastal tribes, on the other hand, the "food fishery" has been reduced to several days of the year when Indians obtain special permits from the Department of Fisheries and Oceans to take fish with commercial gear for home consumption. Most of these groups have entered fully into the commercial fishery and their systems of property rights have been quite differently affected. These groups also tend to perceive the locus of struggle for recognition of aboriginal rights in terms of access to commercial high seas fishing licences and fishing areas (Nishga Tribal Council, Fishery Management Proposal, 1980). This also appears to be the perception of the Department of Fisheries and Oceans (DFO, James, 1984). To understand how these systems were transformed, one must trace the development of early Indian participation in the industry.

INDIANS IN COMMERCIAL FISHING: FISHERS AND SHOREWORKERS

In the traditional system both husband as captor and wife as processor have use rights over the fish. When the Indian family unit entered the commercial fishery, however, the husband and wife were incorporated into different specific relations of production with the canners who purchased the fish, although they continued in their traditional technical roles as fishers and processors.

Indian Commercial Fishers

Canneries, especially in the north, were, in the beginning, almost completely dependent on Indians as a source of labour in both fishing and shorework. Canneries would locate near Indian communities for the explicit purpose of using Indian labour if the communities were near a particularly large salmon run (Gillis and McKay, 1980). In 1887, for example, the Nass and Skeena Rivers salmon were caught almost exclusively by Indians, many fishing as contract wage workers for the cannery on cannery-owned boats (Sessional Papers, 1888). During this time, out of one hundred fishing licences on the Skeena, only forty were held by fishers, the balance being held by canners (Sessional Papers, 1892). However, "about one thousand Indians live by fishing alone in this district" (Sessional Papers, 1890), because many Indians without licences sold fish to those on licensed vessels.

Many of those who did fish with licences refused to pay for them, arguing that fishing was their right. The canners paid for the licences under the Indians' names, and were thus apparently able to control the licences. On the Fraser three thousand to thirty-five hundred Indians fished, of whom forty owned their own licences in their own names. The remaining fishers made less money because they sold at wage rates and used company boats and/or used licences bought by the companies (Sessional Papers, 1892).

Thus both Indians' aboriginal fishing rights and their immersion in a noncash economy contributed to the relations first established between processors and Indian fishers. When fishing became a more expensive enterprise in the 1920s, requiring more investment in engines and vessels, the pattern of this relationship was further accentuated. Indians, forbidden to sell reserve land and therefore without borrowing power, could seldom afford to buy a fishing boat and net outright. However, canners at this time were becoming less interested in direct ownership of the more expensive boats and were willing to lend Indian fishers money toward boat purchases, in return for guaranteed delivery of all their fish. Compared to opportunities for Indians in other industries, fishing was unique, leading a Native Brotherhood spokesman to declare, with reference to the canners: "The fishing industry is the only one to help Indians by financing them and giving them a chance to make money. Any progress we have made is due to their help" (*Vancouver Sun*, April 9, 1952).

In practice, however, relatively few Indians actually benefited from the opportunity to become boat owners. There are several reasons for this. One relates to the contrast between a kin-based community life and an individually based competitive life. The other reason has more to do with the canners' attempt to maintain the Indians in a dependent relationship in order to guarantee continuing supplies of fish.

As the first cannery owner at Alert Bay noted: "It is the hardest thing to induce those people to go fishing for me — they are a happy-go-lucky people" (Laurence, 1951). Work rhythms and work discipline in a pre-industrial society organized by kin obligations and authority of the chief differ of course from the rhythms of industrial production. At the establishment of this cannery in the early twentieth century, the authority of the chief was already weakened and had not been replaced by a similar authority in the industrializing province.

Moreover, the safety net offered by the Indian community and by the Indians' ability to rely partially on traditional subsistence did not create the most favourable conditions for the development of a highly disciplined capitalist workforce, either as "company boat" operators or as independent fishers. As one of the Indian fishers who did become an independent boat owner notes: "Lots of guys lost their boats for not taking care of them, not fishing hard. You had to work very hard to make it. If you didn't work, you never made anything" (Sparrow, 1976:41). Many Indian fishers simply continued indefinitely as partial boat owners and regular workers, but never attained a status whereby they could save money.

Even when a fisher did impose the most stringent commercial work discipline on himself, however, it was apparently still very difficult to become independent:

> I don't want to go work for the companies again because they hold you down for a certain length of time. Although you're making money they don't want to take your payments if they could help it. They try to keep you in

debt, just to have control over you . . . when I paid off my boat, . . . they didn't want to take it. I offered them . . . one thousand one hundred dollars — the balance of my payments. I was only supposed to pay five hundred and fifty dollars a year, that's all. That was the agreement we made when I got my boat. . . I said if you ain't gonna take my money. . .you'll never get no more. That boat is mine. . . . Finally they decided. . . to take whatever money I had owing them and I got my ownership of the boat. Oh, they do that all the time to almost everybody. Try to keep you in debt so they have control over you. (Sparrow, 1976:225)

Although the relationship of indebted vessel owner was one which the canners seemed to prefer with all fishers, Indians had fewer possibilities than others for liquidating their debts because they tended to be in the weakest competitive position on other job markets. "You couldn't find work if you did go out, too much discrimination then, you know. Logging camps and mills was [*sic*] about the only places you could get on" (Sparrow, 1976:122). But even in these cases, being a fisher in summer could create a work disadvantage: "You can't get no work here [Vancouver] hardly. As soon as they find out you're a fisherman, they won't look at you. They know you're not going to go steady. I tried time and again around here in the mills" (Ibid.).

For most Indians living on reserves, competing in job markets outside the fishing industry involved protracted periods of time away from home in a world where one was usually defined as the worker of lowest status and dependability. Logging was the main exception to this state of affairs in earlier years, but when logging became unionized and organized into more semipermanent, family-oriented communities and camps in the 1950s and 1960s, all but local resident Indians were again in weak competitive positions for these jobs (see Chapter 13). The canners were apparently able to exploit the ethnic position of Indians in a way similar to that documented for fish buyers and oppressed ethnic minorities in other parts of the world (Anderson, 1970).

The fishing industry thus became an enclave for Indians where they could acquire credit and job security largely unattainable elsewhere. Long-term relations of patronage developed between cannery managers and individual fishers. The reward for a long and dependable delivery record with the cannery included job opportunities for other family members. Such relationships probably inhibited questioning about the canners' handling of licensing, boat ownership, and payment arrangements, as well as inhibiting the development of entrepreneurship. Many Indian fishers interviewed in our survey did not understand the details of their accounts with the companies and had never questioned the companies about them, although in this they were not alone (see Chapter 8).

Indian vessel ownership appears to have declined from over 50 per cent to about 12 per cent of all vessel owners between World War I and 1983. Records of vessel owners do not allow identification of Indians as a separate group of licence-holders from the post-1962 period, and the records of the early periods fail to distinguish between owners and deckhands. The Department of Indian Affairs estimates there were 11,488 Indian fishers in 1929 (Department of Fisheries, Annual Departmental Report, 1929–30), but it is not clear how many were vessel owners.

The largest number of Indian fishers have always been gillnetters. In 1946, 1,653 Indians held salmon gillnet licences, comprising 20.5 per cent of all salmon gillnet licences (including deckhands, who were licensed at that time). This had declined to 805 or 14.4 per cent by 1962. Indian trollers remained constant at about 10 per cent of the troll fleet (628 to 690 licences) during this period. Indian ownership of salmon seines increased from 37 in 1946 to 51 in 1962, while skippers not owning vessels increased from 119 in 1946 to 122 in 1962. By 1983 Indians held 11.6 per cent of all the salmon gillnet, troll, and combination vessels and 12.5 per cent of all the salmon seine vessels. When the vessels acquired by three Indian tribal councils in 1982 are included, Indians owned 14 per cent of all salmon vessels, as well as 28 per cent of the roe herring personnel licences (Campbell, 1974; DFO, James, 1984).

This slight increase in Indian licences and vessels in 1982 resulted from the creation of the Northern Native Fishing Corporation (NNFC). Through this, and with government funding, three tribal councils purchased 252 gillnet and combination salmon licences with associated vessels.

The purchase, funded by the Department of Indian Affairs and arranged in cooperation with the Department of Fisheries and Oceans, possibly served to balance the effects of the Indian Fishermen's Assistance Program (1967 to 1979), which resulted in the retirement of vessels whose skippers could not afford to upgrade them to pass inspection and the recirculation of their licences into larger newer vessels, especially in the seine sector (McKay and Ouellette, 1978; McKay and Healey, 1981; DFO, James, 1984). Those who benefited tended to be the more successful and relatively wealthy fishers who could afford to make the 12 to 20 per cent qualifying downpayment, thus increasing the distance between the wealthy and the poor, and making it virtually impossible for a young person who had not inherited a boat to get into the industry.

Other forces also influenced the loss of Indian licences and participation in the small-boat sector. Poor salmon runs in 1969 and 1971 influenced processors who normally rented out vessels or had purchasing arrangements with indebted fishers to reduce these commitments. The number of gillnetters rented to Indian fishers declined from 482 in 1968 to 205 in 1971 (Friedlaender, 1975). Some of these fishers had accumulated debts and were considered poor prospects. Their opportunities were further restricted by the 1971 buy-back program that allowed companies to sell older vessels or recombine smaller licences into larger vessels (DFO, James, 1984). The closures of central-coast canneries with attached rental fleets at

Klemtu, Namu, and Butedale during this period also contributed to the trend, stripping at least one Indian village of all its fishing opportunities (Friedlaender, 1975).

Altogether, programs to assist Indians accomplished one of two things in addition to their intended consequences. First, they reinforced Indians' perception of themselves as a special interest group that could expect their situation to become better primarily through lobbying as Indians.[2] Second, some Indians came to perceive that status and wealth differences within the Indian fleet were being accentuated, with some Indians becoming owners of large vessels while others lost even basic employment. (The implications for the formation of Indian political institutions will be discussed in the final section.)

The aboriginal system can influence Indian commercial fishers in at least two yet unmentioned ways. First, a system of informal territorial property rights remains in some areas. Fishers who have traditionally fished an area, or who have kinship-based claims over the resources in an area may as a group exercise informal social control to regulate the access of other fishers to the area (Lando, personal communication). Second, the ownership of vessels has to some extent replaced the titular ownership of fishing sites, and vessel ownership carries with it similar obligations for those Indians who choose to be part of the local status system. Vessels are used as a common resource in many ways by community members, despite the formal definition of private ownership. For example, vessel owners transport other community members to food-gathering sites, on food-fishing expeditions, to feasts, and large community gatherings. Vessel owners may hire deckhands from the community, even if they do not require them, and distribute fish gathered with food-fishing permits to the entire community.[3] The fishing vessel is both an essential economic tool and a prestige symbol. Indian fishers resident in Indian communities direct their fishing activities not solely to competing with other fishers, but to fulfilling obligations as well. On one reserve, a chief and seine boat owner was said to spend $30,000 a year in maintaining his chiefly obligations.

Indian Shoreworkers

Unlike Indian fishers, Indian shoreworkers, in common with all shoreworkers, did not acquire the means of production and had few opportunities for upward mobility in the industry. The main question is whether they were dependent on the fishing industry for their livelihood and therefore required to operate at the convenience of capital. A second and related question is whether shoreworkers perceived themselves primarily in class or interest-group terms or in terms of their common bonds with other Indians. It is difficult to provide a general answer to these questions because of the historical and regional variation in the industry.

In the first few decades, the canneries already resembled factories, but Indians were in a different position than the many other workers who had been forced off the land into the wage labour force. Both the organization and the seasonal nature of the work permitted Indians to retain many aspects of their traditional social and economic arrangements, especially before particular stocks or particular areas were affected by increased commercial pressure. For many of the workers, processing jobs were available only during the six-week period at the height of the season. In certain cases, the timing of salmon runs in different areas permitted Indians to finish the canning season in an urban area and return home in time to process their own winter supply of salmon (Pritchard, 1977; Sparrow 1976:149). As long as the canners were dependent on Indian labour, they accommodated the Indians' traditional social and economic arrangements. Indeed, many canneries allowed Indians to process their own fish at the work site on days off.[4] Hiring patterns, work arrangements, and living accommodations at the early canneries were adapted to Indian labour (see Chapter 2). An Indian "cannery boss" or contractor with special language facility (usually one per company per tribal area) would organize the hiring and housing of entire kin units who moved to the canneries during the two- to three-month season. Whole villages are reported to have relocated during this time. In outlying areas, many of these arrangements continued into the 1960s. In later years, as language facility and personal acquaintance grew between cannery managers and individuals, certain families would be hired regularly.

Indian shorewokers of all ages expressed a preference for fish-plant work over other jobs, even though these jobs often involved the pain associated with tendonitis, carpal tunnel syndrome, and back problems.[5] Their refusal to complain may be partly related to fear of losing high-seniority jobs. Other factors may also contribute to the desirability of shorework jobs for Indians: like fishers, Indian shoreworkers were often unable to obtain jobs in other industries and could exploit family connections in fishing and shorework. Both family structure and traditional year-round activities were less disturbed by this industry than by others. It appears that, even following the period when Indians provided much of the labour for fishing and canning, many of the canners continued to be relatively flexible and accommodating to Indian lifestyles.

From the beginning of the commercial fisheries, there were always a few Indian women, and even families, who worked in the canneries year-round (Sparrow, 1976:79); they made nets and repaired machinery at the cannery in the winter. Later, there was also year-round work in areas such as the west coast of Vancouver Island during booms in the herring and pilchard-reduction industry. For the majority of shoreworkers, however, year-round work was only achieved after one had attained high seniority and was feasible only if one lived in Vancouver or Prince Rupert.[6]

In the case of fish plants in the more isolated areas of the coast such as Ucluelet, Tofino, Bella Bella, Port Hardy, Klemtu, Namu, and Butedale, shorework has never been year-round, but has provided critical employment for the regular workforce in the area. When the processing industry went through a period of centralization in the late 1960s, many Indian shoreworker jobs were eliminated with the closure of most of the remaining outlying fish plants. By 1970, Indians comprised only 1,500 of the 3,700 shoreworkers (Pearse, 1982). These jobs were irreplaceable for people living in isolated reserves, not only because they were traditionally held by Indians, but also because they permitted Indians to retain their relatively traditional lifestyles in the off-season.

The lack of other job opportunities near reserves and a desire to maintain the reserves as cultually distinct groups still make seasonal shorework preferable to urban nonshorework jobs in which Indians must compete at a disadvantage. This preference exists even when shoreworkers on reserves are limited to seasonal and temporary employment.

Virtually all Indian shoreworkers are in the United Fishermen and Allied Workers Union and appear, from our interviews, to maintain a more militant and consistent trade-union stance than do Indian fishers. As well, those interviewed said they had more in common with non-Indian shoreworkers than with Indian fishers in the Native Brotherhood. It was not uncommon to hear an Indian fisher express negative or neutral attitudes about the UFAWU and be interrupted by his shoreworker wife, who wished to express her support for the union.

UFAWU personnel stated in interviews that in 1975, when the formerly unionized Bella Bella plant was decertified so that it could be operated by the Indian-owned Central Native Fishermen's Cooperative, a number of Bella Bella shoreworkers wrote to the UFAWU apologizing for the fact that they were obliged to withdraw from the UFAWU in order to hold a job in the cooperative. The structure of this cooperative required that all shoreworkers be members.

CREATING NEW INSTITUTIONS: COMPETING STRATEGIES AND IDEOLOGIES

In the traditional aboriginal system, concepts of ownership and methods of resolving competing claims over resources were highly developed. Elaborate feasts, ceremonies, naming procedures, and dances accompanied transfers of title to fishing spots. A major focus of political activity, indeed of all social activity from religious celebrations to war, was managing conflicts between owners, or between owners and nonowners, and forming coalitions of people with differing claims over resources. There were grounds for almost any individual to come into conflict with almost any other individual. There were also grounds for coalition. The traditional chief,[7] a hereditary owner of considerable property and title, was expected to be a master at forming coalitions and resolving or preventing conflicts.

The contemporary coastal and lower-river Indian groups of B.C. have lost none of their old concern for rights over resources. The coming of capitalism has made matters of ownership and control more problematic, because the authority of the chief has been eroded and because capitalism has specified property rights in a different way. Indians have responded by creating new institutions through which to manage conflicts between different categories of owners and workers and between commercial fishers and communities. These carry on traditions of mutual support, consciousness of ownership, equitable access, and the managment of group interests by an authority, though in other respects they are new.

The most comprehensive ground for coalition is Indian status and ethnicity itself.[8] However, an Indian might also be a seine boat owner, a small-vessel owner, a rental skipper, a crewman, a shoreworker, or a subsistence fisher. As such, he or she may be drawn into groups including non-Indian workers in the fishery. Should a fisher join with other Indians of all categories in the Native Brotherhood or throw in his or her lot instead with all fishers of one category? Indians are just as subject as other fishers to the question of whether small-boat owners should stand with large-boat owners or join with crew. The UFAWU allows small- but not large-scale boat owners to join, thus forcing a division between vessel owners and between large-vessel owners and crew.

One of the main reasons the Native Brotherhood includes all classes of fishers is that it began as an organization concerned with the promotion of Indian health, education, and aboriginal claims, in addition to improving the Indian position in the fishing industry (Gladstone, 1953; Drucker, 1958). In addition, government programs have tended to treat Indians as a group when devising programs to benefit Indian fishers or when addressing the question of aboriginal title. However, as previously noted, most programs have benefited the upwardly mobile segment of the Indian fleet. As one successful Indian boat owner ruefully observed: "Most Indians can't even get into the economic system, because the bottom two rungs of the ladder are missing," i.e., most Indians cannot raise the down payment required to participate in government programs.

Some Indian fishers who cannot make the leap to the third rung of the ladder and some who cannot go beyond it are attracted to the trade-union movement. Very strong union sentiments characterized at least the first half-century of Indian involvement in the fishery (Gladstone, 1953). On the other hand, those who are making a transition up the ladder, and have successfully exploited access to capital, often prefer to ally themselves with others in the same situation, using their status and ethnic identity as a means of tapping special sources of capital and other government aids.

The Native Brotherhood continues to articulate the definition of Indians as a status group whose interests are not adequately represented by the UFAWU. Indian vessel owners who had borrowed large sums in the 1970s to purchase newer and larger vessels identified themselves principally as owners rather than as workers. They needed to keep their boats working to keep up payments. Strikes over minimum prices and shorework wages were perceived as most beneficial to fishers not owning vessels and to shoreworkers. But the rank and file of the Native Brotherhood, composed of small and less upwardly mobile vessel owners, rental skippers, and crew continue to feel at least some solidarity with the union or with union ideology, and to resist the tendency, often coming from their own leadership, to act only as a status group acting solely in native interests.

This division within the Native Brotherhood, which is submerged at times by an overriding rivalry with and even hostility toward the UFAWU,[9] was particularly apparent at the 1982 annual convention when the brotherhood was formulating its position toward the Pearse Commission report. The leadership had invited Commissioner Pearse to attend the convention and evidently endorsed the majority of his recommendations, especially regarding the creation of Indian Development Corporations and other methods of assisting Indians in boat ownership. During debate of these recommendations, however, more support was shown for the UFAWU position. The UFAWU especially opposed the Pearse recommendation for removing 50 per cent of the fleet via a bidding system, which it felt would penalize the owners of smaller vessels. Indian and UFAWU small-vessel owners alike feared a repeat of the problems resulting from the Davis Plan.

The division among Native Brotherhood members was also apparent during interviews with two leaders in one Indian community that became split in its reaction to the formation of the Central Native Fishermen's Cooperative in 1975. Many of the independent vessel owners who formed Central Native Fishermen's Cooperative were also leaders in the Native Brotherhood, and a few had been leaders in this community as well. The cooperative had been in the planning stage for several years, and the leaders seized the opportunity to begin operations and buy a processing plant that was offered for sale because of the 1975 strike. By responding to this offer and setting up a cooperative at this moment, the cooperativists found themselves with a dilemma. They had supported calling a strike in their capacity as Native Brotherhood members, but could now fish during the strike as new cooperative members. They did fish, while most Native Brotherhood members were on strike, and their actions had a divisive impact on the community.

The rank and file of the Native Brotherhood in this community, who were not free to join the cooperative because of their indebtedness to processing companies, are reported by one community leader interviewed to have perceived the founders of the CNFC as a privileged class who were not helping other Indians. To them the co-op represented betrayal and

desertion by an elitist faction because it involved violating the most important expression of company-dependent fishers' solidarity — the strike.
Perhaps even more importantly, community leaders were seen as not
representing the interests of the entire community. "We sink or swim
together: we're not going to be pulled up one at a time," was the attitude
of one rank and file NBBC member. Clearly the cooperative would help
some Indian shoreworkers in one community, but, they felt, at the expense
of other Indian rental fishers, indebted boat owners, and crew on company seiners.

CNFC members, however, believed the creation of the cooperative
would benefit the entire Indian community and provide more secure
processing jobs in at least one community (which otherwise might have
lost the plant altogether). Later they hoped to build plants in other Indian
communities and hire more Indian shoreworkers. In addition, CNFC included
shoreworkers and crew as members, thus defining itself along Native
Brotherhood lines as the expression of a status group which overrode
other interest-group categories.

In this community, where cooperativists had held leadership positions in both the community, the co-op, and the Native Brotherhood, they
were subjected to disapproval from important segments of the community for several years and did not immediately seek re-election as community leaders. Northwest coast ethnography has richly documented the fact
that to be wealthy and successful is acceptable and honoured in Indian
communities as long as it is accompanied by at least some distribution of
benefits and/or expression of solidarity with less fortunate fellows.[10] The
institutional expression of privilege in the CNFC seems to have strained the
tolerance of some Indian fishers in this community, because it appeared to
them to be based on the principle of exclusion of other status-group
members. A healing of wounds in this situation is possible, however,
because considerable wealth (in one case $20,000) is in fact distributed at
potlatches by owners of large vessels from potlatching families, and the
large vessels are used to transport community members to these communal affairs. Whereas such events may not include the entire community,
the good name of the whole community is enhanced by extension. In
another community where some of the large-vessel owners had moved to
Vancouver a decade earlier, CNFC members continued to make efforts to
hire reserve residents as crew, to potlatch in the community, and to provide services to community residents when they travelled to Vancouver
(Susanne Hilton, pers. comm.).[11] CNFC did not survive the difficult 1980s
and was dissolved in 1984 for reasons that are incompletely understood.
It is not clear whether stresses on CNFC, or within communities associated
with it, contributed to its demise.

Another cooperative venture, though a different model, is the Northern Native Fishing Corporation, formed in 1982 when three Indian tribal
councils purchased the rental fleet of B.C. Packers. Disciplining the rank

and file members allows the corporation to prepare to join individual entrepreneurship with tribal interests.[12]

The corporation has some precedent in the paternalism of the traditional rental fleet and the authority of the traditional northern tribal corporate unit. It was launched with an $11 million grant from the federal government to the three tribal councils for purchase of the B.C. Packers' rental fleet. Fishers still deliver to B.C. Packers and borrow money from the company, but the leaders hope to make a leap beyond this dependence into much greater independence. They hope Indian fishers will become independent vessel owners with new attitudes, skills, and responsibilities consonant with ownership. The corporation works toward this goal by maintaining control of the fishing licences, while attempting to extract a commitment to the corporation from fishers by requiring them to purchase the vessel attached to the licence. Thus, in return for access to commercial fishing privileges, loans, training programs in accounting and vessel maintenance, current NNFC members subsidize the entry of more fishers into the corporation. The purchase of a vessel pays for the construction of newer vessels to replace the 50 (out of 250) inoperable by 1984.

In the eyes of the sponsors of the corporation, the transition to vessel ownership is the critical bridge that must be crossed in order to create a commitment to a self-sustaining institution instead of one that will be continually government funded, and that would replace a paternalistic company with a parternalistic government and leadership structure. Interviews revealed that the Indian tribal councils perceive this transition as necessary for economic as well as psychological reasons. According to managers and the corporation board, if the Indians perceive that they are in fact fully responsible for the success of the corporation, and that it must pay for itself, just as each vessel operator must pay for, repair, and keep accounts on his own vessel, they are likely to see the dovetailing of their personal interests with the interests of the corporation. In purely economic terms, the corporation must extract this commitment from its fishers if it wishes to keep the entire fleet in working condition.

The corporation tries to express the interests of the three tribal groups and their communities, not just the individual interests of fishers as businessmen. In this way, it attempts to meld the interests of the fishers to the group and communities at large, so that the issues of stratification and elite formation within native communities are less disruptive, and so they are perhaps less likely to repeat the history of the CNFC. The tribal councils, through the corporation, are the actual owners of the licences, and maintain control of access to the licences in a manner not unlike the aboriginal system. The tribal councils must therefore also balance the interests of individual fishers with what they perceive as the interests of the corporate group as a whole.

During fieldwork, I observed that commercial fishers were often among the few people in Indian communities who were employed. From the community and tribal perspective, then, the fisher is a relatively privileged individual who has certain obligations toward the group as a whole. From their perspective, the vessel owner should see the vessel and licence obtained through the corporation not as something the government purchased for him or her as an individual, but as something the government purchased that was acquired by the tribal councils as a whole for the development of the larger tribal welfare. One of the goals of the larger group is to maximize employment, i.e., to keep all the licences operating to employ tribal members and to make the transition to individual entrepreneurship. Sale of vessels is used to generate funds to accomplish these other purposes.

The contract with fishers states that they may lose the privilege of access to a licence if they are found negligent in causing damage to a vessel or if they stop production for other than health reasons. In this case, the licence may be awarded, by a decision of elected board members, to another tribal member who is without employment. Fishers are thus encouraged by the corporation to develop entrepreneurial attitudes, but within a structure of responsibility to the larger group. Managers and board feel this is the only exit from the paternalistic trap that has allowed dependency to inhibit initiative and responsibility. It is an exit that affords certain kinds of support (loans against boat purchase, assistance with major mechanical replacements) but attempts to do so in a manner that fosters the rapid transition to greater independence, while at the same time probably weakens the potential for united action with non-Indian fishers.

The corporation does contain an important contradiction of its own, however, because it uses vessel ownership as the driving force of community development. With some groups, such as the Gitksan-Wet'suwet'an of the upper Skeena River (who also constitute part of the corporation), commercializing the food fishery is the major hope for economic development. This involves setting nets in the river some two hundred kilometres inland instead of using vessels at sea. Vessel owners and corporation members in this tribal council constitute only a small percentage of the workforce and fish commercially outside their traditional territories. In addition, the Native Brotherhood (1982) has supported the idea of Indians acquiring territorial fishing rights organized around terminal river fisheries, where salmon would be taken as they enter their "home" territories. Many Department of Fisheries and Oceans policy makers favour terminal salmon fisheries as a superior form of management, reducing some mixed stock problems and diminishing or eliminating overcapitalization of vessels. While some tribal groups or portions of tribal groups work toward terminal fisheries and/or commercializing food fisheries as a form of economic development that could benefit the entire community, the vessel-owning route of the corporation would benefit some Indian groups less, because in some groups far fewer people have access to vessels than to river nets. A potential conflict is thus created even within the same tribal group between members who are working toward vessel owner-ship and members working toward capturing the same fish by different methods many kilometres upstream.

Conclusion

The first purpose of this chapter was to describe the conditions that created the contemporary Indian fishery, focussing on two different sets of forces. First, there are forces that push Indians to act as a group solely in terms of their ethnicity. These include aboriginal, political, and socio-economic organization in its continuing features, the potential political gains from aboriginal title and Indian status, and a perceived common bond in the way Indians have developed special dependencies on proces-sors and suffered similar losses of fishing privileges and position in the fleet. A second set of forces pushes Indians to find more in common with non-Indian shoreworkers or fishers than with all other Indians. These include solidarity with the UFAWU, which has developed among Indian shoreworkers and some Indian crew, rental skippers, and small-vessel owners. Another example is the identification some Indian large-vessel owners have shown with other large-vessel owners who wish to avoid strikes.

The second purpose of the chapter has been to show how Indians have used modern institutions congruent with traditional political forms to attempt to reconcile the conflicts created by these two sets of forces.

NOTES

1. In some cases, weirs or traps were built by the whole community, with no distinction in access. However, in these cases the houses standing at the weir sites, which were necessary for smoking the catch, were owned by individuals or extended families (Suttles, 1960). Thus these individuals indirectly regulated access. There is also varia-tion in whether ownership of fishing sites is vested in the group as a collectivity or in the highest ranked member of the group (Riches, 1979). In either case, access of group members is regulated. Descrip-tions of California Indians' regulation of access to salmon weirs strongly suggests that the ritual specialists who directed the timing of fishing were practising conscious conservation (Swezey and Heizer, 1977). Tlingits in southeast Alaska refused to continue fishing for a cannery in 1907 "for conservation reasons" at a time when their system of property rights to the fish was still intact (Rogers, 1979).

2. The Supreme Court of Canada decision on aboriginal title in 1973 and the subsequent establishment of the Office of Native Claims in 1975 led Indians to expect that at last the government might be prepared to negotiate a settlement for the comprehensive land claims Indians made in the absence of treaties.

3. See Chapter 13, "The Fishing-Dependent Community."

4. Indians adopted European technology for some of their own home processing and also continued many of their traditional smoking and drying techniques.

5. One fifty-year-old non-Indian shoreworker was in awe of the Indian shoreworkers' stoical attitude toward pain, since it made her weep every night during the beginning of the season. Tendonitis and carpal tunnel syndrome result from the repetitive motions of filleting or butchering. Backache can be chronic for graders, who twist the torso and throw with the grading of each fish. One high-seniority grader reported having taken painkillers in order to work.

6. It has not been possible to obtain accurate statistics about where the lower-seniority Indian shoreworkers in these centres live the rest of the year. We do know, however, that many of them, especially in the north, live part of the year on outlying reserves and that shorework is their only employment.

7. The literature on British Columbia coastal Indian groups is so enormous that only a token citation is made here. A classic general overview is found in Drucker (1951). Chiefs' accounts of their world as they see it are particularly revealing: see Ford (1941) and Spradley (1969). A graphic, popular account of Indian-white conflicts over ownership of fish resources in premodern times and destruction thereof in modern times is given by Raunet (1984). See also Cove (1982).

8. See note 7.

9. Drucker (1958) noted this in the earlier days of UFAWU–Native Brotherhood relations and cites as a typical example of rivalry between the two groups the fact that, after the brotherhood had received its formal charter in 1945 under the Societies Act, union leaders lodged an official protest to the Trades and Labour Congress of Canada against allowing any organization chartered under the Societies Act to sign labour contracts. Another source of conflict Drucker notes is that Indian fishers have been less able to tolerate losing the entire fishing season by a strike than have many non-Indian fishers, who had better job opportunities in the off-season. A source of conflict mentioned by Indians today is the welfare fund administered chiefly by the UFAWU because of historical agreements. The UFAWU does not recognize non-Indian members of the NBBC and will not grant them welfare benefits, insisting they must join the union to qualify. NBBC members complain that the UFAWU thus uses the welfare fund to recruit members and

contend that if Indians can join the union, non-Indians should be entitled to join the brotherhood (see debate at 1981 NBBC convention).

10. See note 7.

11. Understanding the redistribution of benefits within Indian communities is beyond the scope of present research. The impression received by the author is that there is a range of situations from the Gitksan-Wet'suwet'en's institutionalized policy of redistribution to one situation (not discussed here) in which no redistribution occurs. It appears, however, that in the majority of Indian communities, the wealthier members have chosen to stay and lend support to less fortunate members.

12. The following analysis is based on multiple interviews between 1982 and 1984 with the corporation managers, on interviews with twelve corporation fishers in three communities who were part of our survey sample, on interviews with and speeches by the tribal-council leaders involved, and on numerous conversations with Indian observers of the corporation.

12

Shoreworkers and UFAWU Organization: Struggles between Fishers and Plant Workers within the Union

ALICJA MUSZYNSKI

Shoreworkers, unlike fishers, have always been a labour force totally dependent on fish processors or labour contractors for employment. As a factory labour force, their wages and working conditions place them within the traditional working class, although seasonal work and low annual wages confine many of them to a marginal economic position.

Historically, certain aspects of fish-plant work have remained relatively constant. The work is often dirty, almost always seasonal, and invariably organized as a factory process. But other aspects have changed. Unionization has succeeded in gaining high hourly wages (B.C. shoreworkers are said to be the highest paid fish-processing workers in the world); security of employment (seniority clauses guarantee workers are rehired from season to season and employed for the duration of each season); increased benefits (e.g., pension and welfare plans); and improved working conditions.

However, it took decades of labour organization and negotiation before these changes were introduced. Fishers organized the shoreworkers, with important consequences for both groups. When fishers went out on strike, shoreworkers closed the plants. By refusing to process catches, the shoreworkers forced processors to negotiate with fishers.

It was only after twenty years of UFAWU organization and struggle that shoreworkers began formulating and fighting for their own demands. During this latter period, divisions among fishers became more acute, increasing their reliance on the shore sector in price negotiations. The increasing militancy and group cohesion of shoreworkers, in conjunction with an increasingly divided fishing fleet, created a context within which processing workers developed a stronger union position. Strike issues are a good indicator of this strength, and the demands articulated by shoreworkers in the 1967 and 1973 strikes illustrate this process.

Until the introduction of collective bargaining, employers, through the Chinese contract system (described in Chapter 3), controlled their plant labour forces. While workers at individual plants might walk out to protest wages or working conditions, generally there was little amelioration in either until union agreements replaced Chinese contracts as the method of organizing labour. Unlike the fishing sector, where a number of labour organizations developed and competed, in shore plants the UFAWU has been the dominant union since 1945, the year in which it was founded. While the Native Brotherhood has also been an important labour organization for native workers, its plant agreements have been patterned on those of the union. Some small plants remained unorganized, and in various ways, tried to prevent UFAWU organizing. But no one would dispute the dominant influence of the UFAWU in the history of B.C. shoreworkers.

Fishers and labour organizers (most of them members of the Communist Party) were the founders of the union. Shoreworker support was crucial to the establishment of a provincewide industrial union, their goal. But here they found a labour force divided by race and gender, receiving very low wages. A stated union principle was equality of all workers regardless of ethnicity or gender, but this principle has yet to be achieved among shoreworkers. The difficulties of achieving equality forms one of the issues explored in this chapter.

This chapter examines the process of struggle *within* the union between fishers and shoreworkers. The argument developed here is that to ameliorate one's own working conditions involves a consciousness that these conditions need changing as well as a commitment to fight for these changes. The great majority of shoreworkers, once organized, were not prepared to struggle for two decades. Until the strikes of 1967 and 1973, shoreworkers were a fairly passive group, following the lead of fishers and supporting their demands. In those two strikes, however, shoreworkers became militant and began to struggle for their own causes. How that militancy was nurtured by UFAWU organizers is the theme of this chapter.

The data are taken largely from one source, the union newpaper, *The Fisherman*. The method is a content analysis, used to trace the changes in perspective from one supporting fishers' interests to a growing awareness of the unique problems faced by the majority of shoreworkers, as well as the emergence of leaders from their ranks. Today, the president, Jack Nichol, is a former shoreworker.

A word of caution is in order. The perspective adopted here focusses on the UFAWU and its internal organization. No attention is paid to other labour organizations in the industry or to events important to the industry but not relevant to the struggles occurring within the union. The justification for such an approach is the importance of the UFAWU in determining relations between employers and their plant labour forces.

PRELUDE TO THE FOUNDING OF THE FISHERMEN AND ALLIED WORKERS UNION, 1933–1945

Organized fishers had little bargaining power if the canners could keep operating during their strikes. While there are early instances of spontaneous strikes by both fishers and shoreworkers, these were largely confined to individual plants where specific issues were disputed. Shoreworkers would support fishers since many of them were related by kinship, and, in turn, fishers would insist that conditions in the canneries be improved. Often, however, this involved no more than an adjustment in pay rates to bring a particular plant in line with others in the vicinity. In the 1930s, the frequency of such spontaneous strikes rose.

The recessions in the 1920s had severe impacts on the provincial fishing industry, resulting in the closure of many plants, and, in 1928 British Columbia Fishing and Packing Company attempted to regain a position of dominance in the industry by buying and closing a large number of canneries. Adversely affected by such closures, fishers and shoreworkers began to escalate their resistance. In 1933, the Fishermen's and Cannery Workers' Industrial Union emerged from a fisher's union to incorporate both groups. Its only concrete achievement, however, was the signing of an agreement in 1935 with the Deep Bay Fishing and Packing Company. The agreement was terminated when the plant burned down two years later. In 1936, the fishers involved in forming that union reorganized themselves into two separate unions, the Salmon Purse Seiners' Union and the Pacific Coast Fishermen's Union. They worked together and eventually amalgamated into the United Fishermen's Federal Union (UFFU). They again sought to incorporate shoreworkers into their organization. When the Trades and Labour Congress refused to give them jurisdiction over shoreworkers, they founded, in 1941, the United Fish Cannery and Reduction Plant Workers' Federal Union, Local No. 89. The UFFU provided financial assistance and shared its office space with Local 89.

Fishers' interests influenced shore-plant organization for the next several decades. Although the mandate of Local 89 was to organize all provincial plant workers, it was not keen on organizing the large, seasonal Chinese and female crews. Shore-plant organizer Bill Gateman noted that the long hours, low pay, poor living accommodations, and poor working conditions characterizing fish canning were directly connected to the "racial question," which hampered cooperation.

> Workers in the fish canning industry consist of Native people (Indians), Chinese, Japanese and white workers. The employers have successfully played one group against the other for the express purpose of keeping them disorganized so they could maintain poor conditions. . ... The employers have utilized the contract system to their advantage in maintaining low rates of pay. Practically all employers use the services of a Chinaman contractor known to the workers as the "China Boss." The "China Boss" contracts at so much a case to can the season's pack and then employs his labor at the cheapest possible rate. There is a consistent conflict between the respective Chinese contractors. One will under bid the other in order to receive the contract, which in turn affects the wage rates. This process has been going on for years

and has worked very successfully in keeping down wage rates. (*The Fisherman,* vol. 3, no. 52, p. 4)[1]

Canners resisted the efforts of union organizers to certify plants. They had a strong self-interest in maintaining a system that allowed them to hire a cheap seasonal labour force. In addition, they were united in their own organization, the Salmon Canners' Operating Committee (the name was changed in 1951 to the Fisheries Association).

With the onset of World War II and increased wartime food production, regional War Labour Boards were established to ensure that industries essential to the war effort were not hampered by labour disputes. The state thus became a third participant in negotiating labour contracts.

The UFFU desired provincial certification. The companies insisted on negotiating agreements for individual plants. The union victory in this set of negotiations hinged on organizing reduction plants for three reasons. First, these plants were lucrative and essential to the industry's war production. Second, they were situated throughout the province, adjacent to the major canneries. Third, and most importantly, these plants were staffed by a small group of skilled workers, most of whom were men of European descent. Unlike cannery workers (a large, seasonal, mixed group), they were a small and unified labour force. Some of them fished and thus had further ties with fishers.

A similar group worked inside the canneries, maintaining the machinery (as did reduction-plant workers since most work in reduction plants involved overseeing machines). Reduction-plant workers and cannery machinemen earned guaranteed monthly wages. This system of payment worked well when packs and/or markets were poor, since a fixed monthly salary was assured. However, the situation created by the war and excellent salmon and herring runs reversed the situation. For example, in 1941, the Kildonan plant on the west coast of Vancouver Island (the plant responsible for forming Local 89), operated twenty hours on some days. Despite increased work, no satisfactory adjustments were made to the fixed monthly salaries. Thus, reduction-plant workers, as well as cannery machinemen, were exremely receptive to union organization, and Gateman found organizing these two groups a relatively easy task. With the intervention of the Regional War Labour Board in 1942, Local 89 was certified as bargaining agent for reduction-plant workers and cannery machinemen. Negotiations centred around monthly guaranteed hours, and straight time was henceforth to be paid for hours worked beyond the stipulated minimum (overtime rates were negotiated in future contracts). Thus, Local 89 was able to achieve provincial certification without having to organize the massive seasonal labour force.[2]

The other sector organized by Local 89 was the workforce in fresh-fish plants, and here women played an active role. However, certification proceeded on a plant by plant basis. It was the inside workers who pressed for the inclusion of women in collective agreements. For example, the labour force at Edmunds and Walker unanimously decided to include its thirty-two women in its agreement because they were performing the same work in the freezer room and on the dressing tables performed before the war by men. However, the bulk of the female labour force was seasonally employed in the canneries.

The Native Brotherhood was the first organization to negotiate agreements for the entire cannery labour force (see Chapter 10 for the history of the Native Brotherhood). In 1943, five companies (ABC, B.C. Packers, Canadian Fish, Nelson Brothers, and Nootka Bamfield) signed agreements with the Brotherhood. Native workers formed the majority of the labour forces in these plants. This was the first time that uniform rates and conditions were established for a whole group of canneries. Women received forty to fifty cents per hour with equivalent piece rates for hand filling salmon and herring and for heading herring, while men received forty to sixty cents per hour (*The Fisherman*, 5[23]:4; 5[25]:2).

THE EARLY YEARS OF UFAWU

In 1945, the UFFU and Local 89 received joint jurisdiction from the Trades and Labour Congress and amalgamated to form the United Fishermen and Allied Workers Union (UFAWU). It was not until 1947, however, that the entire cannery labor force was covered by collective agreements. And for the first twenty years of its existence, the union remained basically a bargaining agent for the fishers, even though its legal status in that respect was constantly challenged under the Combines Act (see Chapter 10). As well, numerous groups of fishers refused to join and challenged the union's right to represent them. The union's legitimacy as an industrial trade union organization therefore rested on its shore-plant membership.[3]

Guaranteed packs came to an end in 1947, obliging the companies to reduce production for a smaller domestic market. At the same time, the process of consolidation and concentration that had begun with the introduction of refrigeration techniques on packers and in plants early in the century picked up speed. By 1949, canning operations on the west coast of Vancouver Island were almost nonexistent. On the Skeena River, the companies contemplated establishing one cannery to process all the fish on a pool basis. As shoreworkers became organized, their numbers shrank. Under Chinese contracts, employers did not worry about the numbers employed, since payment of workers was the responsibility of the contractor. But legislation introduced during the war made this system obsolete. Now companies had to include names of employees on their payrolls. One reaction to government legislation and unionization was to increase mechanization and to concentrate operations in a smaller number of factories.

Women were brought into the master agreement of 1946, but they were given a separate supplement to cover wages and working conditions. The shortened season curtailed their employment in the urban centres, and the closure of canneries in more remote areas reduced their employment opportunities. In 1948, herring canning ended and a shorter season ensued. This severely reduced the duration of seasonal employment for women. As noted in *The Fisherman*: "Not only are women penalized by lower wages while working, but falling markets reduce the amount of work available for women more than for men" (11[15]:1). UFAWU organizers estimated a 50 per cent reduction in the number of women employed between 1944 and 1953 (6,150 in 1944 compared to 3,947 in 1952) (16[18]:1).

Chinese male workers also had a separate supplement before 1949 and afterward were included in the general male categories, categories at the bottom of the wage scale. As Chinese contracts were replaced by union agreements, Chinese male workers were displaced from the industry. The Exclusion Act of 1923 barred further immigration of Chinese into the province. By 1950, these workers were advanced in age. Since employers were now forced to negotiate for individuals rather than groups (the "China gangs"), these men appear to have suffered loss of employment.

THE 1950s

Though the entire shore labour force was organized by 1949, the original structure of work by ethnicity and gender persisted. With declining markets, continuing emphasis on the fishing sector by the total membership, internal strife between union and nonunion fishers, and the strong anticommunist atmosphere of the period, as well as the continuing legal battles over its right to represent fishers, it was possibly beyond the union's powers to eliminate the discrimination inherent in such a structure. While coping with its other burdens, the union was suspended from the Trades and Labour Congress in 1953 because of its communist leadership and orientation. It was not readmitted to the Canadian Labour Congress until 1973.

The controversial legal status of a union representing vessel owners might eventually have caused the demise of the UFAWU but for the legality of its representation of shoreworkers and tendermen. The example of an antitrust trial in Los Angeles in 1947, in which thirteen union fishers and one top union officer were found guilty of conspiracy to restrain trade by attempting to set minimum prices, was used by Canadian processors to press their case. According to the union newspaper, by 1952 American sister organizations were all but decimated and "the UFAWU was standing alone as the single functioning industrial union anywhere in the fishing industries of North America" (44[2]:10).

Though they received relatively few benefits, shoreworkers continued throughout the 1950s and 1960s to support fishers. Part of the explanation for their lack of militancy may lie in the divisions within the shoreworker force itself.

The most militant shoreworkers were reduction-plant workers, cannery machinemen, monthly classified netmen, and fresh-fish and cold-storage workers. Most of them were men, and many were also fishers. They worked closely together and on a permanent basis, often alternating employment between fishing and work inside the plants. They were the group most closely tied, among shoreworkers, to fishers and thus receptive to their organizing efforts. They were also the most skilled workers and could point to disparities between their wages and working conditions with other organized industries in the province. During the 1950s and 1960s, these groups did realize significant gains. Their guaranteed monthly rates slowly decreased to the equivalent of a forty-hour week. They fought for and won multiple overtime rates. They sought parity with their fellow tradesmen in other B.C. resource industries. Although reluctant, the Fisheries Association was willing to negotiate better conditions for them. In fact, this militant sector consistently pressed for percentage increases, which served to widen the gap between wages paid these workers and the general labour and female categories. Ultimately, the Fisheries Association was willing to grant the militants their demands because they represented such a small number of workers. For example, in 1954 the cost of shore requests (a forty-hour week with no reduction in take-home pay) was estimated as representing one-fifth of a cent per one-quarter-pound can.[4]

Union organizers found the Fisheries Association adamant in refusing to grant any wage increases to seasonally employed women and casual labourers, who formed the greater portion of the labour force. These two groups went without any wage increases for three years (1951–1953) and faced very different conditions from the rest of the labour force. Large numbers of women had been hired during the war years to replace enlisted men and to fill vacancies created by expanded production. Women of European descent were hired for urban-based plants to meet these labour shortages (further aggravated in 1942 by the removal of all Japanese from B.C. coastal areas). Guaranteed packs came to an end in 1947, and from that time until the late 1950s, the companies had a difficult time re-establishing their domestic markets and finding new outlets. Thus they had a surplus labour supply and could resist wage demands.

Mechanization, pursued throughout the 1950s, allowed processors to close outlying canneries and consolidate operations in a few key plants. These plants became multiproduct and multiprocessing operations, combining diverse processing techniques and product lines in one geographical location in or near an urban centre. Despite such reductions in their numbers, male casual labourers and cannery women refused to take strike action. In an effort to promote active participation by these groups, the union organized wage conferences in 1952. Each category of workers received proportional representation and elected members from its ranks to participate in union-company negotiations. These categories included cannery men and women, net workers, fresh-fish and cold-storage workers, reduction-plant workers, steam and refrigeration engineers, watchmen, and saltery workers. In each category, there was a further subdivision in collective agreements by gender. At the wage conferences, women were elected according to their numerical proportion in the various categories; for example, none for reduction-plant workers but well over half for cannery workers (14[3]:1). Voting results were compiled on this basis.

In 1952, agreements began to be negotiated on the basis of percentage wage increases because it was recognized that the tradesmen were more dissatisfied with their wages than the other groups. Although the business agent, Alex Gordon, responsible for organizing the shore sector, felt that male general labour and all women required increases, the workers were not prepared to press their demands. "Among this group, however, it was recognized by the wage conference that dissatisfaction with the rate during the 1951 season was not strongly expressed" (14[3]:1). Shoreworkers were thus left with agreements basically unchanged from those of the previous year. The two-thirds majority needed for strike action was not attained (although large differences were recorded among the areas in which voting took place, with the northern areas of Prince Rupert and Skeena River being particularly militant).[5] On the other hand, fishers went out four times that same year. There was a seven-week salmon strike and the herring strike resulted in the complete loss of the season. Fishers responded by advocating separate groupings among themselves. The Native Brotherhood was particularly opposed to strike activity, and gear conflicts between gillnetters and seiners also surfaced. Shoreworkers reacted by becoming even more conciliatory. "[The] attitude of shoreworkers is that every effort should be made to speedily settle 1953 agreements with a minimum of the unpleasantness which characterized 1952 negotiations" (Ibid., 15[7]:1).

Shoreworkers settled the following month with no general wage increase. The Native Brotherhood's opposition to strike activity was reflected in its support of shore-plant negotiations. Fishers' attitudes toward shoreworkers were reflected in *The Fisherman*. "Several fishermen have asked what the proposed increases so far granted to shoreworkers will mean in increased costs to the fishing companies" (Ibid., 15[19]:2). In 1954, both Native Brotherhood representatives and shoreworkers were "unanimous in feeling that the past policy of the shoreworkers in striving in every way possible to settle agreements without stoppage of work should be steadfastly adhered to" (Ibid., 16[18]:1).

Although such a policy did not significantly improve wages in the 1950s, the union and Native Brotherhood worked jointly to rectify racial discrimination against native workers in the industry. Efforts were made to gain better company housing and the payment of transportation costs to and from the plants, and it was mainly native women who received these benefits. Instances of discrimination in the application of seniority rights to Native Brotherhood workers were uncovered in several plants and fought. In addition, in the 1950s, shoreworkers were included under social legislation (Workmen's Compensation, Unemployment Insurance, welfare, and pension plans) and companies established and contributed to pension and welfare schemes.

The union had adopted an extremely discriminatory structure when it acquired certification rights to shore plants. This resulted, in part, from the way in which they initiated unionization, beginning with the most skilled groups on a provincewide basis (rather than, for example, organizing the entire labour force within a plant) (Muszynski, 1984). It was then forced to fight these inegalitarian conditions from within. Racial inequality was tackled first in the 1950s, followed in the late 1960s by attempts to end allocation and payment of work on the basis of gender. As late as 1956, "guarantees, paid transportation, seniority, and board conditions do not apply to Native women at B.C. fish canneries" (Ibid., 18[13]:7). But in that year, equal treatment of native workers was written into the agreement. It was also in that year that the union adopted a "one strike, one settlement" policy. It insisted that all three agreements (fishers, shoreworkers, and tendermen) be settled together.

The companies were quick to respond to this challenge. When the net fishers went out on strike in 1957, the companies closed down the canneries and laid off net workers and monthly paid cannery men. Fishers had to work on their gear, and, with the net workers laid off, the companies hoped to create internal frictions. Fishers and net workers solved this by allowing fishers to work on their own gear and declaring all company gear untouchable by anyone but net workers (Ibid., 19[22]:1). The strike was settled but the 150 monthly shoreworkers remained locked out. A ballot conducted among this group resulted in a 64 per cent vote for complete walkout, less than the two-thirds majority required. The issue went to conciliation, and the shoreworkers lost their pay for that period of time. A split board award ruled that monthly workers were subject to sporadic layoff (Ibid., 20[4]:1).

Another tactic employed by the companies was hiring large numbers of casual employees during the period when a strike vote was being conducted. Thus, in 1962, 53.1 per cent of the workers voted to strike. Within the various categories, however, 74.4 per cent of fresh-fish and cold-storage workers voted to strike. The number of cannery workers voting was inflated by an extra 897 workers hired (as compared to 1959) to cope with the large run. The union organizer insisted these people had little previous experience. By holding the vote at this point in the season, the combined cannery, net and reduction worker vote was only 47.8 per cent (Ibid., 25[25]:1). In 1963, a second general strike occurred. Shoreworkers voted to accept a conciliation award, but the Fisheries Association turned it down.[6] Again the companies hired a large number of workers for one or two days during the voting period, but this time they were unsuccessful in preventing a strike vote. The companies responded by threatening to close down for the season. While shoreworkers and tendermen settled, the net fleet sailed without an agreement. The issues were referred to a one-man arbitration committee that led to the appointment, in 1964, of the federal-provincial committee to examine the entire issue of negotiations over fish prices. It recommended compulsory arbitration.

The threat of compulsory arbitration was effective in dampening the desire of net fishers to strike the following year. The numerous strikes conducted in the previous years had aggravated tensions among fishers. This was reflected in poor attendance at union meetings. The fishers would not have realized any gains in 1964 had not shoreworkers and tendermen voted in favour of cancelling their newly negotiated agreements to support the net fleet's demands. A crisis was brewing, and it erupted two years later.

THE 1967 STRIKE IN PRINCE RUPERT: UFAWU SHOREWORKERS BECOME MILITANT

The process of consolidation and centralization that occurred in the 1950s and 1960s was facilitated by technological developments. The lines were futher mechanized and existing equipment modified to speed the pace of production. Plants in outlying areas were closed. The same volume of fish was processed in fewer plants with a reduced labor force. The introduction of brine systems on packers resulted in the total elimination of tendermen on herring packers. The introduction of air pumps for mechanically unloading fish from packers and fish boats eliminated even more labour. The workforce was further reduced by the introduction of refrigerated packers and the increased use of contracted tug boats and refrigerated trucks. Historically the smallest of the three sectors, by the 1970s, tendermen were fighting for the survival of their trade.

These technological changes also affected the fishing fleet. New refrigeration systems as well as improved engines and electronic equipment influenced the emergence of a big boat fleet. Combination boats involving both the combination of gears — gillnetting and seining — as well as the pursuit of more than one type of fish — salmon and herring — emerged. In addition, the development of the fresh/frozen market encouraged groundfish operations, in which large trawlers emerged on both coasts. The economic slump in the fisheries experienced in the 1950s began to lift in the following decade. While some fishers plowed profits into their vessels and became small capitalists, those on shore began to search for new fisheries to exploit. B.C. Packers began to expand its groundfish processing operations. In 1962, the Weston food conglomerate took control of that firm, and B.C. Packers turned to groundfish operations in the Atlantic as well as the Pacific.

The UFAWU viewed both developments, the increasing importance of the big boat fleet over which it held no jurisdiction and the exploitation of groundfish operations on both coasts by B.C. Packers, with great alarm. In January 1967, the union began a disastrous attempt to organize trawl fleets in the Maritimes. Negotiations ended in a seven-month strike in 1970 against Booth and Acadia Fisheries. The companies responded by closing plants, declaring them uneconomical. The fishers lost an entire season, workers lost their jobs, and the UFAWU was driven out of the area.

At the same time, the union was trying to gain readmission to the Canadian Labour Congress (CLC), which stipulated that the UFAWU would only be readmitted through affiliation with an existing charter member. It contemplated mergers with two of these as a way of gaining admission and also in the belief that the only way to confront the giant conglomerates in the food industry was through a strong national union of food workers. The question was one of retention of autonomy and power in such an association, and the negotiations with other unions came to nought. Nevertheless the CLC readmitted the UFAWU in 1973 on condition that it relinquish its organizational attempts on the Atlantic coast.

On the Pacific coast, the union attempted to extend its jurisdiction to crews working in the trawl and longline fisheries in 1966. The history of the various fishers' unions that became involved is documented in Chapter 10. The Prince Rupert Fishermen's Cooperative became involved, and the resulting dispute tore apart the community of Prince Rupert. On March 23, 1967, the Supreme Court issued an injunction to the Prince Rupert Fishing Vessel Owners Association (PRFVOA) against Prince Rupert shore-workers who refused to handle "hot" fish. On April 3, a coastwide ballot forbade the sending of a UFAWU telegram ordering Prince Rupert shore-workers to handle "unfair" vessels. The strikes were justified in the following terms:

> Refusal by the Association [PRFVOA] to negotiate a trawl agreement and the abrupt lockout of longliners are being linked with the major expansion of the B.C. trawl fishery, reflected in a better than 100 percent increase in landings in the past three years, and plans by major monopolies in the industry to enter the trawler field on a large scale. (Ibid., 30[12]:12)

Three associations were involved. Prince Rupert was the provincial capital of the groundfish industry. Vessel owners in this fishery had organized the PRFVOA and had expanded operations into the net fisheries, constructing a salmon cannery. Until 1966, UFAWU had not concentrated on organizing groundfish crews since the Deep Sea Fishermen's Union held jurisdiction. But in the 1960s the groundfish industry began to look as if it might challenge the traditional salmon-canning industry. The UFAWU thus sought to expand its jurisdiction to meet the changing situation. By organizing the groundfish industry, it also hoped to extend its jurisdiction from the Pacific to the Atlantic. Its policies during this period have been questioned by many in the industry. However, in what follows, attention will focus on the impact of these events on the shoreworkers in Prince Rupert who took the brunt of the strike that occurred there.

The UFAWU had no effective way of preventing the big vessels from fishing. The only way to block their activity was to refuse to unload their catches at the plants, and to refuse to supply them with ice and bait. This proved extremely effective. It also polarized the community. Wives of the big-vessel owners banded together into a group, calling themselves the "Marching Mothers," and held parades and generally waged a campaign against the union. The female membership of the UFAWU responded to the attacks of the "Marching Mothers" by issuing the following statement: "We are the Allied part of the letters. Some of us are wives of fishermen and all of us are shoreworkers" (Ibid., 30[23]:5)

Intimidation and threats of deportation were levied against immigrant and foreign-born shoreworkers. On May 24, a co-op worker issued a broadcast over a local radio station. Part of it stated: "UFAWU trawlers and longliners have declared Rupert Vessel Owners' boats unfair, therefore, we won't touch them or any other 'hot' fish coming from them" (Ibid., 30[18]:4). On June 2, shoreworkers shut down the entire co-op plant, declaring it unfair. Co-op and PRFVOA vessels tried to unload their catches at B.C. Packers' Seal Cove and at Royal Fisheries, with resulting plant walkouts. The UFAWU set up pickets at the co-op. The DSFU retaliated by picketing Seal Cove and Atlin, declaring the UFAWU unfair. Injunctions were invoked. Jack Nichol, the union's business agent, was arrested twice. The president and secretary treasurer were eventually jailed for ten months. The legal suits launched by vessel owners against the union resulted in damages of $250,000 being awarded against the UFAWU.

The dispute ended in July when the DSFU and the UFAWU worked out a jurisdictional agreement. The strike ended without any negotiations made for the shoreworkers who had supported the strike and refused to cross picket lines. Eight employees at Royal Fisheries were rehired on the basis of a thirty-day "cooling-off" period. Their seniority was restored in early September. However, within the PRFCA, an irreconcilable breach had been created. Approximately fifty PRFCA shoreworkers supported the strike, and the union tried to have their seniority rights restored. PRFCA management demanded a thirteen-month "cooling-off" period during which time their seniority would not be recognized. This meant they would lose employment during the fall, winter, and spring. Eventually, a formula was worked out whereby ten displaced workers were reinstated in January 1968, ten more in March, and the rest were to be called in from the list of the "outside group" in the order of their original seniority. They would lose their seniority rights during the summer, but be fully reinstated September 1, 1968. Such was not to be.

In February 1968, the shoreworkers and office employees at the co-op voted to decertify the UFAWU (104 employees voted for decertification). None of the fifty was allowed to vote. Thus, the co-op kept the most militant workers out of the plant, and replaced them with workers who were intimidated by the possibility of a similar fate if they became militant. Royal Fisheries voted to keep the union. The DSFU took over the PRFCA, creating the Amalgamated Shoreworkers and Clerks Union (ASCU). On November 6, 1968, PRFCA employees waged a six-hour strike over the seniority rights of the "outside group," demanding that these be cancelled. They won their strike. Those workers who had supported the UFAWU were thus kept out of the PRFCA labour force. The divisions between PRFCA shoreworkers represented the antagonism between those supporting vessel owners (some shoreworkers were related to boat owners) and those who supported the UFAWU. However, both groups supported fishers. The strike, after all, was over jurisdiction on boats, but it was waged in the plants. Shoreworkers became involved, not on their own behalf, but in support of either the vessel owners or the union. In the end, they were the real losers, as neither the PRFCA nor the UFAWU did very much to help them once the strike of fishers was over. The experience did, however, teach shoreworkers how to fight (in this case, for someone else's cause).

COMPANY UNIONS VERSUS INDUSTRIAL UNIONS

Although the UFAWU has been judged by many to have advocated strike action too often and to have overextended its powers in 1967, there is no question that an industrial union with a wide geographical base exercises considerable power. The historic weakness on the Atlantic coast can be traced to provincial legislation that outlawed the organization of fishers beyond a county base. Many of the shore plants were unionized, but agreements covered specific plants with wide discrepancies among them. Such a situation was ripe for company unions. The ability of shoreworkers to negotiate effectively was further eroded with the growth of the large conglomerates and their entry into the fishing industry. In the event of a strike or militant action by shoreworkers, they simply closed their operations and moved to a less militant area. On the Pacific coast, the ASCU has found it very difficult to negotiate with the PRFCA.

Most of its agreements mirrored those negotiated between the UFAWU and the Fisheries Association. On June 23, 1978, the PRFCA locked out its five hundred employees. The main issue was the protection of seniority rights. The PRFCA offered to settle for a wage ten cents per hour higher than that of the industry settlement, but proposed to replace departmental seniority with seniority based on job function. It sought injunctions against the use of pickets at both its Prince Rupert and Vancouver plants (thirty UFAWU members respected these pickets, although much hard feeling was involved). The B.C. Labour Relations Board upheld the ASCU's right to picket, noting that the union had little or no leverage other than this. The strike was settled seven weeks later. Shoreworkers narrowly voted (58 per cent) to accept the proposal that they had already rejected three times. Attacks against the union's business agent by the Prince Rupert press led to her resignation, after five years of service.

While the companies levied combines charges against the UFAWU and questioned its right to represent fishers in a trade union, they used injunctions to prevent strikes and picketing. This weapon has a long history, one in which the UFAWU has been heavily involved. Not only have union leaders frequently had injunctions issued against them to restrict their activity, they have also led marches to Parliament in order to protest against the drafting of such restrictive legislation.

In summary, during this period of time, the UFAWU was consistently more responsive to the interests of fishers than those of shoreworkers. Within the shore sector, the skilled tradesmen were the militant segment best able to formulate and press for their demands. This included the fresh-fish and cold-storage workers. And here skilled female filleters were to be found, as well as women who did "men's" work. They were paid lower wages, simply on the basis of gender. However, they too became a fairly militant sector. Within the union, the unequal wage structure between men and women had always been recognized and denounced. In the end, women had to take matters into their own hands and convince their male coworkers that parity in wages was to the benefit of all workers in the industry. The Prince Rupert strike showed that women were not a passive labour force, totally subservient to the demands made of them by their employers. The struggle to formulate and win demands for themselves, however, took time — over two decades. But it did come.

"EQUAL PAY FOR EQUAL WORK": THE 1973 STRIKE

The UFAWU has always adhered to the principle of "equal pay for equal work." However, as noted earlier, it incorporated job classifications based on racial and gender distinctions in its earliest negotiated agreements. These distinctions were the basis upon which companies obtained the cheapest labour forces they could find. When the contract system was supplanted by negotiated collective agreements as a system of wage payment, the companies responded by filling their crews with large numbers of women rather than keeping on Chinese workers. World War II introduced a period of expansion in the industry, and the least skilled jobs that developed in the new fisheries were assigned to women. The job classifications previously marked by racial distinctions became distinguished by gender. Job classifications came to be incorporated in two supplements, a blue one listing men's work and a pink one for women. However, racial discrimination did not end, and the efforts of union organizers were initially directed at correcting these abuses, especially those suffered by native women.

The fight for "equal pay for equal work" was waged on two fronts: in labour negotiations and through legislative action. In 1954, the B.C. government passed an "equal pay for equal work" bill. The Fisheries Association flatly refused to write a section into the 1954 agreement simply stating the principle. Like the federal legislation enacted during the war, the provincial bill was ineffective since no provisions for enforcement were included. In 1968, one of the UFAWU general organizers, Mickey Beagle, presented a brief to the hearings of the Royal Commission on the Status of Women outlining the cumulative results of the use by companies of women as a cheap source of labour. An experienced female general fish worker received 9.3 per cent less than an inexperienced male worker and 24.5 per cent less than an experienced male employed in the same category. A fully qualified filleter received $2.34 per hour (and had to pass tests as well as meet production standards to earn this rate), while inexperienced male help received $2.37 an hour. Helen O'Shaughnessy, vice-president for shoreworkers, noted: "Women generally perform much of the key production line work . . . and are affected directly by speed-up and intensification of work loads" (Ibid., 31[14]:9). The report concluded that stronger equal-pay legislation in B.C. was necessary, with enforcement of provisions that the minimum pay for women not be less than the general female labour rate in the same industry. Also, a clear definition of "equal work" was required, since work performed by women could be entirely different from work performed by men, but still be equal in value; that is, it could be equal in training time, skills, hours of work, production, intensity, and necessity to overall production.

> The difference in work performed by men and women is not in the skills [but in] the fact that there is some heavy work performed by men that women are not physically able to do. This does not mean that work performed by women is not equal — it simply means that at times it is not the same. (Ibid.)

In the late 1960s, women became more aggressive in pressing for equal conditions of employment and pay. Many of them were losing their jobs. Plant closures at Klemtu on the Nass River, and on the Fraser and Skeena Rivers, in 1969, resulted in the elimination of an estimated one thousand jobs in shore plants that year. The twelve canneries operating in

1968 were reduced to five by 1971, with a loss of sixteen hundred jobs. In 1978, the salmon cannery at Namu, the only cannery left on the coast between the lower mainland and the northern area, was closed. The native village of Bella Bella was left without its major source of income, a fate also suffered by the village of Klemtu (the cannery had originally been built on the reserve there to take advantage of native labour). Most of these plant closures affected native shoreworkers and fishers more severely than other groups.

As native women became more militant, they also became a less attractive labour force for employers. Just as women had replaced Chinese labourers during World War II, especially in new jobs created by diversification, in the 1970s a similar displacement occurred in a new fishery — roe herring.

The roe fishery represented yet another expansion into new fisheries by the provincial companies. In the 1960s, Japan had overfished its own banks and began searching for new sources of supply. B.C. companies encouraged Japanese interest in provincial waters. The roe was originally extracted from salmon, but the lucrative market lay in herring. Traditional net gear-types could be used, since gillnetters and seiners already captured salmon and herring. Developments encouraged larger vessels for the capture of herring. The real change for labour occurred on the shore with construction of a separate section in the plants for extracting the eggs. Since the herring operation preceded salmon, this could serve to extend seasonal employment for women. Here is where the struggle ensued.

The practice of importing Japanese workers began in the early 1960s when about forty Japanese technicians were imported to work at the Coal Harbour whaling station (jointly owned by B.C. Packers and a Japanese firm). The companies claimed that Canadian workers were not skilled in butchering whale meat for human consumption. The union countered with the charge that the technicians did not teach these skills to the Canadian crew. The operation was short-lived. Similar employment of Japanese nationals occurred in the sea urchin roe fishery on Vancouver Island. The issue exploded with the development in the late 1960s of the lucrative salmon and herring roe industry. In 1969, the union applied to the federal minister of manpower and immigration to prevent the import of workers from Japan to work in salmon roe operations. However, even more technicians were brought in to work in the herring roe fishery. Between 1971 and 1978, an estimated four hundred working permits were granted. Suspicion arose that they were being paid much lower wages than those received by UFAWU members. An important factor in the increasing militancy among shoreworkers was this use of foreign labour at a time when jobs were being eliminated in the industry.

Roe operations began in 1964, first on a limited scale in salmon roe, and then expanded to mass production in the herring roe fishery. Use of a foreign labour force was viewed by the union as one means of trying to introduce a cheap and unorganized workforce into a new industry. Another was the substitution of women for men, allowing the companies to pay a lower base rate. "Until a few years ago salmon eggs were regarded as waste and discarded with the rest of the offal. Whenever the eggs were needed for bait they were pulled out of the fish by men" (Ibid., 34[2]:6). Once a large market was found for the roe, however, only women were placed on these operations. The reasons given were that their fingers were more agile, and, therefore, women were more suited to the work. The women wanted to be paid the same rate as the men who had previously pulled the eggs. The companies refused. The women, although said to be more suited to the work, ended up receiving wages four cents above the base rate for women, or thirty-four cents less an hour than the men's rate.

The herring roe operation expanded early in 1972, when eighty-five boats were engaged. A number of fishers made fortunes, and the companies realized substantial profits. Shoreworkers, however, did not receive commensurate benefits. The operation did provide several weeks extended employment to a large number of women. However, herring roe processing operations were introduced during the life of an existing agreement, and the companies refused to negotiate wages or conditions for either the 1972 or 1973 seasons. In the meantime, herring fishers won a price increase of 140 per cent, and herring tendermen also negotiated for the new fishery, winning substantial increases. This issue was a major factor in fuelling dissatisfaction among shoreworkers. Compounding the problem was the fact that, because it was largely a manual operation, the processing of the herring roe could be contracted out to small, unorganized firms. Many of these sprang up overnight, and the larger firms contracted out operations to them. Fishers were called on by the union to deliver their catches only to organized plants.

By 1972, women were also feeling the effect of negotiations carried on over a number of years based on percentage increases. The increase made that year left women with a rate hike three cents an hour less than that received by men. Women working in fresh-fish operations received an increase six cents less than that given to men. Their base rate in 1972 was $2.89 per hour. Men received a base rate of $3.87 per hour (Ibid., 35[2]:4).

Thus, by 1973, many developments had increased the dissatisfaction among shoreworkers, and they became more militant. Jobs were of primary consideration. Japanese demand created a lucrative fresh/frozen market that diverted salmon from the can to the freezer and eliminated many jobs in the process. No efforts were made to create alternative employment opportunities. New operations, such as roe popping, were designated as women's work with corresponding low rates of pay. Where feasible, they were contracted out to firms that could hire even cheaper labour forces. For example, in 1975, a plant in Tofino on Vancouver Island tried to export its frozen herring to Mexico for processing. Sixty UFAWU shore-workers were laid off when they refused to load 250 tons of fish into freezer trucks. A Japanese company had an investment in this plant and had also purchased one in Mexico. Wages there were $1.20 per hour, compared to $4.50 in Tofino. Legislation was enacted preventing a company from exporting more than 25 per cent of its total landings in a raw form (Ibid., 40[10]:1,12). At home, Canadian Fish was financing the development of machinery that could process the fish and eliminate the large number of workers required in the manual operation.

Women employed in fresh-fish operations also had grounds for dissatisfaction. With the movement of B.C. Packers to the Atlantic coast, groundfish operations at home were pursued only in years when substantial profits could be made, thus creating sporadic employment for those women who had traditionally received the most secure employment. Conversely, the opening of the Japanese market for B.C. fish and fish products meant that fishers were receiving good incomes. Shoreworkers, however, were given little opportunity to participate, unless they were prepared to take strike action, as they did in 1973.

> The union proposes to delete all contract references to women and simply list classifications and the pertinent wage rates. As a union whose shoreworker membership numbers women in the majority, the UFAWU has a responsibility to establish a lead in a fight for equal rights for working women and equal pay for work of equal value. (Ibid., 37[9]:12)

The central issue in 1973 was not wages, but the introduction of a single cannery schedule. The Fisheries Association agreed to merge the two cannery supplements, to list job classifications by labour groups, to standardize rates, and to raise the lower-paid classification by fifteen cents over two years, in addition to any general increase. However, it proposed to do this by employing men at wage rates lower than those they currently earned. "Clearly, the union's aim is to raise the level of the rates paid to women and not to depress male wage rates" (Ibid., 37[10]:11). Turnouts at union meetings were the largest in decades, and the negotiating committee received the strongest membership support in its history. Cannery, net, reduction and watchmen classifications voted 90.5 per cent for strike action, while the fresh-fish and cold-storage sector voted 92.6 per cent in favour. The change in negotiations that year among the fishers, tendermen, and shoreworker sectors was noted. "In the 28-year history of the UFAWU, shoreworkers have been on strike a total of five weeks. Even when they have struck—as in 1959 and 1963—their actions have been closely linked to tieups of salmon net fishermen" (Ibid., 37[14]:8). This time they were going out to meet their own demands. The strike lasted one week, from July 6–15. While fishers capitulated as time went on, shoreworkers and tendermen became more adamant.

A breakthrough was won with this strike, but inequalities were not erased. A differential of seventy-one cents between filleters and male labour rates was narrowed to twenty-one cents. References to male and female net workers were abolished, and, as of April 15, 1974, rates were standardized. In cannery classifications, the forty-four-cent base rate differential between men and women was eliminated over a three-year period. The new wage structure elevated B.C. shoreworkers to among the highest paid workers in a primary food industry in the country, but sex differences in pay rates were only narrowed, not eliminated.

A major discrepancy that remained was in the fresh-fish and cold-storage operation. A one-thousand-hour differential was established in the agreements based on differences between heavy and light work. Those doing heavy work earned $6.95 per hour after one thousand hours. Men were automatically given this rate, while women were locked into a four-hundred-hour rate of $6.34 per hour. In February 1977, Susan Jorgensen, a fresh-fish worker, filed a complaint charging discrimination under Section 6 of the Human Rights Code. Although she performed the heavy work classification in racking fish, she was paid the lower rate. In addition, she estimated that she lost about three months of work each year because she was placed on the female seniority list and could not do the cold-storage work available to men with her seniority. In January 1979, to help strengthen her case, the UFAWU executive agreed to allow the union's name to be added to the complaint. The case dragged on for several years, but Jorgensen won in the end, when a decision was handed down in her favour. The union hoped to establish a precedent and end the discrepant four-hundred-hour and one-thousand-hour rates, establishing the same hourly basis for both men and women.

CONCLUSION

Although gender differences still existed in terms of pay differentials, the 1973 strike demonstrated that a large majority of shoreworkers were not prepared to continue tolerating such a structure. It also demonstrated the conflict between the companies and the union. The companies have always wanted a cheap and available labour force. The Chinese contract system was an ideal arrangement in supplying such a labour force.

Fishers of European descent have a history of struggle with these companies dating from before the turn of the century. A group of them realized, in the 1930s, that to be effective they required an organized shore-plant labour force. They battled with the companies for jurisdiction. Since the struggle occurred during World War II, the state became involved and was forced to make concessions to labour in order to guarantee uninterrupted production for the war.

At this point in time, shoreworkers, apart from a small group of skilled men (some of whom were themselves fishers), were not directly involved in union activity. Thus, the most vocal groups (the skilled male categories) realized gains at the expense of the majority — seasonal cannery workers. What this also did was to perpetuate labour categories based on race and gender, a system developed under the Chinese contracts and incorporated in union agreements. While companies reluctantly agreed to better conditions for skilled workers, they continued to structure labour as they always had done. In fact, remnants of the Chinese contract system, in terms of contractors recruiting labour forces, persist to this day.

The argument put forward in this chapter is that those workers most expoited by such a system would have to struggle for change. By 1973, for the reasons outlined, they were engaged in such a struggle. Struggles had to be waged not only against the companies, but also within the union, where fishers' interests had predominated since its founding. Today, more than at any other point in history, shoreworkers occupy a position of parity with fishers in the UFAWU. That parity is crucial, it has been argued here, to fight the more general discrimination in the industry.

NOTES

1. Further primary-source references to *The Fisherman* list the volume followed immediately by the issue number (in parentheses), followed by the page or pages cited, e.g., volume 13, no. 3, page 6 appears as 13(3):6.
2. According to provincial fisheries statistics, in 1941 there were 7,914 workers employed in fish plants. Of these, Gateman estimated only about 150 were cannery machinemen, with an even lower number of reduction plant workers. Readers who desire more information on shoreworker organization for this early period are directed to Muszynski (1984).
3. There are three sectors in the union: fishers, shoreworkers, and tendermen. Tendermen comprise the smallest numbers and form an intermediate group bearing characteristics of the other two sectors. Tendermen have been a very militant group, but their numbers have been decimated by technological changes introduced on packers, fish boats, and in unloading facilities.
4. The figure was calculated on the basis of three hundred monthly paid men receiving seven months of employment and 12 per cent overtime; six hundred hourly paid men for three and a half months and 10 per cent overtime; and fourteen hundred women for two and a half months and 10 per cent overtime (*The Fisherman*, 16[22]:1). The

monthly paid men were the militant sector. According to these figures, however, they represented only 13 per cent of the labour force.

5. Northern militancy was understandable in view of the fact that at least twelve hundred shoreworkers were out of work by September 9, 1952, because of northern plant closures and a shorter season (*The Fisherman*, 14[28]:2).

6. In just about every round of bargaining, the UFAWU and the Fisheries Association were unable to come to terms jointly and had to apply for conciliation boards to help resolve the differences. The Fisheries Association set the pattern of agreements among the companies, although a number of "independents" bargained on an individual basis with the union.

PART 3

COMMUNITY
AND REGION

13

The Fishing-Dependent Community

EVELYN PINKERTON

This chapter provides a description of one rural isolated fishing-dependent region that is characterized by both underdevelopment and persistent attempts to overcome underdevelopment. Following a brief description of the region as a whole, the first half of the chapter focusses on the community of Tofino, the decline of its fishery, and attempts to sustain a fish-processing facility there. The second half describes the Indian village of Ahousaht, where fishing provides the organizing principle for nearly all social life, in addition to being the major source of employment. In conclusion, the survival strategies and development possibilities of both communities are outlined.

The purposes of the chapter are: (1) to show something of the character of the communities and the extent to which social life and culture have been patterned around fishing; (2) to show what dependence on fishing means in both social and economic terms, and how people in this situation are affected by government policies related to fishing; (3) to recognize the potential that is demonstrated in the way people cope with underdevelopment: the efficiency of a small-scale fishery at Ahousaht, the locational advantages of a fish-processing plant at Tofino, and the degree of human effort mobilized to save the resource and the communities.

THE REGION

Beginning in February 1982, I lived for nine weeks in the region of the west coast of Vancouver Island. This region extends from Winter Harbour to Nitinat, 360 kilometres of coastline (see map, page 7). I concentrated mostly on Tofino, but also spent time in Port Alberni, Ucluelet, and Ahousaht, interviewing fishers and shoreworkers about their work histories, opinions, attitudes, and community relations. As well, I interviewed fish-plant managers, cash buyers, community leaders, local businessmen, local historians, and Department of Fisheries and Oceans personnel.[1] At least seven other fishing-dependent regions on the coast share some of the characteristics of this region, and therefore this one may be considered representative.[2]

For a high percentage of people in this and similar areas, fishing is a skill learned early in life, often to the exclusion of other job skills. For some villages in these areas, particularly Indian villages, fishing is the activity around which life has always been organized, and it serves important functions in distribution, socialization, social participation, and mutual support.

Similar degrees of fishing dependence in these regions have produced similar degrees of vulnerability to fleet rationalization policies and to centralizing processing and services. With the exception of three villages within the eight areas, all of these sites have suffered similar patterns of decline in the number of salmon licences held by local residents (Figure 13.1).

These fishing-dependent villages have also been affected by capital withdrawals from the regions, either directly, when in-plant jobs were eliminated with the closure of a local cannery, or indirectly, when the removal of a cannery or packing services influenced local fishers' ability to continue fishing. A locally sponsored study (Lewis, 1977) suggests that many fishing licences were lost in these areas even before 1969 for this reason. For example, when processing plants and/or packing services were withdrawn from Queens Cove and Nuchatlitz on the west coast of Vancouver Island, small-scale local fishers who could not afford to invest in more sophisticated gear and mobile vessels had to abandon fishing. Part of the reason for the withdrawal of services and processing facilities from such isolated areas was the overexploitation and subsequent decline of local stocks that made packing from these areas less economical, as will be discussed later.

In sum, fishing-dependent regions tend to be exceptionally vulnerable. Villagers often have few job opportunities and marketable skills apart from fishing. Important features of their social organization are often conditioned by fishing. They cannot always protect or maintain local stocks, which may be overfished by the larger and more mobile members of the British Columbia fleet. Local fishers tend to lose fishing licences. Processing plants and packing and collecting services, without which it is difficult for small-scale fishers to operate, have often been removed. The two communities described below are both vulnerable in these ways, though they differ in degree of fishing dependence, in fishing-related social organization, and in apparent destinations.

TOFINO TODAY

Tofino is a district municipality of one thousand located on the Esowista peninsula of the west coast of Vancouver Island at the terminus of the Trans-Canada highway. It is a service centre and transportation link for three outlying Indian reserves accessible only by water or air (Ahousaht, Opitset, and Hot Springs Cove) with a combined populated of 850, and for a scattering of some 50 to 100 settlers around Clayoquot Sound. One

Figure 13.1
Salmon Licences by Community

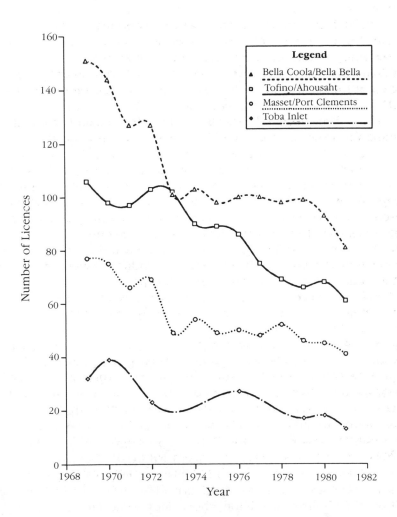

Source: DFO, Licensing data, 1969–1981.

logging camp employing one hundred men is located in Clayoquot Sound, but it does virtually all of its hiring and servicing from the east coast of Vancouver Island.

Tofino services a regional population of two to three thousand people with a twenty-one-bed hospital, a two-hundred-pupil elementary school, a post office, a Department of Fisheries and Oceans office, an Indian

Health office, a Royal Canadian Mounted Police office, a lifeboat station, a local telephone switchboard, and a bank. A small Forest Service office closed in 1984. Except for the school and hospital, however, these institutions employ only one part-time person to three persons each.

Tofino also supports about thirty small businesses, most of which have fewer than ten employees. The majority of these are restaurants and hotels, which depend on the tourists who visit Pacific Rim National Park between Tofino and Ucluelet twenty-five miles to the south during the two-month summer season.[3] Since tourism is seasonal, these businesses need the patronage of Indians coming to and from outlying reserves, as well as nonresident fishers and fish buyers who come during salmon and roe herring season.

There are also a number of businesses with major clients in the fishing industry, such as fuel suppliers, marine services, construction, a marina, and some of the hotels and grocery stores. In a 1980 study conducted by the village office to measure the importance of fishing for Tofino,[4] it was estimated that these enterprises depended on the fishing industry for 50 to 100 per cent of their business, while the tourist businesses were 30 to 50 per cent fishing dependent. The town also estimated that 40 per cent of household heads are directly or indirectly dependent on the fishing industry and that all the 50 to 100 per cent fishing-dependent businesses would cease to operate without the fishing-related industry.

In addition to these businesses, in 1982 there were two fish-processing plants and four fish-collecting stations in Tofino. One very large plant operated by B.C. Packers, with 2.5 million kilograms cold storage and 30 tonnes ice-making capacity, closed in 1983. During the mid-1970s it employed 250 people at the season's height; in the 1980s there were fifty employees at the high point. The other fish company, Canako Sea Products, operates a small geoduc plant with a small freezer and two-tonne ice maker employing up to fifteen people. The fish-collecting stations consist of two small family-run operations based principally in Vancouver (Vancouver Shellfish and Longbeach Shellfish), a third station operated by B.C. Packers, and a fourth, operated until 1984 by Cassiar Packing through its subsidiary Tonquin. All of these stations made ice for fishers and for trucks transporting roe herring, salmon, and other products to Vancouver. The village of Tofino received $81,727 in property taxes from these six establishments in 1980, about 20 per cent of the municipal budget. In addition, local shoreworker and office staff wages circulated many times in the community, employing others as well.[5] The activity of fish processing alone, then, contributes substantially to the local economy.

During roe herring season in February and March of 1982, Tofino was the centre of activity for the entire region. In the following two years Ucluelet was the centre. Fishers trucked their skiffs to Tofino and waited there for the announcement of openings. Fish buyers and major companies flew search missions and manned their radios from Tofino. During this time and in summer salmon season, the town buzzes with activity; restaurants, pubs, hotels, and docks overflow with fishers. At other times, public life is at an ebb, and socializing is confined to family gatherings, nights at the Royal Canadian Legion lounge and the Hotel Maquinna pub, or young people's dances at the hotel. Fishers are the exception: they socialize year-round while helping each other work on their boats at the docks.

In winter the cash flow through the village is restricted; local business-men measure their trade by the cyclic appearance of government cheques. They know they will do some business on "poggie day" (unemployment insurance) and on "pecker day" (family allowance). The social assistance trade is supplied chiefly by people who were formerly employed in fishing.

The political issues that dominated the town in 1982 centred around the question of who would bear the cost of providing water and sewer services for real estate and tourist development, and around persuading the province to place the region's tourist development over the interests of two major forest companies. The companies intended to log Meares Island, a scenic area facing Tofino, much valued in its natural state by the Clayoquot Indians resident at Opitset on Meares Island and by the Ahousahts, who traditionally owned part of Meares Island. The first issue was the chief concern of the older, established fishing families who wished to prevent real estate developers from profiting at the expense of Tofino taxpayers, especially younger families who faced an unhopeful job situa-tion in fishing. As real estate in Tofino began to approach lower mainland values, the older families feared the community would eventually serve as home only to vacationing and retiring urbanites, or the school and hospital professionals. The fears were overpowered, however, by the pressing need for more diversified economic development as fishing opportunities declined in the 1980s.

The preservation of Meares Island was supported by business people concerned about potential damage to tourism, by the Tofino town council, which wanted protection and expansion of its water-supply area on Meares Island, by people favouring maricultural development around the island,[6] by Clayoquot and Ahousaht Indians involved in clamming and trapping on Meares Island, and by "alternative lifestyle" and other local environ-mentalists. At the 1984 Easter celebration organized by the last group, the Indians declared the area a "tribal park" and launched a court action to prevent logging, basing their claim on aboriginal title.[7] Environmental protection and aboriginal title claims unite the community.

The "alternative lifestyle" residents have found an economic niche in Tofino. They came originally as refugees of urban alienation and from settlements in the area, which together became the Pacific Rim National Park in the 1960s. Since the early 1970s they have supplied seasonal or part-time demand for work as chambermaids, shake cutters, tree planters, carpenters, waitresses, clerks, deck hands, and caretakers. They also support themselves on cottage industries such as carving and handicrafts, which are marketed in urban areas. Because their ideology and modest lifestyles permit them to survive with such irregular work, they usually do not qualify for, or do not choose to take, unemployment insurance benefits.

Only about 17 per cent of local residents do receive unemployment insurance benefits. At the January-February "high point" for unemployment in the Tofino-Opitset area in 1982, eighty-eight persons out of a labour force of about five hundred[8] received unemployment insurance benefits; forty of these were fishers, comprising most of the approximately forty-five fishers in Tofino and Opitset. In Ahousaht, only twelve fishers out of thirty-eight vessel owners received unemployment benefits.

During a similar period in 1977 when Tofino had a slightly smaller population, a greater number received unemployment benefits (134), and only 32 per cent were fishers (43 people). The change in overall unemployment benefits and the fact that 1982 was a recession year suggest that fewer people had enough work in 1981 to qualify for unemployment. (To qualify, it was necessary to have worked at least twenty weeks in a previous year, followed by at least ten weeks in the year preceding application.)

Many fishers interviewed in the west coast area and in the fleet at large declared either that they "didn't believe in" unemployment or that they would not apply for benefits unless they had no choice. The total number of fishers receiving unemployment benefits on Vancouver Island and in metropolitan Vancouver increased by 210 and 211 respectively between 1977 and 1982, although the absolute number of fishers in both areas declined. The increase in fishers' use of unemployment benefits is thus a provincewide phenomenon, evidently related to the greater economic difficulties suffered in 1982 compared to 1977. The fact that the increase in use of unemployment benefits has been great in Vancouver and the Vancouver Island region while it has been relatively constant in Tofino suggests that off-season job opportunities have been curtailed more recently in these regions, while they have been a more long-term problem in Tofino. The unavailability of unemployment insurance data prior to 1976 made it impossible to quantify the statement made by many fishers that they did not collect benefits until large-boat payments became current in the 1970s.

HISTORY OF THE REGION:
THE DECLINE OF FIVE FISHERIES

Commercial fishing in Clayoquot Sound consisted of five fisheries, each of which declined: the salmon net fishery, whaling, the pilchard fishery, the herring fishery, and the salmon troll fishery. Whereas the first four fisheries were terminated by the failure to control overexploitation, the troll fishery is threatened chiefly by an international treaty.

Clayoquot Sound was first developed for commercial fishing when a cannery was built on the Kennedy River in 1895 by Thomas Earle and sold in 1902 to a Victoria firm, Beckwith and Brewster. The Norwegian Fraser River gillnetters, who were hired to fish for this cannery, were among the first European settlers in this region. The Clayoquot Sound Cannery held a drag seine licence to intercept salmon entering the Kennedy River and bought salmon from the two or three other drag seine licensees in Clayoquot Sound. Purse seines replaced drag seines about 1920 when commercial net fishing in the river was closed. In addition to three company-owned seiners, about fiften Clayoquot Sound residents acquired and operated seine vessels in the area. This local fleet was composed equally of Ahousahts, Clayoquots, and Tofino residents, who mainly fished sockeye salmon in August and chum salmon in September and October.

World War I brought a high demand for sockeye and chum, which kept eight canneries in operation on the west coast of Vancouver Island until 1921, processing over 84,000 cases in 1920 (Lyons, 1969). By 1919 there was such concern over the reduction of the chum stocks that the fishers and residents of Barkley Sound (immediately south of Clayoquot Sound) demanded an inquiry (Roberts, 1970). This did not occur, and by 1924 there were 277,267 cases of salmon processed in this area (Lyons, 1969). Separate records for the area at later dates could not be found, but production probably increased as demand picked up between 1926 and 1929, and six new canneries opened on the west coast of Vancouver Island (Roberts, 1970). As more nonlocal packers came into the area to buy fish, Clayoquot Canning was eventually forced to close its local operation in 1932.

At the same time, the seine fleet was increasing and more nonlocal seiners began to fish Clayoquot Sound at the height of the season. The fishery inspectors felt that "one great menace to the salmon fisheries of the province, particularly the fall varieties, is the huge increase in the number of salmon purse seines operating." These boats increased from 143 in 1922 to 223 in 1923 (DFO Annual Departmental Review, 1926–27). A partial reason for the increase was pilchard and herring seiners who were permitted to fish for fall chum salmon using smaller nets. The fishers and canners alike expressed such discontent in 1928 that by 1929 the Department of Fisheries devised the Transfer System to limit the effort of the increasingly mobile purse seine fleet. Clayoquot Sound became one of twenty-seven salmon purse seine areas in which the local fisheries guardian was empowered by order in council to shorten the fishing time if more than (in this case) fourteen seiners entered the area. One seine was equivalent to eight gillnets; vessels were required to check in with the local guardian upon entering an area and check out when leaving (ADR 1929–30; interviews).

Although the Transfer System remained in effect for over twenty-five years, it was not successful in saving the chum stocks, which were the traditional staple of the area (Drucker, 1951). The first catch records for the area were made only for seine catches between 1934 and 1949. Five-year average catches in Clayoquot Sound during this period show a dramatic and relentless decline from a 79,198 piece average (1935–39) to a 32,681 piece average (1945–49) (see Table 13.1).

In 1951 nylon gillnets were introduced and gillnet effort in the area began to increase. In 1953 the Department of Fisheries began closing the area for a week or two at a time, but no stock recovery resulted. Since by this time escapements (fish not caught that return to the creek to spawn) as well as catch were recorded, it is possible to estimate more accurately the decline of chum stocks (catch plus escapement) as a whole. Between 1951 and 1968, the total chum stock in Area 24 declined from an average of 134,315 (1951–55) to an average of 36,981 (1965–69).

The most important reason for the reduction of this already-depressed stock to one-quarter of its early 1950s size was the increased efficiency and political pressure of the mobile seine fleet. Seiners became more efficient as the puretic block and the snap purse ring made five to seven sets a day practicable, replacing the former three or four sets. A decade later the drum and modern winch made up to eighteen sets a day possible.

Table 13.1
Chum Salmon Catch and Escapement, 1935–1983:
Clayoquot Sound (Tofino Area)

Year[1]	Catch (in pieces) (5-year averages)[2]	Escapement (pieces) (5-year averages)
1935–39	79,198	n/a
1940–44	59,545	n/a
1945–49	32,681	18,425
1950–54	47,940	86,375
1955–59	33,990	53,205
1960–64	13,990	41,390
1965–69	0	36,981
1970–74	7,352	58,090
1975–79	39,240	59,257
1980–84	24,011	71,340

n/a — not available
[1]For reasons of space, five-year averages are reported. Given the cyclical nature of salmon, this procedure can sometimes distort the true pattern; in this case it does not.
[2]From 1935 to 1949, the catch was by seine only; from 1950 to 1984 the catch was by seine and gillnet. Zero catch from 1965 to 1969 was the result of a fish closure.
Source: Roberts (1970); Department of Fisheries and Oceans.

In 1957 the Transfer System collapsed as a result of increased pressure for fishing time from the mobile seine fleet. By 1964 local chum and other net-caught salmon stocks were so depleted that Clayoquot Sound was completely closed to net fishing. Since then, only an occasional one- or two-day fishery has occurred. Despite this closure, there is little improvement in chum escapements in Clayoquot Sound.[9] The reduction of this fishery constituted a severe blow for local fishers who depended on the fall fishery for a good percentage of their catch. It had been the custom of many local trollers to mount a small table seine on their vessels for the chum season, thus rounding out their fishing season. Overfishing inside Clayoquot Sound also caused local sockeye stocks to collapse after 1969, although closures since then have permitted some recovery (but no fishing).

Whaling was also introduced to the region early, but disappeared more swiftly. In 1905 a Victoria company established whaling stations in Barkley Sound to the south and Kyuquot to the north. The Norwegian-style steam whalers, hiring mostly Norwegian crew on a share basis, brought in five hundred whales between April and August 1908. In a relatively short time, the whale population was radically depleted, and the plants were converted to pilchard-reduction facilities (Scott, 1972).

Meanwhile non-Indian ethnic groups and other fisheries became established. British settlers populated the surrounding islands, while Scots from the Isle of Skye and Japanese Canadians from Steveston settled the area around the turn of the century (Bossin, 1981). In 1917 or 1926[10] the Japanese set up a small fish "camp" in Tofino through their Tofino Cooperative Fishing Association, which purchased troll-caught salmon and transported it to fresh-fish markets in Seattle via their own packer. Some of their thirty-three-member cooperative were Scots, who also took up trolling. This method of capture, which predominated in later years, was thus introduced to the region very early. Small but numerous runs of local chinook and coho salmon, combined with the fall chum fishery and the interception of passing stocks offshore, kept local trollers working much of the year.

The Japanese also established a number of salteries, processing chum salmon and herring for the Chinese market. These were closed when the Japanese were evicted from the coast during World War II (Scott, 1972). The Japanese never re-established themselves in Tofino after the war, perhaps because of a 1947 village resolution forbidding their ownership of land or businesses (Bossin, 1981). Many Japanese attending the Tofino 1982 homecoming celebration were visiting for the first time in forty years.

The west coast of Vancouver Island was the area of the coast where pilchard were most plentiful and where processing of this fish was first established in 1918. Between 1925 and 1948 the "pilchard boom" was the reason for the operation of twenty-six pilchard reduction plants between

Barkley Sound to the south and Kyuquot to the north. Six of these were in Clayoquot Sound. The most notable and long-lived of the west coast plants (Kildonan, Port Albion, Ceepeecee, Hecate, Ecoole) also had attached salmon canneries, and, in the case of Kildonan and Port Albion, later added cold-storage facilities. During these years, the larger plants could be six- or seven-month operations, providing employment to fishers (who often worked in the reduction plants in the winter after fishing season) and to Indian women, who might work a season in more than one cannery if working times overlapped, and to Chinese labour brought from Vancouver. One local historian reports that seventy-five seiners, one hundred tugs and scout boats, and fifty packing scows kept some five hundred fishers and five hundred shoreworkers employed in 1927 (Nicholson, 1965).

Whatever interest the companies operating these plants (B.C. Packers, Canadian Fish, Nelson Brothers, Nootka-Bamfield, Todd) had in fresh, troll-caught fish appears to have diminished during the 1930s. Buying stations in Quatsino and Kyuquot were closed, and Kildonan would only accept fish delivered to the plant. The troll fishery was sustained, however, by the formation in 1931 of the Kyuquot Trollers Association, which acquired its own packers and transported fresh salmon to Seattle until 1955, when a market failure in Seattle caused it to be taken over by the Prince Rupert Fishermen's Cooperative Association. The PRFCA continued to operate the KTA's cold storage facilities at Winter Harbour and Victoria (later moving the Victoria facility to Vancouver), and the buying stations at Kyuquot, Esperanza, Hot Springs, Tofino, and Ucluelet (Hill, 1967).

B.C. Packers redeveloped its interest in troll-caught salmon in 1940 and reintegrated with Edmunds and Walker, a Vancouver firm specializing in fresh fish.[11] Canadian Fish at this time also bought the herring-reduction/cold-storage plant at Port Albion from the Nootka-Bamfield Company and used this plant to purchase trolled salmon until herring reduction ended in 1967 because of fears of overfishing. Both companies set up "troll camps" (supplying ice by packer from their main cold-storage plants and buying fish) all over the west coast of Vancouver Island (Kyuquot, Nuchatlitz, Queens Cove, Esperanza, Hot Springs Cove, Ahousaht, Tofino, Ucluelet). They were followed in the 1950s by competitors Blaine Myers and Malcolm McCallum, both of which were later acquired by B.C. Packers, and, in the late 1950s and 1960s, by Tulloch-Western, which was acquired by North Coast Fisheries in 1964, and was in turn acquired by Canadian Fish and Norpac Fisheries.

After World War II, Tofino fishers returned home with enough capital to buy trollers. As fresh and frozen salmon markets improved in the postwar period, many former salmon seine fishers turned to trolling by the late 1940s. (Some seiners fished fall chum salmon in combination with seasonal herring or pilchard operations.)

The build-up of seiners during the pilchard boom, followed by the herring reduction fishery on which the seiners then concentrated after pilchards disappeared in 1948, also led to drastic overfishing of herring and the termination of the herring fishery in 1967. This left the west coast of Vancouver Island little but the troll fishery. (Halibut had been largely depleted, and groundfish markets had not yet achieved the modest rise that occurred in the late 1970s.) Even the troll fishery was damaged by the seine build-up because chinook stocks, which had flourished in areas of herring concentration (chinook feed on herring), declined sharply after herring stocks were reduced, leading to the closure of highly productive local chinook salmon fisheries and of local troll camps dependent on them (Roberts, 1970).

The local troll fishery survived, however, because it was only partially based on local stocks, and because it was able to diversify by catching sockeye and pink salmon. Much of the fish taken by the large troll fleet that focussed on the west coast of Vancouver Island consisted of passing stocks, intercepted on their way to the Fraser and other Canadian and American rivers. Fraser sockeye and pink salmon stocks were protected and built up by the International Pacific Salmon Commission.[12]

By the 1970s, however, the large companies that had serviced areas where salmon were traditionally trolled were withdrawing. In 1973 Canadian Fish withdrew its last troll camps from the region, and B.C. Packers retained its troll camps only in Winter Harbour, Kyuquot, Tofino, and Ucluelet. Kyuquot camp was closed in 1983.

The last surviving traditional fishery in Clayoquot Sound may be threatened by the ongoing negotiations of the 1985 Canada-U.S. treaty. The general intent of the treaty is to reduce Canadian interceptions of U.S. chinook and coho and to reduce U.S. interception of Fraser River sockeye and pink, thus giving each country greater incentives to improve its own stocks. By 1984, international arrangements for the conservation of chinook salmon established an area-species closure that reduced fishing time on the west coast of Vancouver Island. The 1985 treaty terms limit the chinook and coho catch in the area for the next two years. The chinook ceiling is well below average catches, while the coho ceiling is slightly higher than average. These ceilings will be renegotiated every two years through the new Pacific Salmon Commission. The implied expectation is that local enhancement effort in the next eight years will be so successful that locally produced fish can replace intercepted fish.

From the turn of the century to the 1960s, the Clayoquot Sound area thus witnessed the disappearance of four fisheries: salmon seining, whaling, pilchard reduction, and herring reduction. Even trolling now appears endangered. Trolling, however, can survive under very different conditions from those that existed for the other fisheries. The first four fisheries had been propelled by the local existence of processing and collecting facilities, which required a high-volume production, and which therefore shut down as stocks were reduced. Trolling produces a high-quality fresh or frozen product, which does not require high volume to be viable, either in fishing or in processing. The construction of a road to the west coast of Vancouver Island and the growth of frozen seafood markets would change processing economics and create new opportunities for the area.

THE EXPLOITATION OF
LOCATIONAL ADVANTAGE

The structure of competition among processors for the trolled fish was to change completely in the 1960s, with the advent of a road to the west coast of Vancouver Island. The area had previously experienced some population growth from military personnel who, stationed at the airforce base constructed during the war, remained afterward. But more growth occurred because the logging road connected Tofino to the east coast of Vancouver Island in 1959 (consequent to the development of large-scale logging by major forest companies), and brought regional administrative offices and services into the area. Tofino grew from 89 households in 1951 to 182 households in 1961 to 202 households in 1971. The Canada Postholders directories listed 27 per cent of household heads as fishers in 1961 and 24 per cent in 1971 (Table 13.2). The populations of Tofino and Ucluelet increased steadily and comparably during this period, but the proportion of fishers continued to decrease, at least partially because government and services were increasing in economic importance. A hospital, government services, and tourism became important in Tofino, while logging was more important in Ucluelet.

Once road connections leading to the Vancouver ferry had been established, the area enjoyed an advantage peculiar to fish processing. Had the industry at this point be geared exclusively toward canned markets, the process of urban centralization of production would have continued. However, the expansion of fresh and frozen markets created demands for fish products that had neither the same economies of scale nor the same technological constraints that existed for canned products. Whereas large companies could mass-produce canned salmon more cheaply by transporting it to centralized locations, such was not necessarily the case with frozen salmon, groundfish, shellfish and, in the 1970s, herring roe. (Herring stocks recovered sufficiently after a five-year closure to support a roe herring fishery in the early 1970s: the roe were brined or frozen.)

When the fishery on the west coast of Vancouver Island came to depend chiefly on intercepted troll fishing, it relied on a product delivered at a higher price. This could best be sold on the fresh/frozen markets where salmon commanded a higher price. The large Vancouver canners did not have all the advantages in acquiring and processing trolled fish, which they had in net-caught fish. In fact, the eventual processing strategy of smaller firms became the exact opposite of the large firms' centralization

Table 13.2
Demographic Profile for Tofino and Ucluelet, 1950–1980

Year	Tofino	Ucluelet
1951–51		
• Households	89	158
• Household heads fishing	16 (18%)	56 (35%)
1960–61		
• Households	182	275
• Household heads fishing	50 (28%)	77 (28%)
1970–71		
• Households	202	420
• Household heads fishing	48 (24%)	87 (21%)
1980–81		
• Households	220	560
• Household heads fishing	88 (40%)	n/a

n/a — not available
Source: Canada Census (various years), W.Sinclair (1971), J.Burns (1976), Tofino Village
Census, personal interviews.

and concentration strategy (Chapter 4). These larger firms closed their last outlying canneries in the B.C. central coast area in the 1960s and 1970s and their herring reduction in the 1960s. They declared that they would need a much larger supply or combination of supplies than presently existed on the west coat of Vancouver Island to warrant establishing a plant there. Troll fish alone held less interest for them and, with the exception of Barkley Sound,[13] locally based net fishing had largely ceased in the region.

In contrast, the smaller firms preferred to locate plants near the supply source rather than near markets. By processing the fish at the supply end, and only transporting them to market later, they could realize four locational advantages, as follows.

1. *The capture of quality.* By processing the fish immediately instead of transporting them to Vancouver for processing, a much higher quality product could be achieved. Hence a higher price or a larger share of the market could be realized. Transpacific Fish in Ucluelet, for example, has gradually increased its share of the frozen salmon market through its reputation for quality.

2. *The capture of shrinkage.* By processing the fish immediately, a greater portion of the fish was usable, and there was less weight loss. Since fish are sold by weight, the reduction of the 14 per cent shrinkage or weight loss normally experienced in packed fish would represent considerable savings.

3. *The lowering of transportation costs (based on weight).* The final dressing and processing of the fish at the supply site meant that no excess weight was transported to a distant processing site. Specific orders were often filled on site. This factor proved particularly important when on-board holding techniques improved sufficiently for net-caught fish to be dressed for the frozen market (Appendix B, Pinkerton, 1983a).

4. *The exploitation of underutilized species in small quantities.* Clams, oysters, crabs, shrimp, prawns, abalone, food herring, and several species of groundfish could not be economically harvested or processed unless the plant was near the capture site. Small shellfish plants on the east coast of Vancouver Island had already demonstrated that discarding the shell and processing close to the capture site produced a viable operation. Weight, quality, available quantity, knowledge of fishgrounds, knowledge of markets, and labour flexibility all played into the way a small, local plant with an interest in pocket markets could effectively exploit these species.

What developed in this particular area was thus a regional specialization of production by small firms, which was possible because there were no significant economic or licensing barriers to entry into fresh and frozen processing. Vast capital outlay was not required, because it was possible to rent cold storage in Vancouver for excess product and to purchase relatively small-scale ice-making equipment. In the 1970s, government optimism about increasing supplies fostered a liberal licensing policy for new plants. The success of this strategy had been evident since Seafood Products began operating a small canning and freezer firm at Port Hardy on the northeastern end of Vancouver Island in 1966. This plant acquired troll fish captured in the northern area of both the east and west coasts of Vancouver Island and specialized in quality production. The west coast of Vancouver Island was the only other isolated and plentiful fish-supply area (roe herring, groundfish, shellfish, trolled salmon) of the coast without a plant and that was now accessible by road. The way in which other forces played into the attempt to establish a plant that would exploit local advantages is examined later, when I sketch the most recent stage of the development of fishing and processing in this region.

LOCAL ICE STATIONS AND FREEZER PLANTS

Ice stations were established by local businessmen in Tofino and Ucluelet in 1961, a decade before plants were constructed.[14] These made it possible to truck fish to Vancouver for two cents a pound, rather than pay the packing cost of seven cents a pound. The large companies followed suit with ice stations of their own. By 1981 there were several ice stations in Tofino and in Ucluelet. Four of these purchased for small Vancouver-based companies, two purchased for cooperatives (one local, one the PRFCA), and two purchased for major companies. In this respect, all the companies with ice stations enjoyed the locational advantages of ice plants, and they all transferred the bulk of production (freezing) to the urban centre. Tofino was in a particularly advantageous location, since it was the nearest town with a connecting road to service the entire northwest coast of Vancouver Island (except Winter Harbour to the extreme north, which also had a road connection). It was thus the collection point for a large supply area, where packers from the northern troll camps transferred their fish onto trucks to get into Vancouver quickly.

TOFINO FISHERIES LTD.: HOW BIG CAN A LOCAL PLANT GET?

The first big plunge into building a local processing facility in the area was taken by a third-generation Tofino businessman in 1972. Perhaps to his disadvantage, he began in a period of expanding markets when the federal and provincial governments and even individual banks were willing to make large sums available for the construction of fish plants in isolated areas, especially (in the case of government) if they employed Indians (Foodwest, 1979). In the 1960s and earlier, a firm establishing itself in an isolated region began cautiously, fully aware of the slowness with which processor reputations and loyalties are built with fishers, and the unpredictability of supplies and markets. But the mood of the 1970s was far more optimistic and expansive.

In an attempt to understand why a man experienced in the fish business had built a very large plant and gone into receivership in a few years, only to be eagerly bought out by a large, long-established firm that likewise could not make a profit on the plant, I undertook a detailed investigation of the events surrounding this conscious venture to capitalize on locational advantage. I interviewed two of the three partners in the Tofino company, three of the nine committee members on the special ARDA (Agricultural & Rural Development Agency) federal-provincial team that funded the plant, numerous Tofino residents involved with the plant's operation, managers of Canadian Fish and B.C. Packers, which later purchased the plant, and managers of smaller buying and processing ventures in the area. I read the provincial government file on how the decision was reached not to fund an Indian purchase of the plant after the first failure.

There were differences of opinion about the plant's viability. The Tofino businessman was convinced the plant was operational and was hindered only by delays in funding. Managers at Canadian Fish, a large and long-established firm, were convinced the plant was viable when it made the purchase, but later concluded it was not. Personnel at Special ARDA (the unit focussing on Indian employment) believed the plant was viable and blamed the failure on problems among the partners. Smaller processors and fish buyers who worked in the area termed the plant a "white elephant," too large to be sustained in the long run by local supplies. They pointed to a smaller plant that was later established in Ucluelet as a more appropriate size for the region.

Both Robert Wingen, the Tofino businessman, and one of his partners, Andrew Tulloch, had worked for Canadian Fish for years. Wingen had operated all of Canadian Fish's troll camps north of Tofino on a contract basis since 1962, and since 1965 had run a small oyster- and clam-processing plant next to the family boatworks in Tofino. Because he had already been selling salmon roe from trollers to the Japanese, he began processing herring roe in 1972 with an established market, hiring some forty-five local people, mostly Indians, during the two-month season. Wingen was well versed in using location to best advantage through managing the Canadian Fish troll camps. By living year-round in Tofino and hiring and dispatching packers from there instead of Vancouver, he was able to operate a less costly and more efficient collecting service than otherwise would have been possible. Because of Wingen, Canadian Fish continued to offer this service for some years in locations that other companies had abandoned as too expensive. They discontinued it in 1973 after Wingen became completely independent.

Wingen intended to expand gradually over the next five years and had developed forecasts for the increased capital and labour required for his venture. He planned to process shrimp, crab, oysters, abalone, clams, roe herring, food herring, salmon, salmon roe, and groundfish in quantities large enough to make the operation year-round, claiming that none of these species alone or in fewer combinations would have made a viable operation. Wingen's strategy was to combine many small fisheries, whose supplies he knew well, and which were largely unexploited by Canadians. (U.S. shrimp boats exploited grounds twenty miles off Tofino, and the capture of groundfish by foreign fleets within Canadian waters was later turned into a scandal by the UFAWU.) This was a strategy not favoured by the larger companies, which were interested predominantly in roe herring, salmon, and only some species of groundfish, profitable commodities for which markets were large and established. Wingen also believed there were ways to economically service the Indian communities north of Tofino, from which the large processors were withdrawing packing services. It was in fact largely through frustration with the production and marketing decisions of the large companies that Wingen and Tulloch decided to go into business.

Wingen was considered "competent at anything he touched," and Tulloch, then production manager for Canadian Fish, and formerly operator of his own west coast troll camps, was termed "dynamic." The two had solidified their determination to go independent after their initiative and willingness to take risks had earned Canadian Fish a bonanza on the 1972 Nitinat chum run.

Wingen's original caution about expanding the business gradually was dispelled by Tulloch's optimism and ARDA's interest in funding a $1.6 million plant that would employ a significant number of Indians. Building a larger plant than they had originally intended involved considerable risk for the firm, however, since funding only covered 34 per cent of capital costs, and the firm had to operate through the 1974 slump in groundfish, salmon, and shellfish prices before construction was complete and funding could be obtained. Although Wingen adopted a number of innovative strategies (such as flying fishers on reconnaissance to develop new fisheries and persuading the Department of Fisheries to open small pocket fisheries), and Tulloch brought with him some of Canadian Fish's highliners, the firm was in financial trouble by August 1974. This was exacerbated by the fact that ARDA delayed half of the $500,000 subsidy (the first half was paid in June 1974) because of doubts raised about the operation, and banks refused to pay creditors for fear the firm would go under. Some of these doubts may have been raised to the government by the major processors, who have made no secret of their position that subsidies to small firms constitute unfair competition and increase the problem of provincewide overcapacity in processing (B.C. Packers, 1981). According to several informants, this was "standard industry practice."

Tofino Fisheries offered the plant for sale at $3 to 4 million (including land and vessels) in September 1974. The provincial government considered assisting the Nuu-chah-nulth Tribal Council in purchasing the plant. A provincial study was favourable, but the province abandoned the idea after consultation with the Prince Rupert Fishermen's Cooperative. Canadian Fish did not share the co-op's negative assessment, however, and, after a complex series of biddings and negotiations, purchased the plant from the receiver at $1.9 million in January 1976. Its operation did not prove lucrative, however, and by 1978 Canfisco sent roe herring from Tofino to its Vancouver plant for popping. B.C. Packers purchased the plant "cheap" as part of its acquisition of Canadian Fish's northern operations in 1980. I was informed in 1982 that neither it nor Canfisco had ever made money on the plant. From that time to the present, B.C. Packers has apparently been willing to sell the plant, but not to a competitor. Both the buying of the plant and the subsequent unwillingness to sell it suggest that the company was engaging in pre-emptive buying and/or pre-emptive holding of a facility that might be viable in the right hands (for discussion of pre-emptive buying see Schwindt, 1982).

One very small nonunionized geoduc plant still operates in Tofino. In Ucluelet there are two plants owned by Vancouver-based companies: one is a small salmon-roe herring-groundfish operation. A larger plant operated by Central Native Fishermen's Co-op received some government protection because it employed Indians, but experienced severe supply problems and financial difficulties in the 1980s. By 1985 it was leased from the receiver by a group of former CNFC trollers and operated chiefly through inventory financing from Booth Fisheries, a Seattle firm that sold all its products. (By 1986 McMillan replaced Booth and the plant was unionized.) Operating at 60 per cent capacity for three months (10 per cent for 9 months), it survived by custom processing and purchasing from smaller buying stations. The plants as a group clearly suffered from supply competition once they were all established in the late 1970s.

Whereas the failure of the Tofino plant may have some unique causes, its failure suggests the operation of larger principles when considered in a regional context.

Optimistic predictions about fish supplies and buoyant markets encouraged government funding or assistance to three plants in the Tofino/Ucluelet area between 1972 and 1979. These plants competed against each other as well as the major companies with ice stations in the area for the regional supply of roe herring, groundfish, and trolled salmon. This level of competition and its resulting undersupply to each facility tended to eliminate the locational advantage one or two small- to- medium-sized plants could potentially enjoy in this area.

The species harvested in this area did not require economies of scale in order to be profitable. Large investments became, in the case of Tofino Fisheries, a long-term liability; a smaller investment or a more gradual expansion might have stood a better chance. Large companies may have thought of the operation also in terms of economies of scale, and later learned that a large plant capacity under such competitive circumstances was not warranted.

The volatility and cyclical nature of fish markets (often related to fluctuating supplies as other countries overfish their domestic stocks and seek new suppliers while domestic stocks recover), create high-risk conditions in fish processing. It is particularly inappropriate for government and new firms to assume the existence of large and stable markets, supplies, and fishers' loyalty if they cannot offer financing to them.

A significant opportunity for long-term development in the region may exist for one or two fish-processing plants, but the policy of licensing and offering assistance to three plants in the region, while yet a fourth exists, appears to have weakened the position of all the plants, and largely dissipated the advantages the region could potentially enjoy.

LOCAL EMPLOYMENT IN FISHING

Records of fishing employment in the region are available only after the introduction of licence limitation, and the whole Clayoquot Sound area and adjacent northern area, Hot Springs Cove, residence of the Hesquiat Band, are grouped together. Between 1969 and 1981, salmon licences in the area declined from 106 to 61. The loss of fishing jobs to traditional fishing families is even more severe if one calculates that some of these jobs went to new local residents who left other professions (e.g. teaching) to take up fishing, and who could afford the rising cost of licences because of earnings elsewhere.

In Tofino the fishing labour force was bifurcated into two groups: one that had grown up with fishing and the newcomers, usually people in their thirties or forties. Unlike the newcomers, people who had grown up in fishing, whether young or old, had begun their apprenticeships between the ages of nine and twelve, deck handing with a father, uncle, or brother for five to seven years before acquiring their own boat in their early twenties. Many had not held other jobs, except very sporadically or seasonally, and perhaps 50 per cent had not applied for unemployment insurance benefits until the late 1970s when times became exceptionally difficult.

The majority of fishers interviewed in this category had also learned conservationist attitudes and skills: they viewed conservation practices as enlightened self-interest. They avoided areas that were known to be frequented by juvenile fish at specific times and otherwise fished in ways that minimized damage to rearing areas. They supported the use of barbless hooks (to permit release of immature fish) when this regulation was proposed. Young people found themselves without job opportunities in fishing when they were of age to enter the industry as skippers.

Another phenomenon related to local licence holders can be observed in Figure 13.2. At the same time that salmon licences were declining in the area, other licences (herring roe, groundfish, crab, shrimp, halibut) were increasing. Whereas in 1969 the total number of local licences was only one more than the number of salmon licences, in 1981 the total number of licences included sixty-five licences other than salmon licences. Because the same fisher often owned both a salmon and herring licence, the increase in licences might not indicate an increase in the number of fishers. The separation of fishing privileges that had formerly been included under a single salmon licence also increased the number of licences. (Halibut, for example, was separated from the salmon licence in 1979.) Therefore the larger number of fishing licences does not necessarily indicate a greater number of fishers, since many fishers simply acquired two or three licences. (Some fishers own two herring licences.) About 62 per cent of Indians possess both salmon and roe herring licences, for example. This percentage is higher in isolated Indian villages. In Ahousaht in 1983, forty-six people held sixty-one salmon and herring licences, an overlap of 75 per cent.[15]

The remaining salmon fishers are of course in a different economic position than were those who fished in 1969. A comparison of landed

Figure 13.2
Salmon Versus Other Licences

Area: Clayoquot Sound

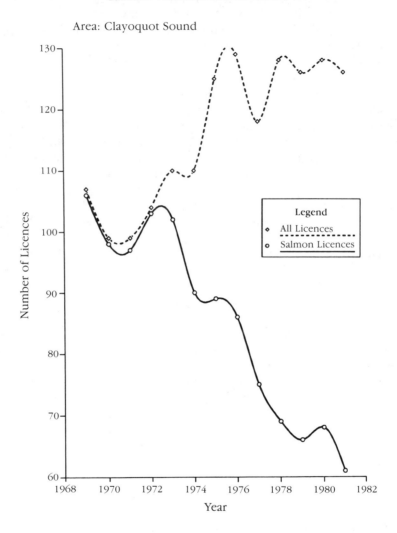

Source: DFO, Licensing data, 1969–1981.

value of catch between these years (Table 13.3) indicates the virtual elimination of low-income marginal fishers who delivered in 1969, as well as the greatly increased differential between the earnings of the lower quartile and the upper quartile of local fishers. The local fleet became more highly stratified with the increase in fishing effort.

Table 13.3
Gross Income Ranges for Local Salmon Fishers:
Selected Years, 1969–1981 (Tofino Area)

Year	No. of salmon licences	"B" category licences[1]	Income range	Median $	Mean bottom quartile	Mean top quartile
1969	106	32	$ 6.00–$ 13,599	4,435	$ 605	$ 8,805
1972[2]	n/a	18	90.00–$ 33,433	4,814	$ 810	$16,188
1976	86	7	$ 97.00–$ 45,498	18,362	$3,705	$27,301
1979	66	2	$817.00–$ 95,432	30,852	$6,069	$53,560
1981	61	3	$ 13.00–$103,625	15,830	$5,200	$53,599

[1]Delivered less than $2,500 worth of salmon.
[2]Although incomplete data exists for this year, the incomes are probably representative of all licences.
Source: Department of Fisheries and Oceans, licensing and catch statistics data, unpublished.

AHOUSAHT: THE STRUGGLE FOR SURVIVAL

While Tofino can, to a limited extent, soften the impact of job losses in the fisheries because of government services and tourism, and partially redistribute the same labour force into these industries, this opportunity does not exist for Ahousaht. Like many Indian communities, Ahousaht is both more dependent on fishing and more vulnerable to programs such as licence limitation than its non-Indian counterparts. As with many present-day Indian reserves that were established on traditional winter village or summer fishing sites, people do not live in Ahousaht because of new job opportunities, but because of old ones. The Marktosis Reserve, as it is sometimes called, has been home to several once-separate tribes (Kelsemat, Ahousaht, Manhousaht) since boat moorage became an important issue in the 1940s. Prior to that time, many people still fished commercially from canoes that could be drawn up on the beach. An Ahousaht elder, born in 1908 in a traditional longhouse on Vargas Island just north of Tofino, declares that he grew up in a world "much like that which existed before contact with Europeans" (Webster, 1983:17, 41). People still pursued the seasonal fishing, hunting, and food-gathering round of activities in a variety of sites, and "it was very difficult for anyone to make a living in the industries run by the Whites."

Nonetheless, during the height of the pilchard boom, some forty to fifty Ahousahts worked in the reduction plants at least part of the year. After the collapse of this fishery in 1948, a number of Ahousahts went out a considerable part of the year to work in logging camps in Kyuquot,

Tahsis, Nootka, and Ucluelet. According to an Ahousaht elder, the conversion of most of these camps in the 1960s to nonresidential camps, drawing labour from family-based company towns, severely curtailed this job possibility for Ahousahts. Those who did continue to find jobs had to contend with a company policy of giving fishers last priority in hiring (Sparrow, 1976). Ahousaht was thrown back upon fishing for its survival.

Ahousaht is not accessible by road, and water transportation to Tofino is across open ocean often rough in winter storms. Hence it is unusually difficult to live in Ahousaht and commute dependably by small craft to a job in Tofino or further afield. It is not surprising, then, that almost half of the Ahousaht band (which totals about one thousand people) live off-reserve. Three-quarters of off-reserve residents live in Port Alberni, Tofino, Ucluelet, and Victoria. The two main reasons for off-reserve residence, according to the Ahousaht Education Study (Cannon, 1980), are lack of employment and lack of educational opportunity.

The most important employment opportunity is fishing, which also provides transportation for the fishing-boat owner that is more dependable than that of a small craft. The drop in on-reserve residence after the 1971 buy-back program is said to have resulted from loss of fishing employment and from the dependence of smaller, less mobile boats on larger boats for packing.

The availability of on-reserve housing has been an important limiting factor, since many people move back to Ahousaht as soon as a house becomes available. Population has been expanding faster than housing. A young person who stands in line for a fishing boat may also leave Ahousaht to make money to buy the family boat at lower than market price, usually from an older male relative who is graduating to a larger boat.

Despite the losses in fishing employment related to the Davis Plan, Ahousaht is the largest of the fifteen Nuu-chah-nulth bands and has the largest number of fishers. Of the eighty-five trollers owned by Nuu-chah-nulth fishers in 1981, thirty-eight were operated by Ahousahts. In addition, some fifteen to twenty unlicensed (contrary to DFO regulations) small trollers — punts, speedboats — called "putters" operate out of Ahousaht, making it also the largest "putter" fleet on the west coast of the island. Only two salmon trolling licences are actually possessed by Ahousaht putters.

The putter fleet fishes close to the village, but the majority of the licensed troll fleet also prefers to fish locally, even if there are larger supplies elsewhere.[16] The Ahousahts consider that their style of fishing is efficient because they spend less time travelling, spend less money on fuel, and because their small, less-capitalized vessels apply fishing effort in a manner that adjusts itself to what local stocks will support.

A larger troller or seiner could be considered more efficient than these small trollers if there were a local supply large enough to support its much larger investment. Even if a larger vessel had to travel great distances to catch enough fish, it could be highly efficient, if fuel were cheap relative to the price of fish. But if fuel prices rise (as they have in the last five years), and if supplies are low and thus openings are less frequent in any one area (as they have been in the last five years), fish prices must be very high for many large, highly capitalized vessels to remain viable operations. The smaller vessels have often been better able than the larger ones to survive low supplies, fluctuating prices, and high fuel costs in the 1980s.

According to one of the licensed putter fishers, his costs are about $4,000 a year on an engine, and another $3,000 on fuel and gear. Depending on how many days of good weather he gets and how many weeks he works, he could theoretically clear up to $16,000 (four days a week for thirty weeks). In a 1977 study, it was estimated that putter fishers' net income ranged between $300 and $10,000 (Lewis, 1977). The variance in these estimates may be because of a combination of seasonal supply of fish and seasonal availability of licensed trollers through whom the unlicensed putters can deliver. At the height of the season, the more highly capitalized trollers who normally deliver fish for the putter fishers have to move farther away from local territories for their fish, and thus cannot service the putters as often. This phenomenon of a little-capitalized, small-boat fleet surviving in an isolated area on low costs and local supply is similar to that at Bella Coola in the central coast region. There, as well, some of the local fleet have eventually become more highly capitalized and more mobile during part of the year in an attempt to try their luck at competing in the larger, more distant openings around the coast. They did so, however, only in the late 1970s and early 1980s when local fish supplies were low. In Ahousaht, this greater capitalization of some of the local fleet seems to have occurred somewhat earlier.

Skippers on the larger trollers (about forty feet) have low opportunity costs in a strictly commercial sense. However, in the off-season, they use their boats to service local needs in several ways. They provide free transportation to guests at community events that occur on a monthly basis (marriages, funerals, memorial services, sports events, and various celebratory feasts and meetings). These events are crucial in maintaining ties with off-reserve relatives and friends and in enabling participation in wider cultural and political events. They provide transportation at cost (covering fuel expenses) to community members for food fishing and commercial clamming trips; clamming is an important source of winter income, and food fishing significantly lowers the cost of living. The distribution of food fish to elderly and disabled people also plays an important role in informal social assistance and in the affirmation of cultural values. In some food fisheries, the band allocates fish to different families and can thus play an important redistributive role to partially compensate for the inequalities in access to job opportunities and job success. Skippers also provide a means to transport freight, beachcombed firewood and other resources, and to conduct hunting trips. These are important inputs into the local subsistence and informal economy.

The licensed trollers also demonstrate an unusually high use of labour per resource unit harvested (Lewis, 1977). That is, the Ahousaht fleet does not catch a higher percentage of fish per boat (in fact, incomes per boat are well below the provincial average on 93 per cent of the trollers), but it does employ a greater number of deck hands per boat than the coastwide average. While the overall troller average is 1.2, in Ahousaht, where the deck hands are usually close kin, it is 2.3. The Ahousahts thus employ more people on fewer resources. As capital investments in fishing boats are used to provide multiple nonmonetary inputs, so capital investment is used to provide more jobs in the community.

Multiple deck-hand hiring serves important functions in simultaneously apprenticing several sons. Inheritance rules dictate that fishing boats and gear are to be shared as equitably as possible in a family. As there are more sons than boats in one family, brothers leapfrog each other into a series of boats, which are passed down in the family as older brothers are able to acquire other boats. The extended family, not the individual, is the unit of survival. The youngest of five brothers in one family, for example, deck handed for nine years on an uncle's boat before owning a boat in partnership with his brother.

Ahousaht fishing careers tend to begin earlier than Tofino careers, sometimes at age eight, and apprenticeships also tend to be long: nine to twelve years is common. The putter troller can also be an apprenticeship route, as well as a retirement route for an older fisherman who makes way for a son to run the boat.

While inheritance of fishing boats and deck-hand jobs is usually confined to the extended family, the unit of information sharing and mutual aid on the fishing grounds is the entire local fleet and community. In the case of Ahousaht, a few friends and in-laws from the neighbouring Hesquiat and Clayoquot bands are also part of this support network.

The major network of communication is the ship's radio, used in intravillage communication (more than the telephone) as well as ship-to-ship and ship-to-shore communication. In fact, the technology that seemed most central to the Ahousaht household besides the stove and the automatic coffee-maker is the ship's radio.

The village community has relevance to fishing in ways more important than information sharing. Perhaps the most important way in which the community acts as a unit is in its shared perception of being disadvantaged as a group by fish-management policies, such as the Davis Plan, and the failure to settle aboriginal claims (Chapter 11). Ahousahts believe that the people who benefited from fleet rationalization were not themselves and that they were disproportionately disadvantaged by the elimination of the marginal fishers. Their marginal fishers had no occupational alternatives and were either forced to leave the community or to go on welfare. This inevitably created more work for the rest of the community, who then had to support them and experience the effects of demoralization, as well. The Ahousaht fleet has declined from a situation where "every person that was able fished: it wasn't the best boats going, but they didn't have to go far," as an Ahousaht welfare officer remarked. By 1977, sixty-five local residents with skippering experience sought entry into fishing but did not have licences (Lewis, 1977).

The community has reacted to government policies as a group, whose survival is threatened. They acted in defiance of the Davis Plan, opposed the Pearse Commission, practised tax resistance, and pressed charges against a multinational logging company that was using a herring-spawning area on their island as a booming ground. Their political ideology unified them, even as their unequal job opportunities divided them. Their sense of common stake mitigated the stratification created by job loss among those who had been unable to retain fishing licences after the Davis Plan and among those who were competing less successfully.

The tension between financial independence and the necessity of supporting other community members was expressed as follows by one fisher: "I bawled him out when I found out he stole my net, but I didn't take it back, because I was doing OK and he was barely making it. It was punishment enough my speaking that way to him." This same fisher publicly berated the author of the Davis Plan for "making criminals of us by depriving our low-income fishermen of licences," and challenged Commissioner Pearse when he was in Ahousaht to correct the situation. As the Ahousahts' chief counsellor put it: "Nobody here thinks their fate is sealed. We all believe in struggle." The "we" in this statement expressed the identification of the whole community with the employed segment, and their sense that as the fate of the fishers went, so went the fate of Ahousaht.

THE LOCALLY RELEVANT FUNCTIONS OF FISHING AND FISH PROCESSING

As is evident from the preceding discussion, fishing has a social significance in Ahousaht that goes beyond employment. Fishing in Ahousaht is close to being a total institution: it involves a complete pattern of living, a comprehensive organization of behaviour. To a lesser extent, this is also true of fishing and fish processing in Tofino. In order to compare the social importance of the fishing industry in these two communities to each other, and potentially to other places, we need to consider the major social functions of fishing-related activities. Warren (1972) identifies five major social functions that are dimensions of community when these functions are organized on a locality basis: production-distribution-consumption, socialization, social control, social participation, and mutual support. In Ahousaht, fishing performs these functions in a manner that has a high degree of locality relevance.

1. *Production-distribution-consumption*. Fishing in Ahousaht is central to the overall organization of all three of these functions. Although much of production from commercial fishing is exchanged for cash and spent on the welfare of the fisher's family and the fishing boat, that vessel and even the labour of the family can also be called upon by the rest of the community at important times. To the extent that there is distribution in the community outside the welfare system (which is now the principal means of support for unemployed families), it occurs through fishing or the fishing industry. Food fish is distributed through corporate band mechanisms. Fishing boats provide the means for distributing transportation, jobs, and access to subsistence goods. The consumption of food fish at feasts symbolizes the identity of the band, the contributions of individual members, and the rights band members exercise over the labour of other members.

2. *Socialization*. Fishing is the chief means by which young men are socialized into work roles and social responsibility toward family and band. Young women participate less than men in commercial fishing but just as much in clam digging, and more than men in fish processing and preparation of feasts. Providing for family and other band members in this way is the most important way Ahousahts achieve social competence.

3. *Social participation*. Being part of the work of fishing and fish processing and being part of the conversation about the entire enterprise, including boat building and maintenance, successful gear, weather, fishing boats, strategies, and financing, are the most important ways a resident of Ahousaht identifies with the community and as part of a social group. Fishing includes old and young alike. Both the apprentices and the elderly can assist in providing food and a little cash by their participation in the putter fleet. All ages also sortie on the boats to dig clams. The ability to contribute fish to feasts is an important form of social participation, as well.

4. *Mutual support*. Aside from the obvious function of mutual support provided by the contributing of food fish, fishing is the activity around which much mutual support is organized. On the fish grounds, this consists of sharing information about the location of fish. It also involves any assistance required by a fisher in distress on the grounds or in need of advice or loan of tools for boat repair. Any member of the Ahousaht community will drop any activity to assist an endangered band member. This mutual support is of course widespread throughout the entire B.C. fleet. Ahousaht differs in that greater attention is paid to the whereabouts and conditions of the local fleet at all times. The Ahousaht fleet may also act as a bargaining unit at times, such as the 1979 herring season, when fishers were able to get a higher price by bargaining together and delivering their fish to one buyer during the most important opening.

5. *Social control.* The obligations entailed by having a fishing boat constitute a form of social control of fishers by the community. Fishing provides the main arena in which community members have rights over one another's labour and even capital. In sum, fishing and related activities in Ahousaht perform crucial social and economic functions that serve to organize much of local life.

In Tofino, fishing also has some degree of locality relevance, but there are more choices and a greater degree of social reference to the outside economic and social world than exist in Ahousaht.

The locally relevant functions of Tofino Fisheries' plant are less easily distinguished, since this plant generally operated according to the world system of markets and finance. To compete successfully against larger companies with ice stations in Tofino, however, the plant needed to adapt to local conditions and use local resources, human and natural, in a relevant fashion. The simple fact of organizing packing and collecting on such a basis enabled Wingen to continue these operations when it had become uneconomical for other nonlocal companies. Wingen attempted to involve fishers in the exploitation of new local fisheries in which they would have a natural advantage. He also organized payment schedules and credit in ways adaptive to local circumstances, permitting local people to manage their jobs more effectively. On the occasion of the receiver attempting to stop plant operations and take possession of all the goods, Wingen organized shoreworkers to picket the plant and prevent seizure of undelivered frozen fish, in order that they might receive wages from the sale of this fish.

PLANS FOR THE FUTURE OF TOFINO AND AHOUSAHT

As the local fleet continues to shrink (three repossessions occurred in 1982 and others have followed), the Ahousahts seek alternative paths of economic development. A corporation that would own the licences and retard the repossession process is being considered (Chapter 11), as is renting licences from the Northern Native Fishing Corporation. The legalization of the putter fleet, which may show a larger catch per unit of effort, would reinforce local attempts at efficiency.

The Ahousahts have practised small-scale salmon enhancement of nearby Anderson Creek since 1980, when they began releasing an average of 20,000 chum salmon fry per year. High school students and band members work with a school project on this enhancement, collecting the eggs from the Atleo River each fall, fertilizing and incubating them in Ahousaht over the winter, releasing them in the spring. It is hoped this will grow into a larger project in which young people become involved in stream clearance, stock inventory, and eventual management of streams in the Ahousaht area. The Nuu-chah-nulth Tribal Council and the Department of Fisheries and Oceans built a larger hatchery on the Kennedy River in 1984, from which "satellite" operations on many nearby streams are conducted; eggs from very small runs can be taken to the hatchery to incubate and the fry released into the stream of origin.

Tofino fishers also have been involved with salmon enhancement, at first through the Thornton Creek Enhancement Society, which organized around the chum hatchery near Ucluelet. This hatchery was originally run by the Department of Fisheries and Oceans, but was later taken over by the local society, which hires a manager. Through the efforts of the Tofino and Ucluelet trollers involved in the hatchery, it was eventually possible to use the hatchery for satellite enhancement of depressed chinook and coho runs in adjacent streams. The department was at first reluctant to support these efforts, since chinook and coho require longer incubation periods than do chum and are therefore more difficult and expensive to raise. By 1984 the society was releasing chinook and coho fry in two creeks in Barkley Sound and two in Clayoquot Sound. They intend to continue the enhancement of three more cycle years in these creeks and also begin the enhancement of chinook and coho in additional creeks in Clayoquot and Barkley Sounds. In the last two years, the Department of Fisheries and Oceans has been highly cooperative and supportive of these plans. By 1985, the Thornton Creek Enhancement Society, the Nuu-chah-nulth Tribal Council, and the Hesquiat Band had released 73,000 chinook and 307,520 coho through various contracts with DFO and Canada Manpower.

The fishers in Ahousaht and Tofino involved in these projects pin their hopes on being able to increase the local supply of fish before negotiations within the Pacific Salmon Commission reduce the interception of the salmon on which they now depend. (The roe herring fishery in this area was closed in 1985 and 1986 to allow stocks to recover.) The main reason for west coast reductions or closures is that a large provincewide salmon fleet concentrates there. However, if local stocks do not recover rapidly enough before 1992, and if sufficient intercepted stocks are not allowed, it will be essential for the west coast area to develop a means by which at least their own fishers' futures can be protected. (The Nuu-chah-nulth fleet took 4.2 per cent of the salmon in Areas 23 to 26 in 1982; by extension, the entire locally resident fleet might take less than 10 per cent of the salmon caught in Area 24 and in the entire region.) In addition, a mechanism could be developed whereby nonresident trollers with a strong commitment to the area could be incorporated into local developmental efforts.

The desirability and probable necessity of making a transition from an interception fishery to a locally supplied fishery suggest that the successful Alaska Regional Aquaculture Association model could be fruitfully attempted on the west coast of Vancouver Island. Alaska responded to historical low salmon catches in the mid-1970s with area licensing and legislation permitting the formation of user-group regional associations for building hatcheries. These created the incentive for fishers in the local area to improve the stocks in that area without having to share the improvements beyond it. (Only licence holders in the area may fish there, although fishers may own licences in more than one area and may buy and sell licences. Licences were originally issued to area residents on a point system.) Fishers licensed in the southern southeast Alaska area contributed 3 per cent of their catch income toward stock enhancement (in this case, coho, chinook, and chums), while the State of Alaska supplied a thirty-year loan to set up hatchery operations. The first loan payment fell due in six years, following the first return from the five-year-cycle fish. But in the southern southeast area, returns to the hatchery through the terminal harvesting of fish and the 3 per cent royalty had already produced enough to make the loan repayment by the fifth year.

Through development of a harvesting plan approved by the Board of Fish, a citizen's policy-making body, the Southern Southeast Regional Aquaculture Association fishers can decide where to harvest the fish — either farther out at sea or terminally — as the fish reach their streams of origin in the smaller Special Harvest Area over which the association has exclusive rights. Thus there is some incentive for fishers to invest in competitive gear and large boats, since they have a mechanism to catch terminally through their association. They are motivated to harvest in the most efficient and cost-effective manner and also to judge which fish give a better return — because of higher quality — when taken farther out.

The Southern Southeast Regional Aquaculture Association has also created a united fishers' force that has considerable voice in regional policy decisions through the Regional Planning Team for the area (deciding what species to enhance and planning the overall development of the area). The association provides half the membership of the Regional Planning Team, and their agreement is necessary for the Alaska Department of Fish and Game (the other half of the planning team) to finalize plans. This has proved a powerful incentive for fishers to work together in planning the development of the region.

The increased predictability of production and the corporate ownership of the product when terminally harvested have created different and more stable marketing arrangements and an opportunity for the group to sell to the highest bidder. A marketing arrangement for chums has developed with Korea, as a result. Formerly this was not possible for individuals, who might be obliged to sell all their fish at lower prices to companies to whom they were indebted (Chapter 6). In their terminal harvest within the Special Harvest Area, the Aquaculture Association charters some of its own fishers to harvest for the group or takes a higher percentage (30 to 50 per cent) of the catch of individuals to repay the loan. But trollers and others who choose to fish farther out also benefit from the increased returns (e.g. six million coho were expected to be released as smolts in 1985, representing an estimated return of 900,000 adults of which 675,000 will be caught in the common-property fishery and 225,000 will be caught in the terminal harvest area). Many fishers keep their traditional arrangements with processors when they harvest outside the Special Harvest Area (Langdon, 1984; W.Griffioen, 1984 & 1985 pers. comm.; SSRAA 1985 Harvest Management Plan; Pinkerton, 1985).

The Alaska Regional Aquaculture Association model suggests a route whereby fishers can achieve greater equity and efficiency in their harvest, while organizing themselves to improve the resource base and participate fully in its management. Such a route allows efficient small-scale harvesters to persist alongside larger harvesters and fosters rather than diminishes the development of the entire region. Such a route empowers fishers, rather than nonfisher outsiders, and creates incentives for existing fishers' groups to work together. And, as Jackson (1984) notes, there is no inherent contradiction between such a user-group development corporation and unions.

West coast trollers may be persuaded to this route by world markets if nothing else in the near future. Norway's expanding production of pen-reared Atlantic salmon, and its ability to put fresh salmon of any size on the market at any time of year had already made inroads into B.C.'s traditional fresh and frozen European markets; it now threatens even their North American markets. In 1983 and 1984 Norwegian salmon were the price leader in the Seattle fresh/frozen salmon market, and approached west coast chinook in actual price (P. Heggelund, 1985 pers. comm.; see also Chapter 7). Atlantic salmon are already farmed on a small scale in the western U.S. and in B.C., and the original half-dozen B.C. farms for Pacific species had increased to thirty by 1984. By spring 1985 a salmon farm in Clayoquot Sound was being operated by a nonresident nonfisher, and applications for pen-rearing sites on the B.C. coast soared in the following year. If fishers do not organize themselves for a vital role in such new developments, these may occur anyway, and in a fashion not necessarily in the best interests of local fishers. Development that will be in the interest of the local communities will, however, require the full support of the state, especially in enabling legislation (orders in council) and in making long-term loans available.

CONCLUSION

In examining the nature of fishing dependence and the socio-cultural and economic effects of policies that disadvantage isolated rural fishing communities, I have simultaneously considered how people in two fishing-dependent communities organize fishing and fish-processing activities to exploit the natural and human resources of the region. Focussing particular attention on fishing in Ahousaht and on fish processing in Tofino, I have shown ways in which the industry in these locations provides both a livelihood and a social fabric for communities. I have also shown that significant opportunities for development of the local fishery exist, within a framework of appropriate local involvement and government support.

NOTES

1. I returned to Tofino for the month of August 1982 with coworkers to administer survey interviews to shoreworkers in Tofino and Ucluelet, and in January 1983 spent two weeks in Tofino and Ahousaht doing survey interviews with fishers. As time permitted, I used the survey interviews as an ethnographic instrument, eliciting other information, as well. Since the interviews were lengthy and demanding, however, ethnographic research was fairly limited during these later visits. I returned for a weekend of final interviewing in April 1984.

2. There are also many geographic areas that are fishing-dependent in a different sense or to a lesser degree: areas such as the Gulf Islands, which have difficult transportation links and few occupational opportunities, are excluded. Areas that have long histories of fishing dependence (such as many areas of the east coast of Vancouver Island) but are within commuting distance of less-dependent areas are excluded. The discussion is restricted, however, to those areas that have an extreme dependence on fishing, that have in the past been culturally defined by fishing, and whose future existence may be actually threatened by changes in fishing policy or the location of fishing-related facilities. These regions are: (1) the Queen Charlotte Islands, including the village clusters of Old Masset/Masset and Skidegate/Queen Charlotte City, (2) the Nass River, including the villages of Aiyansh, Greenville, Canyon City, Kincolith, (3) the isolated north coast villages, including Port Simpson, Kitkatla, Oona River, Hartley Bay, (4) the upriver Skeena villages, including the Hazeltons, Kitwanga, Kispiox, Moricetown, (5) the central coast, including Bella Bella, Klemtu, Bella Coola/Hagensborg, Rivers Inlet, (6) the islands and isolated coast off the northeastern coast of Vancouver Island, including Alert Bay, Sointula, Simoon Sound, Kingcome Inlet, Minstrel Island, Sullivan Bay, (7) the northern Georgia Straits area, including Blind Channel, Stuart Island/Big Bay, Surge Narrows, Refuge Cove, Mansons Landing/Squirrel Cove. A few of these areas have experienced logging booms for periods of time and minor logging for longer periods. Yet logging has been unable to sustain the majority of the population in the long run.

3. The local saying, "You can't make it on tourists alone," is becoming less true in 1984. An $8 million federal grant to supply sewer and water to the entire district municipality — seven kilometres, including the area between the former village of Tofino and the park — will establish the infrastructure for the expansion of tourism.
4. As a submission to the Pearse Commission.
5. The circulation of shorework wages was not studied, but Jeffrey Halpern's (1984) study of local versus capital-intensive tourism in an isolated small town in western Ireland showed that income from small-scale locally owned tourist facilities circulated seven or eight times in the community.
6. Lemmens Inlet on Meares Island has been identified as the most favourable site for maricultural development in the region. If logging occurs, mariculturalists believe mariculture will not be possible.
7. In 1985 this injunction to prevent logging on Meares was at first denied, but an appeal was later upheld on the grounds that logging would destroy evidence of continuous aboriginal use, which the Indians claimed in their negotiations with the government.
8. This is an estimate based on an area population of about one thousand.
9. Between 1976 and 1985 there were five commercial chum fisheries, of one or two days each, taking between 22,000 and 125,000 pieces annually. Whether as a result of this or not, escapements after 1980 remained below 73,000. Low chum abundance in the 1960s was a coastwide phenomenon.
10. Two elderly Tofino residents disagree on the date.
11. Edmunds and Walker had been a part of B.C. Packers up until 1931, at which time they took over all of B.C. Packers' fresh and frozen operations, and worked as a semi-independent company for the next nine years (Bell, 1981). It would appear that the larger company did not wish to divert capital during the 1930s into fresh and frozen operations but that it maintained the option of re-entering this market as it became more profitable in the 1940s.
12. The International Pacific Salmon Commission was a joint Canada-U.S. body, set up under a 1940s treaty, which collected data on the pink and sockeye stocks of the Fraser River and ensured that they were divided equally between the two countries.
13. After 1978, when enhanced stocks were first taken by the net fleet, Barkley Sound supported a growing gillnet and seine fishery, largely assisted by a major chinook hatchery on Robertson Creek, and successful fertilization for sockeye on Great Central Lake. Fertilization of Kennedy Lake in Clayoquot Sound has not succeeded in increasing the sockeye there significantly.
14. An ice station is a modern troll camp, consisting of a small storage shed on a moorage facility with an ice-making machine and road

access. The ice station merely supplied ice to vessels, purchased fish, iced it in small tote containers, and trucked it to brokers, to public freezers, or even to a company's plant in Vancouver.

15. However, in the Tofino area, there are a number of fishers specializing in crab and other species, and to a limited extent the total number of licences in 1981 (126) does reflect greater employment in fishing than indicated by salmon licences. I estimate that seventy-five to eighty licenced fishers-skippers live in the Clayoquot Sound area, including sixty-one salmon fishers. Most of the difference between 80 and 126 is made up by personal herring licences applied to herring skiffs, however (see Figure 14.2).

16. Evidence for the preference for local fishing emerges from both interviews and statistics. The catch of the entire Nuu-chah-nulth fleet is in direct proportion to concentrations of fishers and in inverse proportion to the size of the fishery. In 1982, 41 per cent of the Nuu-chah-nulth catch was off Clayoquot Sound (Area 24), while only 24 per cent was off Barkley Sound (Area 23). Thus 65 per cent was taken in areas 23 and 24, while 91 per cent was taken in areas 23 to 26, covering 158 miles of the west coast of Vancouver Island. In 1982 the Nuu-chah-nulth fleet took less than five per cent of the overall salmon troll catch in Areas 23 to 26, but 7.6 per cent of the catch in Area 24 (calculated from Department of Fisheries and Oceans, James, 1984).

14

Regionalism, Dependence, and the B.C. Fisheries: Historical Development and Recent Trends

Keith Warriner

Essays in this volume have focussed upon the varied aspects of capital and labour in the B.C. fishing industry, their often contradictory goals, and the impact these have on the resource, the subordination of labour and changes in property relations, and the particular influence of class, ethnicity, and status in the social organization of the fisheries. Throughout, we have been concerned with the manner in which capital has subordinated labour. It has been recognized that the roles undertaken, the strategies employed, and the relative successes of these endeavours have been tempered in part by the nature of the resource: by its migratory, seasonal, and perishable qualities and its property status.

There is yet another aspect of capitalist production and organization of labour tied to the fishing resource. This is the development of subordinate relations along regional lines. Regionalism has to do with underdevelopment and the perpetuation of underdevelopment leading to regional subordination, economic disadvantage, and instability. In Canada it is widely acknowledged that pronounced regional divisions exist and that the relative advantage one area has over another has little to do with either "natural" geographic boundaries, or "logical" advantages certain regions enjoy because of a pre-eminence of accessible raw materials or of human capital (Bell and Tepperman, 1979; Campbell, 1978; Clement, 1978; Matthews, 1983; Phillips, 1982; Veltmeyer, 1979).

This chapter examines the development and persistence of regional disparities in British Columbia as they relate to the fisheries. On one level, the regional divisions existing in B.C. fisheries will be identified and examined for the degree to which one area holds an advantage over another. As well, the development of these inequalities will be considered together with an examination of recent trends. An explanation for the persistence of the regional imbalance will be sought.

On another level, the behaviour of fishers, processors, and the state will be examined in light of theoretical explanations provided for regional disparities. Dependency theory is one approach used to account for regional differences both within Canada and among nation states. Here, this theory's applicability to the Pacific fisheries will be considered with regard to the particular harvesting properties of the resource.

A DEPENDENCY MODEL OF
THE B.C. ECONOMY

B.C. has a resource-based economy. The majority of provincial income comes from the extraction and export of raw materials — lumber, minerals, energy, and fish. Even the rapidly developing tourist industry is resource based, deriving as it does from the marketing of natural products. While a sizable percentage of its economic base is concentrated in manufacturing (23.2 per cent of the labour force in 1978, Statistics Canada, 1979; c.f. McCann, 1982: Table 1.1), this overlooks the extent to which manufacturing is simply an extension of resource extraction. Further, this "manufacturing" is for the most part concentrated in low-grade processing of resources for export — fish packing, wood products (mainly sawmilling), paper products (mainly pulp and paper), and primary metals (mainly lead, zinc, and aluminum smelting). Of these principal products, only canned fish can be considered a finished product, and even this is changing as more fish leave the country dressed or in the round, fresh or frozen.

The basic structure of this economy has not changed greatly in nearly one hundred years (Shearer, 1968:14). Certainly, the emphasis has shifted from fishing and agriculture to wood, mineral, and energy products; the rate of exploitation of resources has increased; and new products for extraction have been found. Nevertheless, growth is still tied to economic expansion through increases in these resource-extraction activities. While an elaborate superstructure designed to service the resource base has developed, little true secondary manufacturing activity exists. In B.C., economic growth is tied to a straightforward linear expansion in the rate and diversity of exploitation of primary materials. Thus the pace of development is profoundly influenced by the outside world and the opportunities provided by markets beyond British Columbia borders (Shearer, 1968; McCann, 1982).

Dependency Theory

One of the central hypotheses of "world system" and "dependency" theories is that specialization in the production of raw materials is detrimental to long-term growth prospects of less developed nations (Portes, 1976; Wallerstein, 1974; Campbell, 1978; Cuneo, 1978). The key to understanding the persistence of such specialization is the nature of the subdivision of the world economy into core and peripheral regions (Chirot, 1977).

The core is characterized by early economic development and political centralization facilitating an early policy of expansionism and imperialism. Peripheral states are easily subdued and conquered; their status and economies restructured to the advantage of core societies. The periphery is characterized by an overreliance upon primary resource

extraction and export, boom and bust conditions dependent on international markets, little or no processing of indigenous raw materials, lack
of secondary industrial development, importation of technology and skilled
personnel, and a disadvantageous trade balance with the core. All these
factors, combined with the control of finance at the centre, the draining of
capital from the periphery, and an ideology based upon ethnocentrism
that denigrates regional cultures, statuses and skills, tend to entrench the
disadvantages of the peripheral region and create a vulnerable and unstable
economy.

Within B.C., the urban centres of metropolitan Vancouver and Victoria
represent the core. Situated there are central government offices, most
regional head offices of business firms and financial institutions, specialized health and educational facilities, specialized manufacturing concerns,
and the auxiliary services necessary to the operation of basic industries
distributed throughout the remainder of the province. It is from the core
that the rest of the province, comprising a vast hinterland of scattered
communities and small cities devoted to raw-material production, is
controlled. Thus the periphery or hinterland provides the real economic
basis for the province, while the core contains administrative, political,
and financial control.

This division of power has profound implications for the long-term
growth and prosperity of the hinterland. The draining of capital from the
region by distant corporations and financial institutions, plus a reluctance
by both outsiders and indigenous financiers to invest in an already
underdeveloped and volatile regional economy, tend to thwart many opportunities for further economic expansion. The fate of the region hinges on
its ability to maintain an edge on other resource-abundant competitors
throughout the world. This becomes difficult over time when the high
wage structures of other resource sectors, combined with increasing
unionization, produce an impetus for maintaining high labour costs even
when markets are softening. With resource exploitation, additional investment becomes superfluous in terms of harvesting efficiency. Once the
competitive advantage is low or raw materials exhausted, the economic
viability of the region declines rapidly. In the B.C. fisheries, there is evidence
of such a cycle.

REGIONAL DEVELOPMENT IN THE
B.C. FISHERIES:
THE RISE AND FALL OF THE
CANNERY TOWNS

The coastline of British Columbia is, comparatively speaking, an unpeopled
one. Earlier in this century, a typical Union Steamship vessel serving the
coastal communities could make sixty-five stops between Vancouver and

the Nass River in northern British Columbia. Now the Union Steamship line is gone and so, too, are most of the communities and people, many of whom were dependent upon fishing. Gone are places like Seaside, Longview, Vancouver Bay, St. Vincent Bay, Bliss Landing, Roy, Shoal Bay, Toba Inlet, Forward Harbour, Butedale, Port Essington, Boswell, and dozens of other places that once supported a cannery, a fish camp, or a few fishing boats. True, places remain—some still dependent to a degree on fishing—but the economic vitality of the coast has now shifted from fishing and fish processing to lumber and minerals.

In time it will shift again, perhaps to tourism or offshore energy development. While the overall population of the coast has grown, it has not kept pace with population increases generally in the province and has fallen well short of its earlier promise. Of the population now distributed along the coast outside of the southern lower mainland area, the majority is located in a few lumber or mineral-dependent towns — Nanaimo, Courtenay, Campbell River, Kitimat, Port Alberni, Powell River—or in the northern transportation centre of Prince Rupert.

A smaller proportion of the coastal residents relies on fishing; a much smaller proportion on fish processing (Table 14.1). While once coastal communities outside of the Fraser (lower mainland) district supported over sixty canneries (1924–25), there are now five. Of these, three are in the Prince Rupert district, and two remain along the approximately 4,500 miles of coast between the Fraser and Skeena Rivers. In the early 1890s, about 2,700 plant workers were employed outside the Fraser district, representing about 36 per cent of all plant workers. Their numbers grew to a peak of 6,200, or 86 per cent of all shoreworkers, in 1925 to 1927. After that, their numbers declined as plants were closed. Between 1953 and 1955, the eight remaining north-coastal and Vancouver Island plants averaged only 1,500 workers annually, or about 42 per cent of all plant workers. Today the levels are even lower.

Over much of the coastline, especially along the central mainland and the northwest coast of Vancouver Island, the remaining settlements are, for the most part, native villages that existed prior to the European presence in British Columbia. Significant population declines have occurred here too. Prior to the arrival of Europeans, the population of the British Columbia native tribes is estimated to have been as high as 125,000 (Duff, 1964:39). Today it is 57,000, this representing a partial recovery from a low of 23,000 in 1927 (Pearse, 1982:174).

The decline of the fishing-dependent coastal community from about 1930 signals the gradual elimination of a traditional way of life in this region, in addition to closing off employment opportunities for coastal British Columbians. Nevertheless, the process is not unusual. The rush for spoils, be it for furs, gold, or fish, leads to casualties: some abandoned cabins, a few gravestones, or some discarded machinery bearing faint

Table 14.1
Shoreworker Labour Force for Regions
for Selected Years[1]

Years	Shoreworkers		
	Fraser district	Other districts	Total
1881–1883	1,417	522	1,939
	(73.1%)	(26.9%)	(100.0%)
1885–1887	1,299	554	1,853
	(70.1)	(29.9)	(100.0)
1891–1893	4,682	2,681	7,363
	(63.5)	(36.4)	(100.0)
1911–1913	2,249	5,361	7,611
	(29.5)	(70.4)	(99.9)
1914–1916	2,162	4,639	6,801
	(31.8)	(68.2)	(100.0)
1917–1919	1,570	6,093	7,663
	(20.5)	(79.5)	(100.0)
1925–1927	1,001	6,235	7,236
	(13.8)	(86.2)	(100.0)
1928–1930	942	5,988	6,930
	(13.6)	(86.4)	(100.0)
1931–1933	972	4,007	4,980
	(19.5)	(80.5)	(100.0)
1934–1936	1,460	4,836	6,296
	(23.2)	(76.8)	(100.0)
1943–1945	2,113	3,953	6,066
	(34.8)	(65.1)	(99.9)
1948–1950	1,914	2,187	4,102
	(46.6)	(53.3)	(99.9)
1953–1955	2,092	1,495	3,587
	(58.4)	(41.6)	(100.0)

[1]Regional allocation is unavailable for some years and is based upon the proportion of canneries operating in that region.
Source: SP 1881–1883, 1885–1887, 1890–1892, etc., through 1930; Lyons, 1969; DFO base-book Table 1.

evidence of a briefly flourishing commercial past. While such a process is familiar, the particulars in fishing and fish processing are distinctive, in large part because of the nature of the resource.

From an early terminal fishery concentrated around the mouth of the Fraser River, the elusive quality of fish, combined with increasing competitiveness among processors and fishers, led to the rapid deployment of the industry up the coast. Figure 14.1 charts the number of operating canneries over time for both the Fraser region and the remainder of the coastline. During its early development, the fishing industry operated mainly on the south coast. The number of operating canneries in the Fraser district reached a peak of forty-nine in 1901. This was followed, however, by a period of rapid decline offset only by the tendency of processors to "gear up" every fourth year in response to the dominant run of the Fraser sockeye. At the same time, rapid and sustained growth in the incorporation of the processing plants occurred outside the Fraser district.

After 1917 the pre-eminence of the Fraser passed, and the canneries of the north coast, especially along the Skeena and Nass Rivers, became very productive. Still, this could not be sustained, and after 1931, these plants, too, began closing, eliminating in the process the economic base for many coastal communities. Upon entering the post–World War II era, the number of Fraser River canneries had again grown to outnumber those of the coastal hinterland, although by this time, the number of processors from both regions had dipped to pre-1900 levels. The decline continues today.

What factors led to the rise and decline of these coastal operations? While many things influence capture and processing — including the demands of a dominant socio-economic system and world markets — in fishing it has been the nature of the resource that above all has determined the regional patterns just reviewed. Fish are mobile and, once caught, highly perishable. Unlike other resources in which harvesting rights for a particular region can be formally acquired, then used as a means of guaranteeing access to the resource, such property rights have little point in the fisheries where the fish can be easily captured outside the boundaries of a designated area. In the B.C. fisheries, this condition more than any other has defined the method of harvesting.

Three main factors explain the collapse of the Fraser district as the focal point of the B.C. fishing industry after 1913 — competition, resource depletion, and habitat destruction. Until 1901 competition among processors located on the Fraser was intense, as shown by the rapid increases in the number of canning operations during this period. Between 1889 and 1892 licence limitation helped to restrict new entrants to this competition, since in addition to limiting the number of canning operations outright, a significant number of fishing licenses were tied to the canneries themselves, and inability to acquire more licences effectively curtailed the opportunity to begin new processing operations.

Enforcement of these restrictions was not without problems. The cannery owners — acting as an informal syndicate from 1888 onward — expressed the view that the supply of fish was inexhaustable (Sessional Papers, 1891), and by 1892 had so actively opposed the continued curtailment of licences that the restrictions were lifted. It was the opinion

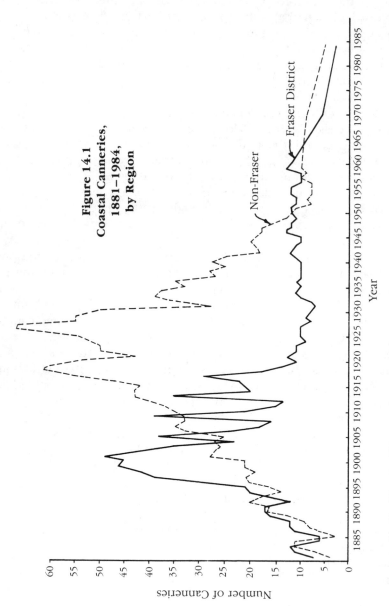

**Figure 14.1
Coastal Canneries,
1881–1984,
by Region**

Non-Fraser

Fraser District

Number of Canneries

Year

Source: Sessional Papers, 1880–1936; Lyons, 1969: Appendix

of the 1892 Commission on the B.C. Fishery that the canners' opposition to licence limitation was the result of existing operators endeavouring to gain oligopsonistic control of the Fraser harvest (Sessional Papers, 1893). In any case, with the relaxation of limitation in 1892 there was a rapid increase in fishing activities and processing. Licences increased from 765 in 1892 to 1,965 in 1893, the combined fishing and processing labour force went up from 3,458 to 8,342 during the same period, and the number of plants continued to increase, reaching 49 in 1901 (Sessional Papers, 1893, 1894, 1902). A decline in the intensity of this competition was only brought about by the amalgamation of twenty-nine of these operations (forty-two along the entire coast) by the B.C. Packers group in 1902 (Lyons, 1969).

A second source of competition for the Fraser canners came from American firms. By 1887 overfishing had led to the near collapse of the king salmon runs on the Columbia and Sacramento Rivers. Consequently, interest turned to the interception of the Fraser-spawned sockeye salmon returning via Puget Sound. Figure 14.2 depicts the rapid growth of this industry in comparison to the Fraser pack. Beginning in 1887, the American pack grew rapidly to rival and then exceed the combined pack of the Fraser canneries. In addition, the use of traps and seines (which were prohibited on the Fraser), allowed the operators to economize on fishing effort and to undercut the price of Canadian products with American and British buyers (Sessional Papers, 1902).

A third basis of competition on the Fraser concerned the number of fishers. This ultimately led to further pressure on the resource and to changes in fishing practices and in the relationships between fishers and cannery operators (Stacey, 1982). Following the unsuccessful attempt to limit fishing licences on the Fraser to five hundred between 1889 and 1892, the fishing grounds became crowded. With the relaxation of restrictions, licences increased from 765 in 1892 to a peak of 3,725 in 1898. As a result, fishers were forced from the Fraser into the Gulf of Georgia. The open waters of the new fishing grounds necessitated changes in vessel design and gears. By the early 1890s, the sturdier, more seaworthy Collingwood boat had been firmly established in the area, and a new type of gillnet, less likely to be seen by fish in the clearer gulf waters, was in use. Both these innovations were costly and led to greater capital expenditures for the fleet than when it was located in the protected Fraser waters. At the same time, the enhanced competition brought on by the growth of the fleet decreased the size of catches per unit of effort (Stacey, 1982:15).

The growth in the fleet also led to fundamental changes in relations between fishers and processors. Previously most fishers were employed by processors and operated cannery-owned vessels. With the cancellation of licence restriction, licences became available to all bona fide British

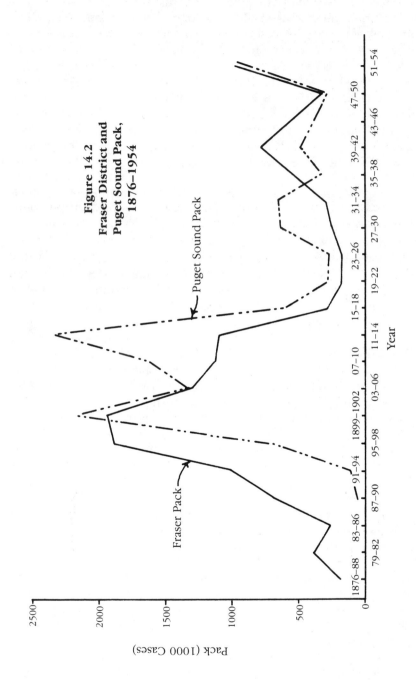

Figure 14.2
Fraser District and
Puget Sound Pack,
1876–1954

Source: Sessional Papers, 1930; Report of the Provincial Fisheries Department, 1956:K10.

subjects and the predominance of the company-owned boat gave way to the independent operator. During the final year of licence limitation, 1892, independent fishers operated 348, or 45.5 per cent of all the Fraser River fishing vessels. By 1894 independent ownership of vessels increased to 1,057, or 65.7 per cent of the district fleet, and by 1898, there were further increases to 2,820 vessels, or 75.7 per cent of the fleet. This important step in the transition of fishers from wage labour to independent, competing commodity producers carried with it all the associated risks inherent in such production, principally reflected in the need to make a profit.

The high level of competition between the Fraser-district processors and fishers, combined with American capture of fish, placed heavy pressure on the resource. From about 1900, the Fraser River pack began to show signs of significant decline (Figure 14.3). The final factor contributing to an already aggravated situation occurred in 1913 when a rock slide in the Fraser Canyon caused by railroad construction blocked the run of returning salmon from reaching the spawning grounds during this "dominant" year in the four-year sockeye cycle. Although some recovery in stocks was achieved by the 1940s, the Fraser River was never again able to reach its former abundance.

The result of this collapse, in terms of redeployment of fishing and processing effort, can be seen in Figure 14.3. In response to heavy competition and declining Fraser packs, the focus of the industry quickly shifted northward, to the Skeena, Nass, and Rivers Inlet regions. From the period 1915 to 1930, the bulk of fishing and processing activities was concentrated here, and these areas flourished. While a detailed breakdown of the labour force by region is unavailable, an estimate of the peak year, 1927, based on the number of fishing licences (5,049) and canneries (50) in the region, puts it at 6,880 for fishers (including crews) and 5,425 for shoreworkers (Table 14.1; Sessional Papers, 1927–28). This total represents 58.7 per cent of the 20,965 people employed in both sectors of the industry during this year.

In terms of the dependency model of regional development, this new concentration is unique. While the loss of competitive advantage from one region inevitably leads to the emphasis on resource exploitation being shifted elsewhere, it need not always require such a heavy reinvestment of new capital. Typically, in the extraction of more durable resources, it is possible to transport the resource to the core region for processing (excepting the possibility of state intervention or highly attractive investment terms encouraging development). This happens for example with wood and minerals in British Columbia. With the exception of some minor processing in preparation for export, the finished product—furniture, high-grade paper, steel — is produced far from where the resource is extracted. This is done in order to take advantage of economies of scale

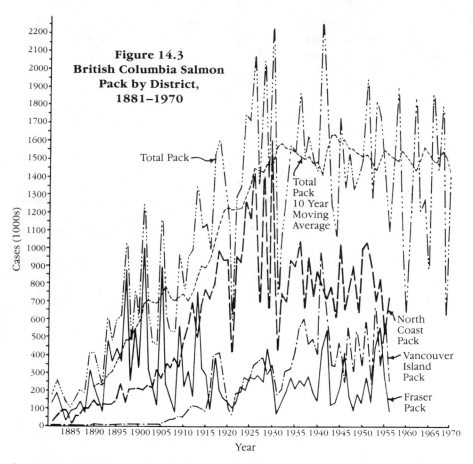

Figure 14.3
British Columbia Salmon
Pack by District,
1881–1970

Source: Sessional Papers, 1880–1930; Lyons, 1969: Appendix.

during manufacture. Also, it is efficient to have such production centrally located at the core to serve as a distribution centre for raw products shipped from surrounding regions. As a result, there is not a great likelihood of a heavy influx of manufacturing capital for advanced processing of materials to take place at the point of extraction for most raw resources. However, in the B.C. fisheries between 1917 and 1930, the central and northern coastal regions were an exception.

The basis for the anomoly is the perishable nature of fish. Once captured, fish must either be eaten or preserved within a short time. At the time, the available refrigeration technology made it uneconomical to transport fish caught in northern waters to sourthern processors. This, combined with the depletion of southern stocks and the requirement to process fish near to the point of capture, forced the relocation of both processors and fishers. The result was that, for a period, the north and central coastal region held the promise of long-term prosperity and economic stability. But it did not last.

A disastrous season in 1931 signalled severe reduction in stocks along the central and northern coast. Like the Fraser, this region now slipped into relative decline (Figure 14.3). While still important, the pack of the north and central coast averaged 198,980 cases for the fifteen-year period 1931 to 1945, 20 per cent less than during the 1916 to 1930 period. More telling, however, is the response of processors to this second, partial collapse. As indicated in Figure 14.3, more processing effort was directed after 1931 to the region of Vancouver Island. Between 1931 and 1945, the annual pack of this region increased to 461,600 cases from an average of 275,430 cases between 1916 and 1930, an increase of 67.7 per cent. Largely as a result of this, the overall pack from all west coast Canadian salmon fisheries declined only slightly after 1930. Between 1916 and 1930 this total Pacific pack averaged 1,520,110 cases, while from 1931 to 1945 the average was 1,487,380 cases, a decline of just 0.9 per cent.

The results of all this regional redeployment of processing effort are depicted by the line representing the ten-year moving average for the total west coast pack from 1881 to 1970 in Figure 14.3. This line smooths out the pronounced pack-size fluctuations caused by both the migratory and seasonal properties of fish and interceptions by foreign fishers. In this way, the figure depicts the level of long-term stability in the industry.

As shown, the British Columbia processors experienced sustained growth until about 1930, followed by high and—despite all the fluctuations —quite stable yields thereafter. This was achieved in spite of the problems just described: competition, pressure on the resource, and the decline in the Fraser runs because of habitat degradation. Further, it appears that much of this success is attributable to the behaviour of processors and fishers who, within a short time following declines in yield of one area, were able to rapidly shift the productive focus to a new region.

What is absent so far in this account is the economic and social impact on an area once capital withdraws. Such a redirection of capital can be more easily weathered in an urban area such as Vancouver because of its more diverse economic base. By contrast, many residents of remote coastal communities were wholly dependent upon fishing and fish processing for their livelihoods and found it impossible to survive this decline in regional employment.

Finally, let us examine the role of fishers throughout all of this. While the closing of a community's cannery may have eliminated two hundred jobs, it did not abruptly end the means of livelihood for fishers based in the area. Opportunities to fish remained, and processing facilities could be sought out. With improved freezing and cold-storage facilities aboard tenders and fishing vessels, the transportation of fish over greater distances became possible. Hence there was additional reason for further processing concentration, and, by the early 1950s, this sector had once again become primarily urban based in the centres of Vancouver and Prince Rupert.

Between 1930 and 1950 fishers acted to buffer the decline of the west coast, rural-based, fishing economy. Even though the number of coastal canneries and their associated labour forces dropped rapidly after 1930, there were, nevertheless, gradual increases in the number of fishers (Figure 14.4). Along the north and central coast, no major decline in the number of fishers occurred until about 1940, by which time the number of operating canneries in the region was half the number in operation ten years earlier. On Vancouver Island, increases in the number of fishers continued until the late 1940s, at which time a postwar surge of new, mainly Vancouver-based entrants drove the number of fishers to an all-time high. From this point, the numbers of Fraser-based fishers remained quite stable at between 3,500 and 4,000 throughout the late 1970s.

Conversely, from about 1950, the population of fishers located outside the Fraser district declined rapidly, even falling for a time below that of the Fraser region, prior to experiencing a moderate recovery in the mid-1970s. What happened to the individuals so affected is unclear. Perhaps some became involved in logging and mining as these districts shifted their economic emphasis. For example, the growth in the number of fishers in rural districts from about 1974 is attributable largely to increases in the size of fleets attached to the lumber and mining towns of Vancouver Island (Warriner and Guppy, 1984). Other fishers no doubt were forced out of the region altogether and resettled in Vancouver. So, for awhile, fishers avoided contributing further to the rapid decay in the coastal fishing economy, but in time their participation in the industry shifted and became much more urban based.

THE EXISTENCE AND PERSISTENCE OF REGIONAL INEQUALITIES AMONG FISHERS

We can now go on to examine current conditions. This investigation is limited to fishers, since, outside Vancouver and Prince Rupert, very little remains of the processing sector. For this purpose, Department of Fisheries and Oceans catch and licensing files from 1967 through 1981 are used to compare the economic circumstances of fishers across various regions. Here there are two goals. First, the investigation will serve to demonstrate the degree to which disparities exist between urban-based fishers and their rural counterparts. The point is to identify the ways in which hinterland participants in this industry derive less benefit and face greater financial disadvantage because of regional locale. Variables such as landings, debt, and capital investment will be examined across various regional categories.

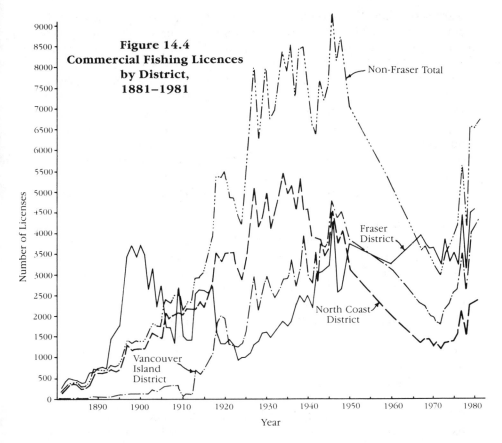

Figure 14.4
Commercial Fishing Licences
by District,
1881–1981

Non-Fraser Total

Fraser District

North Coast District

Vancouver Island District

Number of Licenses

Year

* For period 1949 to 1966 licence figures by region are only available for 1950 and 1960.
  This accounts for the lack of more pronounced fluctuation in licences in the graph for
  this period.
Source: Sessional Papers, 1880–1930; W. Sinclair, 1971: Table 1.4, Graph 1.1; Department
  of Fisheries, 1958: Table 1.

Next, we examine how changes within the industry during recent
years have affected this regional imbalance. Some disproportionate effects
may well be expected because the past fifteen years have been turbulent.
Commencing in 1969, the Davis Plan, which restricted licensing of the
salmon fleet, had a pronounced impact on the industry, contributing to its
current disorder. During the first thirteen years of this limitation, 1969 to
1981, the size of the salmon fleet decreased by nearly 27 per cent. At the
same time, there was an expansion in the fishing capacity of the fleet as
well as a real dollar increase of more than 300 per cent in the combined
average value of vessels and licences. Overall, the number of vessels of all
types, especially multiple gear boats, increased. Salmon seine vessels

increased from 390 in 1969 to 552 in 1981, with consequent increases both in the average size of catches and the number of sets that could be performed each fishing day. In addition, further concentration of the processing sector took place so that by 1981, nearly 53 per cent of all fish production was accounted for by just three urban-based firms. Finally, an explosion in demand for herring roe between 1972 and 1979, combined with an excellent salmon harvest in 1973, contributed to an optimistic mood among fishers that stimulated further fleet expansion and escalation in the market value of vessels (McMullan, 1984; Chapter 6).

Several summary indicators of regional conditions that currently exist among fishers are provided in Table 14.2. Here, variables representing gross earnings from fishing based upon landings from several fisheries are reported between 1967 and 1981, together with vessel characteristics and debt. The table entries are annual averages over this fifteen-year period of each of these variables subdivided by community type.[1]

Community type is classified as follows. Two types of rural communities are identified: remote, representing small, isolated communities, often largely dependent on fishing and without significant service sectors or industrial differentiation; and towns, representing more populous hinterland communities generally dependent upon large-scale mineral or log processing, with more services and acting as hinterland distribution centres. Places such as Masset, Bella Coola, Tofino, and Kitkatla are all categorized here as remote (but Esperanza is not because of its close location to the logging town of Tahsis), and communities such as Nanaimo, Campbell River, and Port Hardy are categorized as towns. These two rural community types of varying size are compared to the coastal urban centres, Prince Rupert and the Vancouver-Victoria metropolitan areas.[2]

As Table 14.2 shows, there are significant differences associated with being rural fishers. During this fifteen-year period, fishers from remote communities and coastal towns averaged less in terms of the gross value of landings from all fisheries, with the exception of the least remunerative of these, shellfish. Overall, fishers in remote locations made landings valued at 53 per cent of those from Prince Rupert and 64 per cent of those from Vancouver-Victoria. At the same time, fishers from towns had overall earnings that were 67 per cent those of Prince Rupert and 81 per cent those of Vancouver-Victoria-based fishers.

Part of the reason for the differences in earnings is that rural vessels make fewer sales. Whether this is because of a lack of fishing effort or less sophisticated vessels is not known. Rural boats tend to be smaller, have less displacement and less power, and they are on average older than urban boats. As a result, their market values are also significantly lower than urban boats. Nevertheless, they tend to be as well equipped in terms of electronic gear.

The average debt carried by rural vessels is also less than that of urban boats. However, when combined with the lesser value of their

Table 14.2
Mean Landings, Vessel Characteristics, and Debt by Community Type

Variable	Type of Community				Totals	
	Remote	Town	Prince Rupert	Vancouver-Victoria	x̄	(N)
Value of landings						
Totals $[a]	9.24	11.70	17.34	14.36	13.50	95,006
Salmon $	8.32	10.64	11.02	10.91	10.56	80,614
Herring $	8.02	14.01	21.42	18.77	16.47	11,153
Halibut $	3.71	10.28	18.85	25.68	18.91	6,497
Ground $	11.05	8.28	79.86	32.18	35.26	1,470
Other $	1.43	1.07	1.81	1.24	1.27	55,091
Price/lb.	0.58	0.53	0.45	0.47	0.50	95,006
Total sales	92.9	84.3	104.52	95.8	94.01	95,006
Total to boat $	1.09	0.99	1.21	0.93	0.99	94,109
Vessel features						
Boat $[a]	17.79	23.42	24.04	27.19	24.90	94,954
Length[b]	28.76	29.70	32.76	33.05	31.79	94,982
Net tonnage[c]	7.94	9.47	10.05	11.35	10.42	90,610
Horse power	98.63	108.73	129.64	139.89	127.13	95,006
Age[d]	18.38	18.32	16.38	16.80	17.26	92,099
Electronics[e]	2.76	2.97	2.54	2.88	2.85	95,006
(N)	11,572	20,136	10,507	52,788		95,006
Debt[f] (1970–1978)						
Debt $[a]	5.96	6.61	9.99	7.29	7.23	38,676
Equity	13.53	17.53	17.97	20.86	18.86	38,676
Debt/Boat $	0.32	0.25	0.29	0.26	0.27	38,676
(N)	5,016	8,877	3,864	20,919		38,676

[a]All values standardized to 1971 constant dollars and reported in thousand dollars.
[b]Feet.
[c]Imperial tons.
[d]Years.
[e]Sum of electronic devices.
[f]1970–1978. Debt not reported for 1967–1969, 1979–1981.

vessels, this still means that the equity rural fishers have in their vessels is less, on average, than that for urban fishers. More telling still is the ratio of debt to vessel market value. Fishers from remote regions carry proportionally higher debt than any group while earning the least. On the other hand, rural fishers from resource towns located predominantly along the southern mainland and east coast of Vancouver Island enjoy the smallest ratios of debt to boat value.

A final issue to be considered with respect to Table 14.2 concerns returns to investment. While figures on operating expenses, travel, and interest on debt are unavailable on a regional basis, a rough indicator of profitability is represented by the ratio of total landings to vessel value.

This ratio is highest for Prince Rupert–based boats but is followed by those from remote settlements and towns and is least for Vancouver-Victoria-based vessels. As well, on the basis of prices paid per unit sold, rural fishers receive more per pound of fish. Thus, while the unavoidable conclusion is that fishers from rural areas receive less income from fishing than their urban counterparts, there are also indications that they may receive greater incremental returns on the basis of invested capital.

DISTRIBUTION OF GEARS

There is also the question of whether some of these regional disparities can be accounted for by the differences in the fishing practices of rural- and urban-based fishers. Some assessment of this is provided in Table 14.3, which compares the kinds of fishing performed both for the salmon fishery and all fisheries by community type. Here it can be seen that rural-based fishers in the salmon fishery are more likely to participate in the troll rather than net fishery. This accounts in part for the greater unit price per pound they are paid for their catch. Also, seine vessels tend to be located in urban areas, thereby inflating to a degree the average landings reported for such areas. Fishers from remote settlements are more likely to engage in combination gillnetting and trolling. Their extra expenditures on gears help to explain their disproportionately high debt loads. Finally there has also been a tendency for proportionally more landings from isolated areas to be reported exclusively for the herring fishery. Since fishers from such areas also tend to operate combination vessels in the salmon fishery, these practices may indicate efforts to alleviate their disadvantaged economic circumstances to a degree, or they may simply be reflections of certain opportunities that are offered by particular regional locales. Overall, however, while some differences in the fishing practices of rural- and urban-based fishers can be discerned (Table 14.3), any conclusions drawn must be tentative, and there are few clues in this evidence to account for the majority of pronounced regional disparities described earlier.

REGIONAL TRENDS

Changes in the regional allocation of benefits over the period from 1967 to 1981 are detailed by community type in Table 14.4. Examination of this table reveals, first, that regional inequalities persist. Thus, for example, during the period 1979 to 1981 remotely based rural fishers still averaged incomes from all landings that were $6,610 (1971 constant amounts) per year less than Vancouver-Victoria-based vessels and $13,910 less than Prince Rupert–based vessels. Thus the basic fact of regional inequality has not changed.

Table 14.3
Gears by Community Type

	Type of Community				Total (N)
	Remote	Town	Prince Rupert	Vancouver-Victoria	
Salmon fishing					
Gillnet	20.0%	23.3	57.2	52.3	43.2
	(1,920)	(3,806)	(5,191)	(24,038)	(34,955)
Seine	1.8	6.7	8.2	9.8	8.1
	(177)	(1,087)	(744)	(4,524)	(6,532)
Troll	64.2	58.4	23.9	25.9	36.8
	(6,159)	(9,134)	(2,170)	(11,906)	(29,769)
Combinations:G-T	12.7	10.8	9.8	11.2	11.2
	(1,223)	(1,768)	(893)	(5,149)	(9,033)
Other	1.2	0.8	0.8	0.7	0.8
	(114)	(136)	(69)	(301)	(620)
Total	99.9	100.0	99.9	99.9	100.1
(N)	(9,593)	(16,331)	(9,067)	(45,918)	(80,909)
All fisheries					
Salmon only	72.9	68.9	72.0	76.2	73.8
	(8,440)	(13,874)	(7,564)	(40,239)	(70,116)
Salmon — other	9.2	11.4	12.3	9.5	10.2
combinations	(1,067)	(2,302)	(1,297)	(5,053)	(9,719)
Herring only	11.4	9.6	8.2	6.7	8.0
	(1,325)	(1,927)	(861)	(3,528)	(7,641)
Other only	5.2	8.9	4.9	6.0	6.4
	(603)	(1,794)	(519)	(3,170)	(6,086)
Other	1.2	1.2	2.5	1.5	1.5
combinations	(143)	(241)	(267)	(799)	(1,450)
Total	99.9	100.0	99.9	99.9	99.9
(N)	(11,578)	(20,138)	(10,508)	(52,789)	(95,012)

At the same time, some differences have become less pronounced. For example, between 1970 and 1978 (the only period for which debt information is available), the debt load of rural-based fishers increased at a faster rate than for urban-based fishers. From 1976 to 1978, the debt of remote fishers averaged 3.04 times what it was from 1970 to 1972, while for Vancouver-Victoria-based fishers, it had increased 2.51 times. At the conclusion of this period, the average debt of rural fishers was still less than that of urban fishers, but they were catching up.

Table 14.4
Trends, 1967–1981, by Community Type

Variable	Type of Community				Total
	Remote	Towns	Prince Rupert	Vancouver-Victoria	
All landings—					
Average value					
1967–1969	4.96[a]	5.02	10.21	9.40	8.24
	(.60)[b]	(.61)	(1.24)	(1.14)	(1.00)
1970–1972	6.42	7.12	12.28	11.70	10.47
	(.61)	(.68)	(1.17)	(1.12)	(1.00)
1973–1975	8.58	11.57	18.87	14.95	13.88
	(.62)	(.83)	(1.36)	(1.08)	(1.00)
1976–1978	11.83	15.43	21.76	17.75	16.72
	(.71)	(.92)	(1.30)	(1.06)	(1.00)
1979–1981	12.28	15.76	26.19	18.89	17.96
	(.68)	(.88)	(1.46)	(1.05)	(1.00)
Number of					
vessels					
1967–1969	709	1,172	813	3,906	6,600
	(10.7)[c]	(17.8)	(12.3)	(59.2)	(100.0)
1970–1972	571	1,049	770	3,567	5,957
	(9.6)	(17.6)	(12.9)	(59.9)	(100.0)
1973–1975	730	1,204	611	3,332	5,877
	(12.4)	(20.5)	(10.4)	(56.7)	(100.0)
1976–1978	973	1,583	640	3,469	6,665
	(14.6)	(23.7)	(9.6)	(52.0)	(99.9)
1979–1981	876	1,704	667	3,321	6,569
	(13.3)	(25.9)	(10.2)	(50.6)	(100.0)
Salmon vessels					
1967–1969	699	1,146	783	3,777	6,405
	(10.9)[c]	(17.9)	(12.2)	(59.0)	(100.0)
1970–1972	560	1,014	741	3,452	5,767
	(9.7)	(17.6)	(12.8)	(59.9)	(100.0)
1973–1975	613	1,021	517	2,874	5,025
	(12.2)	(20.3)	(10.3)	(57.1)	(99.9)
1976–1978	696	1,143	495	2,712	5,046
	(13.8)	(22.7)	(9.8)	(53.7)	(100.0)
1979–1981	618	1,104	475	2,434	4,631
	(13.3)	(23.8)	(10.3)	(52.6)	(100.0)

Continued

Table 14.4-continued

Variable	Type of Community				Total
	Remote	Towns	Prince Rupert	Vancouver-Victoria	
Total landings to vessel value					
1967–1969	0.81[d]	0.61	0.94	0.77	0.77
	(1.31)[e]	(.78)	(.91)	(0.91)	(.95)
1970–1972	0.76	0.61	0.85	0.69	0.70
	(1.09)	(.63)	(.55)	(0.58)	(0.66)
1973–1975	0.81	0.74	1.13	0.87	0.86
	(1.18)	(1.30)	(1.38)	(1.42)	(1.38)
1976–1978	1.37	1.39	1.44	1.21	1.30
	(2.32)	(2.74)	(2.31)	(2.29)	(2.41)
1979–1981	1.43	1.30	1.88	1.13	1.29
	(3.33)	(3.30)	(4.05)	(2.89)	(3.20)

[a]Thousand dollars — all amounts are in 1971 constant dollars.
[b]Proportion of raw total.
[c]Percentage of fleet, 3-year average size.
[d]Ratio.
[e]Standard deviation.

The same can be said for the market value of vessels. Here it is unclear whether this is the result of improvements made in the quality of the rural-based fleet, or is due simply to the escalating book value of vessels in an unstable market. For either reason, while rural fishers continued to earn less than urban ones, their gross earnings from landings from 1979 to 1981 were closer to the industry average. For example, between 1967 and 1969, fishers on the north coast earned 63 per cent that of the overall industry average, and those along the west coast of Vancouver Island, along the mainland, and east coast of the Vancouver Island corridor earned only 60 per cent of this average. From 1979 to 1981, however, these proportions had improved to 68, 76, and 86 per cent, respectively, for each of these regions.

The most significant change occurring over these fifteen years, with possible consequences for the coastal fishing economy, is the size of rural fleets. During this time, there was an apparent reversal in the long-standing pattern of increasing urbanization of the fleet. This urbanization trend, begun about 1950 following relocation of the processing sector to a primarily south coast industry, appears to have been arrested by about 1972. From this point on, the overall number of vessels has gradually increased in rural areas, with corresponding losses occurring in Vancouver and Prince Rupert.

For the salmon fleet alone, during the period when the Davis Plan policies of licence limitation were enforced, there were drops in the number of vessels associated with all community categories. These decreases, however, occurred at a slower rate for rural areas than for urban centres. Thus from 1979 to 1981, the number of salmon vessels associated with remote communities had dropped by an average of eighty-one boats (11.6 per cent), but the proportion of the overall salmon fleet located in such areas had increased from 10.9 to 13.3 per cent. For towns this percentage increase was from 17.9 to 23.8 per cent, while for the Prince Rupert and Vancouver-Victoria areas the respective proportions of the fleet had dropped to 10.3 and 52.6 per cent from 12.2 and 59.0 per cent from 1967 to 1969.

While these changes in local fleet sizes may benefit the rural-based fishing economy, they must also be viewed with caution. Though absolute decreases in the number of vessels associated with the Pacific region's salmon fishery have occurred almost everywhere, and in urban areas at a faster rate, any losses to the fleets of isolated communities are more serious than corresponding decreases in urban centres. The economies of such places are near the threshold level for survival and, unlike urban economies, less able to withstand further losses to economic inputs in any form. Hence decreases in local fleets will have disproportionate consequences for such communities.

Further, there are already signs that the reversal that has taken place in the urban concentration of vessels will be short-lived. It is perhaps not surprising that it occurred when it did. Throughout North America at this time, general migratory patterns suggested a rural revival (Fuguitt et al., 1981; Beaujot and McQuillan, 1982; Statistics Canada, 1978, 1980: 5,11). These overall demographic trends appear now to have reverted to patterns of increasing urbanization. As well, during that period, the Pacific fisheries were being buoyed by the combination of good salmon harvests and the unexpected heavy demand for herring roe. In a way it was like the "old days" where the mood prevailed that there were fortunes, or at least a decent living, to be made from fishing, and fishers may have felt ready to risk an uncertain rural location. In the early 1980s, spurred by a financial crisis in the industry, the mood was bleak and once again signs existed of a new downturn in rural participation rates.

Since 1979 there has been a general decline in the number of vessels associated with rural locations, with the exception of towns. These decreases are especially evident among the most remote areas of the coast — the north and the west coast of Vancouver Island. Where gains have occurred they have been in places providing at least some opportunities for supplemental employment — the lumber and mineral-resource communities of the central mainland and east coast of Vancouver Island — or in Prince Rupert and Victoria where the pattern of declining fleet sizes has been reversed. Thus, in this small way, history has been repeated.

CONCLUSION: REGIONAL IMPACTS AND
CURRENT FISHING POLICY

What has been demonstrated from this analysis is the manipulation and exploitation of the coastal fishing-based economy for short-term economic benefit. Both processors and fishers are responsible.

The coastal fishing economy has risen and fallen on the basis of the prevalence of fish and the economic tide. This has happened in close accord with a model of regional dependency in which the current conditions of entrenched underdevelopment and economic disadvantage affecting the rural coast have resulted from actions taken at the urban core. In comparison with other industries, there have been differences in the ways in which this long-term dependence and lack of development have come about. Nevertheless, the principal driving force behind the brief economic heyday of the rural coast followed a period when, because of south-coast competition and resource depletion, the interior and mainland coasts and the west coast of Vancouver Island were perceived as having an apparent inexhaustible abundance of fish and strong markets. When the fish were depleted and markets dwindled, the southern urban core asserted itself once again.

More recently, during a brief optimistic period spurred by good salmon harvests and a burgeoning new herring fishery, there may have been a fleeting coastal revival, but by 1985 this vanished as the industry experienced its bleakest time. The urban focus of the industry appears once again to be at the forefront. As far as the processing sector is concerned, this fact is incontrovertible.

The state has not been indifferent. In its statements, the ministry has asserted, and continues to assert, that policies should not further impinge upon — should even preserve — these isolated coastal economies (Pearse, 1982:5; W. Sinclair, 1971; Minister of Fisheries, 1968). Nevertheless, little in the way of direct action has ever been undertaken. When policies have been implemented, there has been little regard for the regional consequences. The Davis Plan is an example. The fact that this policy of restricting entry to the salmon fishery did not at the time exacerbate the further rapid decline in the size of rural fleets has much more to do with the prevailing ecological and market conditions than any direct motive on the part of the state to intervene.

There are other policies planned or being implemented in the Pacific fisheries that will directly affect the regional balance. Not unreasonably, the strong focus of these programs concentrates on the preservation of the resource through the curtailment of fishing effort. The current policy of placing restrictions on the troll fishery, which commenced with the 1984 season, exemplifies this. The troll fishery occurs mainly along the outside waters of Vancouver Island and along the interior coast. It is disproportionately conducted by rural-based fishers (Table 14.3). Because of their isolated locales, such fishers have fewer opportunities for supplemental employment from other resource sectors or urban industries. Fishers from such regions also have disproportionately low fishing incomes (Table 14.2). A policy to restrict salmon trolling, however necessary on ecological grounds, will have adverse consequences in terms of enhancing existing regional disparities.

Buy-back, the plan to reduce the size of the salmon fleet, is another policy that may have adverse consequences on a regional basis. An across-the-board reduction by 50 per cent in the size of the fleet, as may be proposed, will directly affect rural fishing economies to a much more severe extent than urban economies, simply because of their greater vulnerability and dependence upon fishing. If buy-back is employed rather to remove those vessels most severely handicapped by debt and low fishing returns, it is also likely to eliminate a disproportionate number of nonurban vessels since such vessels earn less, and there are signs of their debts having escalated most rapidly (Table 14.2). If buy-back is used to restrict certain gears, the rural-based fleet will only gain if this limitation is applied to the net fishery, an unlikely circumstance given current depleted spring salmon stocks and the comparative abundance of pinks and sockeyes.

This is not to say there is no hope of government aid in this matter. The state's current program of salmonid enhancement may in time directly benefit certain isolated native communities. A policy of area licensing could also provide some incentive for resettlement by fishers in rural areas, aid in the preservation of local fish stocks, defray the further rapid capitalization of vessels and equipment that now occurs because of competition from the more efficient urban fleet, and help to curtail fishers' operating costs. In time areas of the coast may even benefit from the development of mariculture and ocean ranching or collective harvesting arrangements among fishers.

Nevertheless, such speculations on how new policies will overcome current problems are idyllic and ignore the past. The history of this coast is one of repeated exploitation and entrenched inequalities, leading to the current state of disadvantage and restricted opportunity. One cannot hope to turn this around easily. In consideration of all that has occurred, a more likely scenario would be one in which the search for some new economic advantage — tourism, ocean mining, or energy exploration — would lead to another shift in the regional resource emphasis. What will follow is uncertain, but it would not be surprising if a cycle similar to that which has already occurred in the Pacific fisheries were to be repeated.

NOTES

1. While the data presented here on earnings and returns on investment are suggestive, it is necessary to acknowledge that they do not properly represent actual net returns from fishing effort. To do this it would be necessary to take into account actual operating costs in combination with operator priorities and expended labour. These data are unavailable on a regional basis. Nor, of course, does this analysis reflect the possibly beneficial, but less tangible, social and psychological rewards often felt to be associated with a rural lifestyle. It is likely, of course, that the averages on gross earnings reported here are also generally reflective of greater profitability in the majority of cases, but the analysis of the role of operator choice and priorities remains to be done.

2. In a second method of classification, all rural communities were grouped according to three major coastal areas — the north coast and Queen Charlotte Islands, the west coast of Vancouver Island, and the southern interior coast and east coast of Vancouver Island. In each case, these hinterland areas can be compared with a nearby urban centre — Prince Rupert, Victoria, or Vancouver. These comparisons, of region as well as community size, essentially reproduce those on community size alone. In nearly all cases, the earnings advantage falls with the urban centre.

Conclusion

15

Uncommon History

PATRICIA MARCHAK

We have considered the history, the role of the state, the nature of capital, the characteristics and organization of labour, and the nature of communities in the west coast fisheries. Let us conclude with a review of significant developments in the period since 1968.

The Davis Plan was introduced with the stated intention of conserving the stocks. Excess vessels were to be bought out by the state and the number of fishers was to decline with decreasing availability of licences. The ideology of "common property," emphazing its tragic outcomes, was enunciated during the debates of the period. The union had long argued in favour of limited entry to the fisheries, but opposed the Davis Plan, pointing out that the licensing of vessels rather than individuals and the saleability of licences would lead to overcapitalization. The union was alone in opposition. Small-vessel owners in general approved of the Davis Plan. For those wanting to leave the industry, the buy-back provisions appeared attractive, and for others, the possibility of reduced competiton and increased returns for the catch seemed appealing.

The Fisheries Association, representing processors, was the chief supporter of the plan. The firms at this point had no vested interest in increasing the size of the labour force either in processing or fishing; in fact, quite the reverse. Vessels owned by companies became saleable under the buy-back provisions. Such action would prolong the longevity of the industry without depleting the processors' control of sufficient supplies for existing markets. And the plan might have the additional benefit of reducing claims by fishers in concert with shoreworkers and reduce the membership of the UFAWU.

Licences became private property as required under the Davis Plan, and, with limitation on their numbers, steadily increased in market value. In a stable or declining industry, an upper level might have been achieved for these values. Without an expanding market, there would have been limited incentive for vessel owners to invest in new gear or upgrade properties. However, the market conditions abruptly changed in the early 1970s when Japanese merchants sought new sources of the major dietary component of their population, because Japanese fishers were subjected to international restrictions on their property rights.

Japanese companies became major components of the west coast fisheries between 1972 and 1979, primarily as merchants. Their internal competition caused a dramatic escalation of salmon as well as herring roe prices. Langdon (1983:68) notes that between 1975 and 1977, Canadian seafood exports to Japan increased from $39 million to $142 million, $109 million of this coming from B.C. The increase reflected escalating prices rather than quantities since the tonnage in 1977 was actually less than in 1975. The peak year was 1978 when Japanese companies, such as Mitsubishi and Marubeni, and numerous smaller firms in the Tokyo fish market paid as much as $3,000 cash a ton at sea for roe herring and sidestepped the processors.

The immediate effect of this extraordinary development was an influx of capital into the fleet. The same government that had argued on behalf of conservation and fleet reduction now intervened in fish harvesting with backing for loans to vessel owners and boat construction companies. A vast overcapitalization of the fleet was the result.

In addition, the federal and provincial governments became financial supports for upgrading processing facilities, which were made necessary by the demand for a new product (roe herring), higher quality salmon, and the entry into the processing sector of new firms catering solely to the Japanese market. These new firms sold roe herring and fresh or fresh-frozen salmon rather than canned salmon and had relatively low start-up costs. They included two native Indian cooperatives (Pacific North Coast Native Co-operative near Prince Rupert and the Central Native Fishermen's Co-operative at two plants on the central coast and Vancouver Island west coast), both aided through Japanese capital and federal government support. B.C. Packers continued to be the dominant firm, selling both its traditional canned and frozen salmon and roe herring, but it now had to compete for raw fish supplies with the new entrants. In addition, because the Japanese offered cash on the grounds for fresh roe herring and salmon, independent fishers were able to circumvent the processors altogether.

An investigation in 1978 revealed that of sixty-one companies buying and processing fish in B.C., twenty-nine had some foreign ownership; seventeen had 20 per cent or more. Twelve of the seventeen had Japanese ownership (the remaining five were American) (Proverbs, 1978). The Japanese firms improved existing technologies and obliged traditional firms to increase capitalization (Langdon, 1983:58–71).

When Japanese capital began to displace American or supplement and compete with Candian in the fish-processing sector, two major industry interest groups sought government controls: the traditional processors led by B.C. Packers and the UFAWU representing shoreworkers. There were newspaper stories about Canadian firms being unable to compete with the enormous Japanese merchant houses, rumours that the Japanese government was providing unfair supports to its export-import firms, claims that the Japanese quality-control supervisors placed in plants with sales contacts to the Japanese merchant houses were displacing Canadians, and demands from the UFAWU to reduce the amount of raw-fish sales at sea (cited in Proverbs, 1978). The studies of Japanese investment conducted under DFO sponsorship in 1978 and 1980 were responses to these concerns, the 1978 study concluding that "Japanese interests have the potential to gain control over B.C.'s fish resources" (Proverbs:3).

It should be noted that foreign investment in the fishing industry was in fact no different from foreign investment in all other Canadian industries. In 1968, the Watkins Report detailed the degree of American ownership in manufacturing and resource industries, and though that report sparked the development of a nationalist movement involving some sectors of small national business and much of the intellectual and artistic communities, it was strongly condemned by both international capital via companies in Canada and international unions via national branches. At that time, these groups argued that their interests were advanced by foreign investment; indeed, they argued sanctimoniously that any interference with the free flow of capital was tantamount to national subversion.

The excessive capacity of the canning plants threatened the traditional processors with early demise if the Japanese could not be stopped. Similar fears in Alaska had already led to proposed legislation in the United States to limit joint ventures between U.S. fishing companies and foreign owners of fish-factory vessels (Proverbs, 1978). The cash sales of raw fish threatened both processors and their labour force. In short, the free flow of capital was perceived in this single instance to be a threat: thus the demand that it be subject to state interference.

Concern with conservation by this time had disappeared. The Davis Plan had fallen apart. Buy-backs of vessels had occurred, but the total fleet capacity was infinitely greater than it had been in 1968, because owners — rationally responding to increased opportunities and the incentives offered by government through subsidies and loans — upgraded and enlarged their properties. The issue of 1978 was not whether the fish could survive the capture, but whether Japanese investors should be permitted to compete successfully with established firms. The DFO studies were initial responses to this concern, preparing the way for controls on Japanese investment. Presumably action would have followed had not market events made it unnecessary.

In 1979 the herring roe market in Japan was beset by a consumers' revolt. The de-escalation of B.C. prices was met by strike action that succeeded in reducing sales throughout the season. As well, salmon runs were low that year. These conditions led to declines in market viability for smaller firms and reduced earnings for larger ones. The major Japanese-owned firm (Norpac Fisheries) ceased operations. B.C. Packers recorded its first major loss, and the American-owned Canadian Fish Company sold its Prince Rupert plant and equipment to B.C. Packers. The Japanese market subsequently revived somewhat, but between depleted supplies in B.C., increasing sales from Alaska, and other sources in Southeast Asia coming on stream, together with new supplies from fish farms in Japan and increasing levels of meat consumption there, it became evident that the boom would not recur. It was a unique period.

The history of the 1970s is filled with state interventions designed — but unable — to "save" the traditional processors in the face of new competition created by state intervention at another level. The DFO, left to struggle alone with the problem of conservation, was obliged to shift its attention to regulation on the grounds. It shortened the periods during which fishing was permitted, yet while its officers were turned into a police force supervising fishers, the government simultaneously ensured that fishing capacity would increase.

The fall of the Japanese market re-created the concerns of the 1960s, except that by this time the industry had such overcapacity and such a high level of indebtedness that bankruptcy loomed for a very high proportion of the remaining participants. B.C. Packers, though now recording losses, came out of the fray with greater control of both raw and processed fish markets than before. It could afford to reduce its services to fishers and sell off its remaining vessels. With competition almost eliminated, and with its own processing facilities outside Canada becoming more profitable, the dominant firm relied on its virtual monopsony to obtain as much of the remaining fish supplies as it could profitably market. Its competition since that time has consisted not of processing firms in B.C., but rather the independent selling capacity of vessel owners, whose boats are now equipped with freezers and who can by-pass processors, and farmed frozen fish from Norway, Sweden, and Japan, which has successfully penetrated the domestic and American markets. A survey of B.C. processors in 1983 indicated that investors were not earning a "satisfactory" return, that their fish prices were too high, that they could not pass on costs in higher selling prices, and that some of the companies were in serious financial difficulty (Gordon, 1983).

While investment in a single vessel is still technically possible for a single individual, the cost of vessel and licence has so escalated since 1968 that individual owners are more often than not in debt to banks, processors, and other lenders. In other industries the escalating cost of equipment has often signalled the beginning of the takeover by large capital, the phase-out of small capital, and the increasing direct subordination (in contrast to indirect controls via commodity pricing) of labour. However, given the overall crisis in the fishery, the already marginal profitability of processing, and the range of economic considerations suggested above, this has not occurred.

How can we explain these contradictory state behaviours? Either the federal state has two unacquainted hands from which it delivers opposing policies, or its policies of 1968 and throughout the 1970s were made on behalf of the same general interests but each was dictated by an entirely different set of concerns. This second explanation rests on the assumption that the state in both instances moved on behalf of the dominant processors whose immediate interests altered because of an external threat to their dominance.

While the conservation argument had a basis in reality — declining fish stocks — it can also be understood as a form of legitimation; it appeared to be in the general, rather than in the particular, interest. It was in the general interest in the sense that a genuine decline in stocks would have impeded and eventually prevented any accumulation taking place. But the speed with which the policy was abandoned suggests that its legitimation function was rather more important in 1968 than its conservation function. DFO personnel, whose task is defined in terms of professional criteria rather than in terms of long-run accumulation potential, end up being the one group that takes conservation seriously: as far as they are concerned, conservation is what defines their employment.

In 1978, the state was able to act on behalf of existing capital in the processing sector while also acting, or appearing to act, on behalf of a strongly unionized shoreworker labour force. It was not acting on behalf of small-vessel owners, yet in order to facilitate the demand for raw-fish supplies by the processors, it became involved in financing the expansion of this fleet. This was the means by which continuing accumulation could occur.

Though the state at the federal level has agonized publicly over excess capacity since 1978, appointing the Pearse Royal Commission and stating its intention to introduce much tougher licensing and management procedures, its purchase of B.C. Packers' rental fleet in 1982 on behalf of northern native bands did not lead to a decrease in capacity. Thus, its extension of special halibut licences to native fishers, and the case of exclusive tribal rights to a fishery, oblige us to consider the state's other important concerns, one of which is native land claims in British Columbia.

INDIAN LAND CLAIMS

The Pearse Commission recommended a much greater role for native Indians in the fisheries. The Boldt decision in the United States had awarded half the west coast fisheries to Indian bands, and this clearly had an impact on Canadian Indian expectations and demands. Numerous Indian briefs to the commission suggested that Indian bands be given special allocations of fish on the grounds both that many subsisted off the resource (the food fishery) and that Indian alternative employment, especially in remote regions, was limited. A proposal to give half or more of all fishing rights to native groups was entertained by the federal government during the debates following publication of the Pearse Commission report. In 1985, the federal government announced an $11 million "economic development program" to provide financial aid to ease the debt load of native vessel owners and to enable native fishers using rental vessels to purchase their own boats. The minister of Indian and Northern Affairs (not the minister of Fisheries and Oceans) argued that the program would provide an economic bridge pending settlement of the fisheries component of comprehensive Indian claims (*Vancouver Sun*, July 9, 1985, A3).

Whatever the merits of this proposal for native peoples, it is not consonant with the idea of an overriding concern for conservation of the resource through limitation of access unless other radical changes are also made. It ensures that the banks will be relieved of outstanding debts and useless repossessed vessels, and inceases the number of participants with fishing capacity. To make sense of such inconsistency, we have to suppose that concern with native claims take precedence over concern with resource conservation, and that, at this moment in Canadian history, the Department of Indian and Northern Affairs takes precedence over the Department of Fisheries and Oceans.

How would this be explained except with reference to the potential threat of numerous land claims by native groups in B.C. to forest and other industrial conerns? Prior to this announcement, the B.C. government became a central participant in a court case regarding Meares Island, a small region off the west coast of Vancouver Island where two major forest companies laid claim to tree-harvesting rights. The provincial government made it clear that aboriginal rights were impediments it would not tolerate. But Indian Affairs is a federal jurisdiction, and the federal government is caught between the industrial claims of B.C. forest companies, the intransigent stance of the provincial government, and the pressing need to "solve" the problem of conflicting native claims. It cannot solve these by shoving them aside because by 1985 there were numerous interest groups and much public sympathy supporting native claims. Settlements elsewhere (Alaska and James Bay) together with the publicity surrounding the Berger inquiry on the proposed Mackenzie Valley Pipeline in the early 1970s have given these claims greater credibility.

The federal government could grant fishing rights by way of avoiding settlements potentially far more damaging to the interests of forestry companies. It is within the federal government's jurisdiction to fundamentally alter the status of claims in the fishery, as the U.S. federal court did in the Boldt case, and this precisely because the right to fish is not common property. However, the other participants in the commercial fishery would not idly watch their stakes eroded, the more so since they have been nurtured in the common property ideology.

Assuming that salmon would continue to be processed in some form in the foreseeable future and that the native fishers would continue to sell raw fish, an increase in native resource rights and vessel ownership would be potentially beneficial to the major processor. It would, most particularly, reduce the bargaining power of the union. Federal government support for native fisheries since 1982 has already contributed to conflict between these two groups. At the same time, if all native groups did not benefit equally, purchases on behalf of some groups would affect relations between bands in different territories. If some processing capacity were added to extended fishing rights, small processors and shoreworkers would be adversely affected. Any arrangement that provides differential benefits to one group would offend others, and the numerous associations that have attempted to bargain in good faith through the minister's advisory council are understandably upset at the prospect of special rights being accorded to native fishers.

CONCLUSION

There is no ideal solution to these issues. This is not a situation involving good guys and bad guys: rather it is one in which numerous groups with competing interests are trying to solve an unresolvable contradiction. The contradiction is in the property rights of the fishery: the provincial government has formal ownership of land and resources; the federal government claims formal ownership of the fish and of the right to allocate fishing licences; private individuals with licences are obliged to compete for capture; and captured fish, now as commodities, are private property. The whole situation is further complicated by the ideology of common property and by the efforts of federal conservation officers to save the fish in the name of common property. The complexities mount when the private property interests of other industries and other users of the fish habitat are brought into the picture.

While there is no ideal solution, there is one hopeful sign in the consultation and discussion process that followed the Pearse Commission. Fishers, whether unionized or independent, seiners, gillnetters, or trollers, native or non-native, have tried to work out their common interests and resolve their internal conflicts. They have demonstrated remarkable good will toward one another and a serious commitment to solving the problems. What they need now is an equally serious commitment by governments to a genuine consultation process.

Appendices

Appendix A

Notes on Sample Survey Methodology

This methodological appendix describes the sample surveys of fish processing workers, fishers, and fishing crew members. Material from these surveys is used most extensively in Chapters 8 and 9, although information obtained via these intensive interviews serves as a backdrop for many of the ideas discussed by various authors. The overall research project relied upon a variety of other sources (e.g., key informant interviews, archival material, official statistics) that are not discussed in this appendix.

SAMPLING

In order to obtain interviews with as wide a cross section of people working in the industry as possible, we used survey research methods to conduct formal, structured interviews with industry participants. The people we interviewed were randomly selected from lists provided to us by many different groups.

For shoreworkers, we relied on lists supplied by both the UFAWU and the PRFCA. In addition, we supplemented these lists with a few other names obtained from a select set of companies that did not employ union labour or that were new to the industry. The interviews were conducted in people's homes in the Lower Mainland, Prince Rupert, Port Edward, Tofino, and Ucluelet in May to October of 1982.

A letter of introduction was hand delivered to each of the shoreworkers selected, and an interviewer returned two or three days later to arrange an interview time. Of the 236 shoreworkers selected from our original list, we had improper addresses or other technical problems in 39 cases. Of the remaining 197 people, we completed 140 interviews.

For fishers and crew, we relied on the following organizations to supply us with lists of names from which we could draw random samples: UFAWU, PRFCA, PTA, PGA, NTA, Native Brotherhood, and the Prince Rupert Guild. We interviewed fishers and crew in the Lower Mainland, Prince Rupert, Tofino, Ucluelet, Bella Bella, Bella Coola, Ahousaht, Port Edward, and Hazelton from the fall of 1982 to the early spring of 1983.

Once again we added names to our lists if we learned that certain groups of people were systematically excluded (this was important mainly among Indian fishers in remote communities). However, it presents an important limitation to our sampling since we cannot be certain that we used a complete cross section of fishers to draw our sample. Fishers not appearing on any official list were excluded unless by chance we recognized during our fieldwork that some particular segment of the fishing community was unrepresented. In addition, all of the lists were dated to varying degrees. We relied on 1982 lists for all organizations, but much of our interviewing was done early in 1983. While these lists were the most recent available, they were undoubtedly biased against new members. However, these caveats aside, we believe our sample is as representative of industry participants as is possible given the constraints noted above.

Out of a total of 182 names of fishers randomly selected for our sample, we contacted and completed interviews with 132. Of the crew members we selected for interviewing ($N = 96$), we were only able to complete fifty-three interviews. Once again all of the interviews were conducted in the individuals' homes (although on two occasions this meant on a boat). All but a few of the interviews with fishers were done by either the two research associates who had worked with fishers for many years or by three fishers whom we hired and trained as interviewers. (Although no women fishers were randomly selected, we did complete informal interviews with three women—one a skipper and two crew. The material from these interviews is not included in our survey-research analysis.)

SURVEY MATERIAL

The formal interviews for shoreworkers averaged about fifty-three minutes in length, with the longest taking somewhat over two hours. For fishers, the interviews averaged approximately two hours and forty-five minutes, with the longest stretching into a six or seven hour discussion (with dinner!). We have chosen not to reproduce the actual questions because of length (the shoreworkers' interview schedule is twenty-two pages long, while the questions for fishers run to sixty-five pages).

As in all surveys, the quality of the data is uneven. Two areas were particularly open to systematic error. Attitude questions were generally well received by shoreworkers, but for fishers greater resistance existed (especially among older respondents). Also for some fishers, questions about incomes and expenses were poorly answered. While some people gave us access to their tax records, others could only provide wild guesses as to their costs or incomes.

No comparable survey efforts have been undertaken in the last few years, and hence no equivalent standards exist to determine the precise quality of our survey data. Nevertheless, as reported in Chapters 8 and 9, when indirect comparisons to tax data, accountant data, and survey work from the 1970s are undertaken, the information we gathered is consistent with these alternate sources, giving us even more confidence in the survey data. In short, while precise measures of the reliability and validity of the survey data are hard to come by, those comparisons we have undertaken demonstrate that our evidence is at least equal in quality to that currently available.

Appendix B

Papers and Publications

Work completed additional to this book:

Guppy, N.
1983   "The Roe-Herring Bonanza and its Impact on the Structure of the
        B.C. Fishing Industry." Paper presented at the Canadian Political
        Science Association meetings, University of British Columbia,
        21 pages.
Guppy, N.
1984   "The Transformation of Social Relations in the B.C. Commercial
        Fishing Industry." Paper presented at the Western Association of
        Sociology and Anthropology meetings, Regina, Saskatchewan,
        39 pages.
Guppy, N.
1986   "Property Rights and Changing Class Formations in the B.C. Com-
        mercial Fishing Industry." *Studies in Political Economy,* vol. 19,
        pp. 59–81.
Hayward, B.
1981   "The B.C. Salmon Fishery: A Consideration of the Effects of Licens-
        ing." *B.C. Studies,* no. 50, pp. 39–51.
Hayward, B.
1984   "The Co-Op Strategy." *The Journal of Canadian Studies,* vol.
        19:1, pp. 48–64.
Marchak, P.
1982   Keynote Address to the Western Fishermen's Federation, Victoria,
        15 pages.
Marchak, P.
1983   "Does Consultation Matter?" Opening Address to the Western
        Fishermen's Federation, February, Victoria, 20 pages.
Marchak, P.
1984   "Introduction" to Special Edition of *Journal of Canadian Stud-
        ies,* vol. 19:1.
Marchak, P.
1984   "Political Economy of Renewable Resources." In Michael S. Wit-
        tington and Glen Williams (eds.) *Canadian Politics in the 1980's,*
        2nd edition. Toronto: Methuen, pp. 137–154.
McMullan, J. L.
1983   "Financing Fishermen in British Columbia." Paper presented at
        the Canadian Political Science Association meetings, June, Univer-
        sity of British Columbia, 43 pages.

McMullan, J. L.
1984   "Contradictions in State Policy and the Crisis in the B.C. Fishing
         Industry." School of Community and Public Affairs, Montreal,
         Concordia University.
McMullan, J. L.
1984   "State, Capital and Debt in the B.C. Fishing Fleet, 1970–1982." *The
         Journal of Canadian Studies,* vol. 19:1, pp. 65–88.
Muszynski, A.
1984   "The Organization of Women and Ethnic Minorities in a Resource
         Industry: A Case Study of the Unionization of Shoreworkers in the
         B.C. Fishing Industry, 1937–1949." *The Journal of Canadian Stud-
         ies,* vol. 19:1, pp. 89–107.
Muszynski, A.
1984   "Class Formation and Class Consciousness: The Making of Shore-
         workers in the B.C. Fishing Industry." Paper presented at the B.C.
         Studies Conference, February 18, University of British Columbia.
Muszynski, A.
1984   "The Fish and Ships Project and its Various Dimensions, including
         Shoreworkers." Tradition and Transition, Reports of Recent Field-
         work, cosponsored by the University of British Columbia Museum
         of Anthropology and the Department of Anthropology and Sociol-
         ogy, March 30, 42 pages.
Muszynski, A.
1986   "Class Formation and Class Consciousness: the Making of Shore-
         workers in the B.C. Fishing Industry." *Studies in Political Economy,*
         vol. 20, pp. 85–116.
Muszynski, A.
1986   "The Creation and Organization of Cheap Wage Labour in the B.C.
         Fishing Industry." Ph.D. dissertation (Sociology), University of
         British Columbia.
Muszynski, A.
1986   "Race and Gender: Structural Determinants in the Formation of
         B.C.'s Salmon Cannery Labour Forces." Paper presented to Canadian
         Association for Rural Studies, University of Manitoba, June. Forth-
         coming publication in *Canadian Journal of Sociology,* 1987.
Pinkerton, E.
1981   "Regulating the West Coast Fishery: Managing the Commons or
         the Community?" Submission to the Commission on Pacific Fish-
         eries Policy, North Vancouver.
Pinkerton, E.
1983a   "The Dressed and the Undressed: Competing Processor Strategies
         in the Market for Raw Salmon." Paper presented at the Canadian
         Sociology and Anthropology Association meeting, June 4.

Pinkerton, E.
1983b "Undercapitalization in a British Columbia Salmon Fleet: the Case for Area Management." Paper presented at International Congress of the Anthropological and Ethnological Sciences, August, Vancouver.

Pinkerton, E.
1983c "All you ever wanted to know about Oncorhynchian Oligopsony." Paper presented to the Department of Anthropology, Riverside, University of California.

Pinkerton, E.
1984 "Intercepting the State: Dramatic Processes in the Assertion of Local Co-Management Rights." Paper read to Society for Applied Anthropology, Toronto, March. Forthcoming publication in B. McCay and J. Acheson (eds.) *The Question of the Commons: Anthropological Contributions to Natural Resource Management.* Tucson: University of Arizona Press.

Pinkerton, E.
1985 "Cooperative Management of Local Fisheries: the Route to Development." Paper read to Society for Economic Anthropology, Warrenton, Virginia, May.

Warriner, K.
1983 "From Urban Centre to Isolated Village: Distributional Patterns of Change in the B.C. Fishing Industry, 1967–1978." Paper presented to the Canadian Political Science Association, Learned Societies Conference, Vancouver, 38 pages.

Warriner, K. and Guppy, N.
1984 "From Urban Centre to Isolated Village: Regional Effects of Limited Entry in the British Columbia Fishery." *The Journal of Canadian Studies,* vol. 19:1, pp. 138–155.

Bibliography

The material cited below is organized into five separate parts. Reference to the proper section is essential to locate the material cited in the text.
1. Books, Articles, Theses, and General Resource Materials.
2. Government and International Organizations: Documentation, Legislation, and Reports.
3. Newspapers, Magazines, Radio.
4. Companies and Fishers' Organizations: Reports, Newsletters, Correspondence.
5. Personal Communication.

PART 1

Books, Articles, Theses, and General Resource Materials

Alexander, David
1976   "The Political Economy of Fishing in Newfoundland." *Journal of Canadian Studies,* vol. 11, no. 1.

Allen, G. C.
1970   "Japan's Place in Trade Strategy." In Hugh Corbet (ed.) *Trade Strategy and the Asian-Pacific Region.* University of Toronto Press.

Andersen, Raoul
1979   "Public and Private Access Management in Newfoundland Fishing." In R. Andersen (ed.) *North Atlantic Maritime Cultures.* The Hague: Mouton, pp. 299–336.

Anderson, E. N., Jr.
1970   *The Floating World of Castle Peak Bay.* Washington, D.C.: American Anthropological Association.

Apostle, Richard, Gene Barrett, Anthony Davis, and Leonard Kasdan
1985   "Land and Sea: The Structure of Fish Processing in Nova Scotia." Preliminary Project Report, Gorsebrook Research Institute for Atlantic Canada Studies, Halifax, St. Mary's University.

Bain, Joe S.
1956   *Barriers to New Competition.* Cambridge: Harvard University Press.

Barrett, Gene
1979   "Underdevelopment and Social Movements in the Nova Scotia Fishing Industry to 1938." In R. Brym and J. Sacouman (eds.) *Underdevelopment and Social Movements in Atlantic Canada.* Toronto: New Hogtown Press, pp. 127–160.

1981 "The state and capital in the fishing industry: the case of Nova Scotia." Paper presented at the Annual Meetings of the Canadian Political Science Association, Halifax, Nova Scotia.

Barrett, Gene and Anthony Davis
1984 "Floundering in Troubled Waters: The Political Economy of the Atlantic Fishery and the Task Force on Atlantic Fisheries." *Journal of Canadian Studies* 19(1), pp. 125–137.

Beacham, T. D.
1984 "Distribution of Landings of Pacific Salmon by the Commercial Fleet in British Columbia, 1951–1982." Canadian Manuscript Report of Fisheries and Aquatic Science, no. 1,784.

Beagle, Mickey
1980 "Segregation and Discrimination in the West Coast Fishing Industry." *Canada and the Sea,* vol. 3(1). Association of Canadian Studies.

Beaujot, Roderic and Kevin McQuillan.
1982 *Growth and Dualism.* Toronto: Gage Publishing.

Bell, David and Lorne Tepperman
1979 *The Roots of Disunity.* Toronto: McClelland and Stewart.

Bell, F. H.
1981 *The Pacific Halibut: The Resouce and the Fishery.* Anchorage: Alaska Northwest Publishing Company.

Berkes, Fikret and D. Pocock
1983 "The Ontario Native Fishing Agreement in Perspective: A Study in User-Group Ecology." *Environments,* vol. 15(3), pp. 17–26.

Bossin, Bob
1981 *Settling Clayoquot Sound.* Heritage Series, no. 33. Victoria: Province of B.C.

B.C. Research
1980 "Current Status and Potential Impact of Commercial Salmon Ranching in the Northeast Pacific Region." B.C. Research Industry Information Report, no. 1A. Vancouver.
1982 "Private Sector Involvement in Pacific Salmon Enhancement." B.C. Research Industry Report, no. 7. Vancouver.

Brox, O.
1972 *Newfoundland Fishermen in an Age of Industry.* St. John's: Institute for Social and Economic Research.

Burnett, Nicolas
1981 "Why US Torpedoed Sea-Law Talks." *Winnipeg Free Press,* April 22. Reprinted.

Burns. J.
1976 *Fisheries Impact on Small Coastal Communities,* Interim Report. Vancouver: Department of Fisheries and Oceans.

Buzan, Barry
1976 *Seabed Politics.* New York: Praeger.

Buzan, Barry and Barbara Johnson
1975   *Canada at the Third Law of the Sea Conference: Policy Role and Prospects.* Occasional Paper Series, no. 29, Law of the Sea Institute. Kingston: University of Rhode Island.

Buzan, Barry and Danford W. Middlemiss
1977   "Canadian Foreign Policy and the Exploitation of the Seabed." In Johnson, B. And M. Zacher (eds.) *Canadian Foreign Policy and the Law of the Sea.* Vancouver: University of British Columbia Press.

Campbell, Blake
1974   "Licence Limitation in the British Columbia Fishery." Vancouver: Department of Fisheries and Oceans.

Campbell, B. and D. Buchanan
1955   "Economic Survey of Salmon Fishermen in B.C." Markets and Economics Service. Vancouver, Department of Fisheries and Oceans.

Campbell, Kenneth
1978   "Regional Disparity and Interregional Exchange Imbalance." In D.Glenday, H. Guindon, and A. Turowetz (eds.) *Modernization and the Canadian State.* Toronto: Macmillan, pp. 111–131.

Cannon, G. Harry
1980   "The Ahousaht Education Study." Ahousaht Band Council. Vancouver: Clare Educational Development.

Carmichael, Alfred
1891   "Account of a season's work at a Salmon Cannery." Windsor Cannery, Aberdeen — Skeena. Vancouver: University of British Columbia Special Collections.

Carrothers, W. A.
1941   *The British Columbia Fisheries.* Toronto: University of Toronto Press.

Chirot, Daniel
1977   *Social Change in the Twentieth Century.* New York: Harcourt, Brace, Jovanovich.

Ciriacy-Wantrup, S. V. and R. Bishop
1975   "Common Property as a Concept in Natural Resource Policy." *Natural Resources Journal,* vol. 15(3), pp. 713–27.

Clark, Scott
1982   "Machines or Fish: Control and Development in a B.C. Indian Community." Paper presented at the Annual Meetings of the Canadian Ethnological Society, Vancouver.

Clarkson, Stephen
1981   *Canada and the Reagan Challenge, Crisis in the Canadian-American Relationship.* Ottawa: Canadian Institute for Economic Policy.

Clement, Wallace
1978   "A Political Economy of Regionalism in Canada." In D. Glenday,
        H. Guindon, and A. Turowetz (eds.) *Modernization and the Cana-
        dian State.* Toronto: Macmillan, pp. 89–110.
1983   *Class, Power and Property.* Toronto: Methuen.
1984   "Canada's Coastal Fisheries: Formation of Unions, Cooperatives,
        and Associations." *Journal of Canadian Studies,* vol. 19(1),
        pp. 5–33.
Conley, John
1985   "Relations of Production and Collective Action in the Salmon Fish-
        ery 1900–1925." In R. Warburton and D. Coburn (eds.) *Essays in
        Working Class Formation in British Columbia.* Vancouver: Uni-
        versity of British Columbia Press.
Connelly, P. and M. MacDonald
1983   "Women's Work: Domestic and Wage Labour in a Nova Scotia
        Community." *Studies in Political Economy,* no. 10, pp. 45–72.
Copes, Parzival
1979–   "The Evolution of Marine Fisheries Policy in Canada." *Journal*
1980   *of Business Administration,* vol. 11(1/2), pp. 125–148.
Cove, John
1982   "The Gitksan Traditional Concept of Land Ownership." *Anthro-
        pologica,* vol. XXIV(1), pp. 3–17.
Cruickshank, Don
1982   *Fleet Rationalization.* Fleet Rationalization Committee Report.
        Vancouver.
Crutchfield, J. A.
1975   "An Economic View of Optimum Sustained Yield." In P. M. Roedel
        (ed.) *Optimum Sustained Yield as a Concept in Fisheries Manage-
        ment.* Washington: American Fisheries Society.
Crutchfield, J. A. and Giulio Pontecorvo
1969   *The Pacific Salmon Fisheries: A Study of Irrational Conservation.*
        Baltimore: Johns Hopkins University Press.
Crutchfield, J. A. and A. Zellner
1962   *Economic Aspects of the Pacific Halibut Fishery.* Fishery Indus-
        trial Research, 1(1), U.S. Department of the Interior.
Cuneo, Carl J.
1978   "A Class Perspective on Regionalism." In D. Glenday, H. Guindon,
        and A. Turowetz (eds.) *Modernization and the Canadian State.*
        Toronto: Macmillan, pp. 132–156.
Davis, Anthony and L. Kasdan
1984   "Bankrupt Government Policies and Belligerent Fishermen Re-
        sponses: Dependency and Conflict in the Southwest Nova Scotia
        Small Boat Fisheries." *Journal of Canadian Studies,* vol. 19(1),
        pp. 108–124.

Deschenes, B. M.
1984   *Study of Marine Casualty Investigation in Canada.* Ottawa: Ministry of Transport.
Drucker, Philip
1951   *The Northern and Central Nootkan Tribes.* U.S. Bureau of American Ethnography, Bulletin 144. Washington, D.C.
1958   *The Native Brotherhoods.* Washington, D.C.: Bureau of American Ethnology.
Duff, Wilson
1964   *Indian History in British Columbia.* Anthropology in British Columbia, Memoir no. 5. Victoria: Queen's Printer.
Finkle, Peter
1975   "Canadian Foreign Policy for Marine Fisheries: An Alternative Perspective." *Journal of Canadian Studies,* vol. 10, pp. 10–24.
Fisher, Robin
1977   *Contact and Conflict: Indian-European Relations in British Columbia, 1774–1890.* Vancouver: University of British Columbia Press.
Fisheries Association
1983   *A Status Report on the Commercial Fishing Industry of British Columbia.* Vancouver, B.C.
Foodwest Resource Consultants
1979   See B.C., Ministry of Environment.
Ford, C. S.
1941   *Smoke From Their Fires: The Life of a Kwakiutl Chief.* New Haven: Yale University Press.
Fraser, G. A.
1977   "Licence Limitation in the British Columbia Salmon Fishery." Technical Report Series. Vancouver: Department of Fisheries and Oceans.
1979   "Limited Entry: Experience of the British Columbia Salmon Fishery." *Journal of the Fisheries Research Board of Canada,* vol. 36, no. 7.
Frecker, J. P.
1972   "Militant and Radical Unionism in the British Columbia Fishing Industry." M.A. thesis, Dept. of Political Science, University of British Columbia.
Fredin, R. A.
1980   "Trends in North Pacific Salmon Fisheries." In W. J. McNeil and D. Himsworth (eds.) *Salmonid Ecosystems of the North Pacific.* Corvallis: Oregon State University Press.
Friedlaender, Michael
1975   *Economic Status of Native Indians in British Columbia Fisheries, 1964–1973.* Technical Report Series, Vancouver.

Fry, M. G.
1975 "Canadian Diplomatic Initiatives: the Law of the Sea." In M. Fry (ed.) *Freedom and Change: Essays in Honour of L. B. Pearson.* Toronto: McClelland and Stewart, pp. 136–151.

Fuguitt, Glenn V., Daniel Lichter, and Calvin Beale
1981 *Population Deconcentration in Metropolitan and Nonmetropolitan Areas in the United States, 1950–1975.* Population Series 70–15. Madison, Wisconsin: Applied Population Laboratory, University of Wisconsin.

Garrod, Steve
1984 "The Social Organization of Labour in the B.C. Salmon Canning Industry, 1981–1900." Unpublished paper, Anthropology and Sociology, University of British Columbia.

Gillis, D. J. And Will McKay
1980 *"Economic Development Planning for the Kwakiutl District."* Report prepared for the Kwakiutl District Council, Port Hardy, B.C.

Gladstone, Percy
1953 "Native Indians and the Fishing Industry of British Columbia," *Canadian Journal of Economics and Political Science*, vol. 19(1), pp. 20–34.
1959 "Industrial Disputes in the Commercial Fisheries of British Columbia." Unpublished M.A. thesis, University of British Columbia.

Gladstone, Percy and Stuart Jamieson
1950 "Unionism in the Fishing Industry of British Columbia." *The Canadian Journal of Economics and Political Science,* vol. XVI(2), pp. 146–171.

Gordon, Clarkson
1983 *Survey of the British Columbia Fish Processing Industry: Financial Results, 1978–1982.* Canadian Industry Report of Fisheries and Aquatic Science, no. 146, Ottawa.

Gordon, H. Scott
1954 "The Economic Theory of a Common Property Resource: The Fishery." *Journal of Political Economy,* vol. 62, pp. 124–142.

Granatstein, J. L.
1975 *Canada's War.* Toronto: University of Toronto Press.

Greene, Stephen and Thomas Keating
1980 "Domestic Factors and Canada–United States Fisheries Relations." *Canadian Journal of Political Science,* vol. 13(4), pp. 731–750.

Gregory, H. and K. Barnes
1939 *North Pacific Fisheries.* San Francisco: The American Council of the Institute of Pacific Relations.

Guppy, Neil

1983   "The Impact of the Roe Herring Fishery on the Organization of the B.C. Fishing Industry." Paper presented at the Learned Societies, Canadian Political Science Association, Vancouver, B.C., June.

1985   "Property Rights and Changing Class Formations in the B.C. Commerical Fishing Industry." *Studies in Political Economy,* vol. 19, pp. 59–81.

Halpern, J.

1984   "The Impact of Capital-Intensive Tourist Development on a Small Western Ireland Town." Paper presented at the Meetings of the American Anthropological Association, Denver, Colorado.

Hardin, G.

1968   "The Tragedy of the Commons." *Science,* vol. 162(3) pp. 1243–1248.

Harper, Janet

1982   "Indian Fisheries Management in the State of Washington." Mimeograph, Vancouver.

Hawthorn, Harry B. (ed.)

1966   *A Survey of Contemporary Indians of Canada: A Report on Economic, Political, Education Needs and Policies.* Two volumes.

Hayward, Brian

1981a   "Development of Relations of Production in the British Columbia Fishing Industry." M.A. thesis, University of British Columbia.

1981b   "The B.C. Salmon Fishery: A Consideration of the Effects of Licensing." *B.C. Studies,* no. 50, pp. 39–51.

1984   "The Co-op Strategy." *Journal of Canadian Studies,* vol. 19(1), pp. 48–64.

Healey, M. C.

1982   "Multispecies, Multistock Aspects of Pacific Salmon Management." In M. C. Mercer (ed.) *Multispecies Approaches to Fisheries Management Advice.* Canadian Special Publications, Fisheries and Aquatic Sciences, no. 59.

Hilborn, Ray and Max Ledbetter

1983   "Determinants of Catching Power in the B.C. Salmon Purse Seine Fleet." Manuscript, Animal and Resource Ecology, University of British Columbia.

Hill, A. V.

1967   *Tides of Change.* Prince Rupert: Prince Rupert Fishermen's Cooperative Association.

Hollick, Ann L.

1974   "Bureaucrats at Sea." In A. Hollick and R. Osgood (eds.) *New Era of Ocean Politics.* Baltimore: John Hopkins University Press, pp. 1–74.

Jackson, Ted
1984   *Community, Economic Self-Help, and Small-Scale Fisheries.* Communications Directorate, Department of Fisheries and Oceans, Ottawa.

James, H.
1983   "Monopoly Relations in the Canadian State: 1939–1957." Unpublished Ph.D. thesis, University of British Columbia.

Jamieson, Stuart M.
1968   *Times of Trouble: Labour Unrest and Industrial Conflict in Canada, 1900–1966.* Task Force on Labour Relations, Study no. 22. Ottawa: Privy Council.

Jamieson, Stuart M. And Percy Gladstone
1950   "Unionism in the Fishing Industry of British Columbia." *Canadian Journal of Economic and Political Science,* vol. XVI(1), pp. 1–11.

Johnson, Barbara
1977   "Canadian Foreign Policy and Fisheries." In B. Johnson and M. Zacher (eds.) *Canadian Foreign Policy and the Law of the Sea.* Vancouver: University of British Columbia Press.

Johnson, Barbara and Frank Langdon
1976   "Two-Hundred-Mile Zones: The Politics of North Pacific Fisheries." *Pacific Affairs,* vol. 49(1), pp. 5–27.

Johnson, Barbara and Mark Zacher
1977   *Canadian Foreign Policy and the Law of the Sea.* Vancouver: University of British Columbia Press.

Johnston, Douglas
1965   *The International Law of Fisheries: A Framework for Policy-Oriented Inquires.* New Haven: Yale University Press.

Kanter, R. M.
1977   *Men and Women of the Corporation.* New York: Basic Books.

Kirby, Michael
1982   *Navigating Troubled Waters: A New Policy for the Atlantic Fisheries.* Task Force on Atlantic Fisheries. Ottawa: Minister of Supply and Services.

Ladner, Edna G.
1979   *Above the Sand Heads.* British Columbia: D. W. Friesen and Sons.
1980   *Mosaic Fragments.* From the memoirs of T. Ellis Ladner (1871–1958). Limited edition.

Lane, Barbara
1978   "Federal Recognition of Indian Fishing Rights in British Columbia: the Babine Barricade Agreement of 1906, the Fort Fraser Agreement of 1911, and the Port St. James Agreement of 1911." Mimeograph prepared for the Union of B.C. Indian Chiefs, Vancouver.

Lane, Robert and Barbara Lane
1978   "Recognition of B.C. Indian Fishing Rights in Formal Treaties with
       the Indians and with Foreign Governments." Mimeograph pre-
       pared for the Union of B.C. Indian Chiefs, Vancouver.
Langdon, Frank
1983   *The Politics of Canadian-Japanese Economic Relations, 1952–
       1983.* Vancouver: University of British Columbia Press.
Langdon, Steve
1977   "Technology, Ecology, and Economy: Fishing Systems in South-
       east Alaska." Unpublished Ph.D. dissertation, Stanford University.
1982   "Managing Modernization: A Critique of Formalist Approaches to
       the Pacific Salmon Fisheries." In John Maiolo and Michael Orbach
       (eds.) *Modernization and Marine Fisheries Policy.* Ann Arbour.
1984   "Commercial Fisheries in Western Alaska: Implications for State
       Fisheries Policy." Paper presented at the Western Regional Science
       Association Meetings, Monteray, California.
Larkin, Peter
1977   "An Epitaph for the Concept of Maximum Sustained Yield." *Journal
       of the American Fisheries Society* 106, pp. 1–11.
1979   "Maybe You Can't Get There From Here: A Foreshortened History
       of Research in Relation to Management of Pacific Salmon." *Journal
       of the Fisheries Research Board of Canada* 36, pp. 98–106.
1980   "Pacific Salmon: Scenarios for the Future." The McKernan Lec-
       tures in Marine Affairs. Seattle: University of Washington.
Laurence, Joseph C.
1951   "A Historical Account of the Early Salmon Canning Industry in
       British Columbia, 1870–1900." Mimeograph, Vancouver: Univeristy
       of British Columbia.
Lewis, Michael, et al.
1977   Omeek Fishermen's Submission to the 1978 Sinclair Report. West
       Coast Information and Research. Port Alberni, B.C.
Locke, John
1960   "Two Treatises on Government: Second Treatise on Civil Govern-
       ment." In *Social Contract: Essays by Locke, Hume and Rousseau.*
       London: Oxford University Press.
Logan, R. M.
1974   *Canada, the United States, and the Third Law of the Sea Con-
       ference.* Canadian American Committee. Ottawa: C. D. Howe
       Research Institute [in conjunction with the U.S. National Planning
       Association].
Lyons, C.
1969   *Salmon: Our Heritage.* Vancouver: Mitchell Press.
Marchak, Patricia
1981   *Ideological Perspectives on Canada* (2nd ed.). Toronto: McGraw-
       Hill Ryerson.

1984   *Green Gold: The Forest Industry in British Columbia.* Vancouver: University of British Columbia Press.

Maril, R. L.
1983   *Texas Shrimpers: Community, Capitalism, and the Sea.* College Station: Texas University Press.

Marlatt, D.
1975   *Steveston Recollected: A Japanese-Canadian History.* Provincial Archives of British Columbia, Aural History Series, Victoria, B.C.

Martin, B.
1982   "The 1900 Fraser River Fisherman's Strike: A Reinterpretation." Unpublished paper, Anthropology and Sociology, Vancouver: University of British Columbia.

Matthews, Ralph
1983   *The Creation of Regional Dependency.* Toronto: University of Toronto Press.

Maud, Ralf
1978   *The Salish People, the Local Contribution of Charles Hill-Tout,* vols. III & IV. Vancouver: Talonbooks.

McCann L. D. (ed.)
1982   *A Geography of Canada: Heartland and Hinterland.* Scarborough: Prentice-Hall Canada.

McDonald, Jim
1985   "Trying to Make a Life: The Historical Political Economy of Kitsumkalum." Ph.D. dissertation, Anthropology and Sociology, University of British Columbia.

McDonald, Robert A. J.
1981   "Victoria, Vancouver and the Economic Development of British Columbia, 1886–1914." In Peter Ward and Robert McDonald (eds.) *British Columbia: Historical Readings.* Vancouver: Douglas and McIntyre, pp. 369–395.

McEachern, D. B.
1975   *Highlights of British Columbia's Herring Fishery, 1972–1974.* Technical Report Series no. 7-75-30. Department of Fisheries and Oceans, Pacific Region.

McGonigle, M. and M. Zacher
1977   "Canadian Foreign Policy and the Control of Marine Pollution." In B. Johnson and M. Zacher (eds.) *Canadian Foreign Policy and the Law of the Sea.* Vancouver: University of British Columbia Press.

McKay, Will
1975   *An Investigation of Foreign Influence in the B.C. Fish Processing Industry.* Vancouver: Environment Canada.

1982   "The Effects of Fuel Price Increases on the Coast and Earnings of Fishing Enterprises, 1978–1986." Prepared for the Native Brotherhood of British Columbia. Vancouver.

McKay, Will and Julie Healey
1981   "Analysis of Attrition from the Indian Owned Salmon Fleet, 1977–1979." Mimeograph prepared for the Native Brotherhood of British Columbia and the Department of Fisheries and Oceans.

McKay, Will and Ken Ouellette
1978   "A Review of the British Columbia Indian Fisherman's Assistance Program, 1968/69–1977/78." Vancouver: Will McKay Consultants Ltd.

McMullan, John
1984   "State, Capital and Debt in the British Columbia Fishing Fleet, 1970–1982." *Journal of Canadian Studies,* vol. 19(1), pp. 65–88.

Macpherson, C. B.
1978   *Property: Mainstream and Critical Positions.* Toronto: University of Toronto Press.

McRae, Donald
1980   "Canada and the Law of the Sea: Some Multilateral and Bilateral Issues." *Canadian Issues: Canada and the Sea,* vol. 3(1), pp. 161–174.

Miller, M. L. and J. C. Johnson
1981   "Hard Work and Competition in the Bristol Bay Salmon Fishery." *Human Organization* 40(2), pp. 131–139.

Miller, Phillip
1978   "A British Columbia Fishing Village." Unpublished Ph.D. dissertation, University of British Columbia.

Molson, C. R.
1974   *Foreign Ownership in the Canadian Fishing Industry.* Ottawa: Environment Canada.

Morrell, Mike
1985   "The Gitksan and Wet'suwet'en Fishery in the Skeena River System." Final Report, Gitksan-Wet'suwet'en Fish Management Study, Gitksan-Wet'suwet'en Tribal Council, Box 299, Hazelton, B.C.

Muszynski, Alicja
1984   "The Organization of Women and Ethnic Minorities in a Resource Industy: A Case Study of the Unionization of Shoreworkers in the B.C. Fishing Industry, 1937–1949." *Journal of Canadian Studies,* vol. 19(1), pp. 89–107.
1986a   "Class Formation and Class Consciousness: The Making of Shoreworkers in the B.C. Fishing Industry." *Studies in Political Economy,* vol. 20, pp. 85–116.
1986b   The Creation and Organization of Cheap Wage Labour in the B.C. Fishing Industry. Unpublished Ph.D. dissertation, Sociology, University of British Columbia.

Needler, A. W. H.
1979 "Evolution of Canadian Fisheries Management Towards Economic Rationalization." *Journal of the Fisheries Research Board of Canada* 36, pp. 716–724.

Newman, Peter
1975 *The Canadian Establishment.* Toronto: McClelland and Stewart.

Nicholson, George
1965 *Vancouver Island's West Coast, 1972–1962.* Victoria: Moriss Publishers.

Norr, J. L. and K. L. Norr
1978 "Technology and Work Organization in North Atlantic Fishing." *Human Organization,* vol. 37(2): 163–171.

North, George
1974 *A Ripple, A Wave: The Story of Union Organization in the B.C. Fishing Industry.* Revised and edited by H. Griffin, Vancouver: Fisherman Publishing Society.

O'Connor, James
1973 *The Fiscal Crisis of the State.* New York: St. Martin's Press.

Offe, Claus
1984 "The Theory of the Capitalist State and the Problem of Policy Formation." In L. Lindberg, R. Alford, C. Crouch, and C. Offe (eds.) *Stress and Contradiction in Modern Capitalism.* Lexington: D.C. Heath.

Ostrom, V.
1975 "Alternative Approaches to the Organization of Public Proprietary Interests." *Natural Resources Journal,* vol. 15, pp. 763–789.

Pearse, Peter
1980 "Property Rights and the Regulation of Common Fisheries." *Journal of Business Administration,* vol. 11 (1&2), pp. 185–209.

Pearse, Peter H. [Commissioner]
1982 *Turning the Tide: a New Policy for Canada's Pacific Fisheries.* Royal Commission on Pacific Fisheries Policy. Vancouver: Ministry of Supply and Services.

Pearse, P. H. and J. E. Wilen
1979 "Impact of Canada's Pacific Salmon Fleet Control Program." *Journal of the Fisheries Research Board of Canada,* vol. 36, no. 7.

Philips, Richard H.
1971 "The History of Western Seining." *National Fisherman,* vol. 52:5.

Phillips, Paul
1982 *Regional Disparities.* Toronto: James Lorimer.

Pibus, Christopher
1981 "The Fisheries Act and Native Fishing Rights in Canada: 1970–1980." *Faculty of Law Review.*

Pineo, Peter
1983   "Stratification and Social Class." In M. M. Rosenberg, W. B. Shaffir, A. Turowetz, and M. Weinfeld (eds.) *An Introduction to Sociology.* Toronto: Methuen.

Pinkerton, Evelyn
1983   "The Dressed and the Undressed: Competing Processor Strategies in the Market for Raw Salmon." Paper presented to the meetings of the Canadian Sociology and Anthropology Association, Vancouver, B.C.
1985   "Co-operative Management of Local Fisheries: A Route to Development." Paper presented to the Society for Economic Anthropology, Warrenton, Virginia.

Plourde, Charles
1975   "Conservation of Extinguishable Species." *Natural Resources Journal,* vol. 15, pp. 791–797.

Portes, Alejandro
1976   "On the Sociology of National Development: Theories and Issues." *American Journal of Sociology* 82, pp. 55–85.

Pritchard, John
1977   "Economic Development and Disintergration of Traditional Culture Among the Haisla." Unpublished Ph.D. dissertation, Anthropology and Sociology, University of British Columbia.

Proverbs, Trevor
1978   *Foreign Investment in the British Columbia Fish Processing Industry.* Vancouver: Economics and Statistical Services, Pacific Region, Department of Fisheries and Oceans.
1980   *An Update on Foreign Investment in the British Columbia Fish Processing Industry.* Vancouver: Department of Fisheries and Oceans.
1982   *Update to Foreign Investment in the British Columbia Fish Processing Industry.* Vancouver: Department of Fisheries and Oceans.

Ralston, Keith
1965   "The 1900 Strike of Fraser River Sockeye Salmon Fishermen." M.A. thesis, University of British Columbia.
1968   "Patterns of Trade and Investment on the Pacific Coast, 1867–1892: The Case of the British Columbia Salmon Canning Industry." *B.C. Studies,* vol. I, pp. 37–45. Reprinted in P. Ward and R. A. J. McDonald (eds.) *British Columbia Historical Readings.* Vancouver: Douglas and McIntyre, 1981; pp. 296–305.
1976-77 "John Sullivan Deas: A Black Entrepreneur in British Columbia Canning." *B.C. Studies,* no. 32.

Randall, Roger L.
1950   "Labor Agreements in the West Coast Fishing Industry: Restraint

of Trade or Basis of Industrial Stability?" *Industrial and Labor Relations Review,* vol. 3(4), pp. 514–544.

Rank, Dennis A.
1982   "Assessment of the Community Economic Development Program." Vancouver.

Raunet, Daniel
1984   *Without Surrender, Without Consent: A History of Nishga Land Claims.* Vancouver: Douglas & McIntyre.

Reid, David J.
1973   *The Development of the Fraser River Salmon Canning Industry, 1885–1913.* Economics and Sociology Unit, Northern Operations
_____ Branch, Pacific Region.
1975   "Company Mergers in the Fraser River Salmon Canning Industry, 1885–1913." *Canadian Historical Review,* vol. LVI (3). Reprinted in P. Ward and R. A. J. McDonald (eds.) *British Columbia: Historical
_____ Readings.* Vancouver: Douglas and McIntyre, 1981, pp. 306–327.
1982   "Consequences of the Davis Plan." Report to the 1982 UFAWU
_____ Convention.
1982   "The Small Business Development Bond Program." Vancouver: Department of Fisheries and Oceans.

Riches, David
1979   "Ecological Variation on the Northwest Coast: Models for the Generation of Cognatic and Matrilineal Descent." In P. Burnham and R.Ellen (eds.) *Social and Ecological Systems.* New York: Academic Press.

Ridler, N. B.
1984   "Socioeconomic Aspects of Sea Cage Salmon Farming in the Maritimes." *Canadian Journal of Fisheries and Aquatic Science* 41, pp. 1490–1495.

Roberts, Leslie
1957   *Clarence Decatur Howe.* Toronto: Clarke Irwin.

Roberts, Margaret
1970   *The Status of Chum Salmon Stocks of the West Coast of Vancouver Island, 1934–1968.* Technical Report, no. 3, Pacific Region, Vancouver.

Rogers, G. W.
1979   "Alaska's Limited Entry Program: Another View." *Journal of the Fisheries Research Board of Canada* 36, pp. 783–788.

Rohner, R.
1967   *The People of Gilford.* Ottawa: National Museum of Man.

Rounsefell, G. and G. Kelez
1938   *The Salmon Fisheries of Swiftshore Bank, Puget Sound and the Fraser River.* U.S. Department of Commerce, Bureau of Fisheries Bulletin, no. 27, Washington, D.C.: Government Printing Office.

Schmidhauser, John R.
1976   "Whales and Salmon: The Interference of Pacific Ocean and Cross-national Policy Making." In D. Walsh (ed.) *The Law of the Sea*. New York: Praeger, pp. 144–171.

Schwindt, Richard
1982   *Industrial Organization of the Pacific Fisheries*. Vancouver: Commission on Pacific Fisheries Policy.

Scott, Anthony
1955   "The Fishery: The Objectives of Sole Ownership." *Journal of Political Economy*, vol. 63, pp. 116–124.
1979   "Development of Economic Theory on Fisheries Regulation." *Journal of the Fisheries Research Board of Canada* 36, pp. 725–741.

Scott, R. Bruce
1972   *Barkley Sound: A History of the Pacific Rim National Park Area*. Victoria, B.C.

Shaffer, Marvin
1979   *An Economic Study of the Structure of the British Columbia Salmon Industry*. Vancouver: Salmonid Enhancement Program, Department of Fisheries and Oceans.

Shearer, R. (ed.)
1968   *Exploiting our Economic Potential: Public Policy and the British Columbia Economy*. Toronto: Holt, Rinehart and Winston.

Sinclair, J., B. Hale, and D. Karjala
1984   "Health and Safety for Fishermen and Shoreworkers on the Pacific Coast of Canada." *Canadian Special Publication Fishery and Aquatic Science,* no. 72.

Sinclair, Peter
1984   "Fishermen of Northwest Newfoundland: Domestic Commodity Production in Advanced Capitalism." *Journal of Canadian Studies,* vol. 19(1), pp. 34–37.

Sinclair, Sol
1960   *Licence Limitation — British Columbia: A Method of Economic Fisheries Management*. Ottawa, Department of Fisheries and Oceans.
1978   *A Licensing and Fee System for the Coastal Fisheries of British Columbia*. Vancouver, Department of Fisheries and Oceans.

Sinclair, William F.
1971   "The Importance of the Commercial Fishing Industry to Selected Remote Coastal Communities of British Columbia." Mimeograph, Vancouver.

Sorokin, Pitirim A.
1964   *Social and Cultural Mobility*. New York: Free Press.

Sparrow, Leona Marie
1976   "Work Histories of a Coast Salish Couple." M.A. thesis, Anthropology and Sociology, University of British Columbia.

Spradley, James
1969   *Guests Never Leave Hungry: The Autobiography of James Sewid, A Kwakiutl Indian.* New Haven: Yale University Press.

Stacey, Duncan A.
1978   "Technological Change in the Fraser River Salmon Canning Industry, 1871–1912." M.A. thesis, History, University of British Columbia.
1982   *Sockeye and Tinplate: Technological Change in the Fraser River Salmon Canning Industry, 1871–1912.* Victoria: British Columbia Provincial Museum.

Steinberg, Charles
1974   "The Legal Problem in Collective Bargaining by Canadian Fishermen." *Labour Law Journal,* October, pp. 643–655.

Suttles, Wayne
1960   "Affinal Ties, Subsistence, and Prestige among the Coast Salish." *American Anthropologist,* vol. 62, pp. 296–305.

Swezey, A. and R. Heizer
1977   "Ritual Management of Salmonid Fish Resources in California." *The Journal of California Anthropology,* vol. 4(1), pp. 6–29.

Tennant, Paul
1982   "Native Indian Political Organizations in British Columbia, 1900–1969: A Response to Internal Colonialism." *B.C. Studies,* no. 55, pp. 3–49.

Thomas, R. J.
1982   "Citizenship and Gender in Work Organization: Some Considerations for Theories of the Labor Process." In M. Burawoy and T. Skocpol (eds.) *Marxist Inquiries: Studies of Labor, Capital, and State.* Chicago: Chicago University Press.

Thompson, P. et al.
1984   *Living the Fishing.* London: Routledge & Kegan Paul.

Underwood, MacLelland and Associates
1976   See Government of Canada, Department of Environment.

University of British Columbia, Dept. of Economics
1980   "The Pacific Halibut Fishery." Unpublished manuscript prepared for the Economic Council of Canada.

Veltmeyer, Henry
1979   "The Capitalist Underdevelopment of Atlantic Canada." In R. Brym and J. Sacouman (eds.) *Underdevelopment and Social Movements in Atlantic Canada.* Toronto: New Hogtown Press, pp. 17-36.

Wallerstein, Immanuel
1974   *The Modern World System — Capitalist Agriculture and the Origins of the European World-Economy in the Sixteenth Century.* New York: Academic Press.

Walters, Carl
1977   "Management Under Uncertainty." In D. Ellis (ed.) *Pacific Salmon: Management for People.* Western Geographic Series, University of Victoria.

Ward, Peter
1978   *White Canada Forever: Popular Attitudes and Public Policy Toward Orientals in British Columbia.* Montreal: McGill–Queen's University Press.

Warren, Roland
1972   *The Community in America.* Chicago: Rand McNally.

Warriner, K. and N. Guppy
1984   "From Urban Centre to Isolated Village: Regional Effects of Limited Entry in the British Columbia Fishery." *Journal of Canadian Studies*, vol. 19(1), pp. 138–156.

Watkins, Mel
1968   *Foreign Ownership and the Structure of Canadian Industry.* Report of the Task Force on the Structure of Canadian Industry, Privy Council Office.

Webster, Peter
1983   *As Far As I Know: Reminiscences of an Ahousaht Elder.* Campbell River Museums and Archives, Campbell River, B.C.

West Coast Information and Research Group
1980   *Nuu-chah-nulth Tribal Council Forestry Study*, vol. I. Report prepared for the Nuu-chah-nulth Tribal Council, Port Alberni, B.C.
1981   "Native People and Fishing: The Struggle for Survival." Prepared for the Nuu-chah-nulth Tribal Council, Port Alberni, B.C.

Wilson, W. A.
1971   "The Socio-economic Background of Commercial Fishing in British Columbia." Fisheries Service, Pacific Region. Vancouver.

PART 2

Government and International Organizations: Documentation, Legislation, and Reports

A. Municipal Government

Village of Tofino
1980–81 Unpublished Census.

B. Government of Alaska

1980 *Alaska Catch and Production Commercial Fisheries Statistics.* Statistics leaflet, no. 28–33. Juneau, Alaska: Alaska Department of Fish and Game, Commercial Fisheries Division.
1980 Department of Fish and Game Statistics

C. Government of British Columbia

1. Department of Agriculture
   1979 *The Salmon Industry in British Columbia.* Victoria: Queen's Printer.
2. Ministry of the Environment
   1958 *The Commercial Salmon Fisheries of British Columbia.* Statistical Basebook Series, no. 3. Resources Branch.
   1979 *Financing in the B.C. Fishing Industry.* Prepared by Foodwest Resource Consultants for the Marine Resources Branch.
   1980 "A Permit and Licence Guide for the Prospective Mariculturist."
   1981 "Fisheries Mariculture: A Mariculturist's Perspective."
   1983 "Market for Farmed Salmonids in British Columbia." Prepared by Ference and Associates for the Marine Resources Branch.
3. Ministry of Industry and Small Business Development
   1976 *B.C. Economic Activity, Review and Outlook.*
   1980 *B.C. Economic Activity, Review and Outlook.*
4. Legislation
   1947 Industrial Conciliation and Arbitration Act
   1954 Labour Relations Act
   1959 Trade Unions Act
   1968 Workers' Compensation Act
   1973 B.C. Labour Code
5. B.C. Supreme Court
   1980 *"Couture* v. *Hewison, Stevens, and Nichol."* *Western Weekly Reports* 2: 136–148.

D. Government of Canada

1. General
   1952 House of Commons. *Debates,* vol. 4, 3379–80.
   1953 House of Commons. *Debates,* vol. 4, 3533–42.
   1982 Restigouche Accords between Government of Quebec and Restigouche Indian Band.
   Various Sessional Papers. Volumes and pages as noted in text.

2. Commission Reports and Submissions
   1922 Commission to Investigate Fisheries Conditions in British Columbia. B.C. Fisheries Commission Report and Recommendations. Ottawa: King's Printer.
   1956 Royal Commission on the Commercial Fisheries of Canada. Prepared by the Department of Fisheries of Canada and the Fisheries Research Board.
   1982 Royal Commission on Pacific Fisheries Policy, Final Report. *Turning the Tide: a New Policy for Canada's Pacific Fisheries.*
   1981– Briefs Presented to 1982 Royal Commission by:
   1982 British Columbia Packers
        Gitksan-Carrier Tribal Council
        Kitimaat (Haisla) Village Council
        Native Brotherhood
        Original "B" Fishermen's Association
        Pacific Coast Fishing Vessel Owners' Guild
        Pacific Coast Salmon Seiners Association
        Pacific Coast Trollers Association
        Pacific Gillnetters Association
        Sanderson, John
        Salmonid Enhancement Task Group
        United Fishermen and Allied Workers Union
3. Department of Environment (also Environment Canada)
   1976 "Policy for Canada's Commercial Fisheries."
   1976 "Competitiveness and Efficiency of the B.C. Salmon Industry." Unpublished report prepared by Underwood, McLellan and Associates and Edwin Reid and Associates for the federal Salmon Enhancement Program, Vancouver.
4. Department of Fisheries and Oceans (previously Department of Fisheries; Ministry of Fisheries).
        Annual Reports of the Fisheries Branch (also known as Departmental Reports).
        Annual Statistical Review.
        Annual Salmonid Enhancement Program (SEP) Reports.
        Annual Fish Products Exports of B.C.
   1949 *Employment in Canadian Fisheries.* Preliminary survey, 2 volumes, prepared for International Labour Office.
   1968 "New Regulations for B.C. Salmon Fishing Industry."
   1977 *The B.C. Salmon Industry: Survey of Economic Studies.* Underwood McLellan and Associates; Edwin Reid and Associates.
   1980 "The Road to 1995 — A Blueprint for Western Fisheries Development." Vancouver.
   1983 "Summary of Minister's Advisory Council Meetings." Session II, January 17–21, Vancouver.

1983 "Farmed Atlantic Salmon Industries in Norway and Scotland, 1983." Vancouver.

1984 "A New Policy for Canada's Pacific Fishery." Vancouver.

1984 "Pacific Fisheries Policy Options."

1984 James, Michelle. "Native Claims, Data Collection, and Impact Assessment." Vancouver.

1984 "The B.C. Fishing Industry: The Year 2005." Conference of minister's advisory council, with members of the Department of Fisheries and Oceans and outside resource consultants. Yellowpoint, Vancouver Island, February 6.

5. Department of Indian and Northern Affairs
Regional Indian Fishermen's Assistance Board

1982 "A Review of the British Columbia Indian Fisherman's Emergency Assistance Program." Vancouver.

Indian Aquaculture Task Force

1983 "An Indian Aquaculture Development Program." Ottawa.

6. Labour Canada (also Department of Labour; Ministry of Labour)

1947 "Collective Agreements in the Fishing Industry in Canada." *The Labour Gazette,* vol. 47(10).

Annual Corporations and Labour Union Returns Act.

Annual Reports of Division of Marine and Fisheries.

1984 Employment Injuries and Occupational Illnesses. Ottawa.

7. Statistics Canada
Various Years *Census of Canada.* Ottawa.

1979 *Labour Force Annual Averages, 1975–78.* Ottawa.

1980 *Perspective Canada* III. Ottawa.

1985 *Consumer Prices and Price Indexes.* Ottawa.

8. Legislation
1906, 1914, 1927, 1952, 1970, 1981 Fisheries Act
1970 Canada Labour Code
1970, 1971, 1972 Unemployment Insurance Act
1974, 1975, 1976 Combines Investigation Act
1981 Constitution Act
1982 Restigouche Accords (Quebec and Restigouche Band)

E. *Joint Government — Federal and Provincial*

1964 Federal-Provincial Committee on Wage and Price Disputes in the Fishing Industry, Summary Review.

F. *Supreme Court of Canada*

1978 "*B.C. Provincial Council et al.* v. *B.C. Packers Ltd., Native Brotherhood of B.C., et al.*" *Western Weekly Reports* 1: 621–630.

G. Government of Japan

Annual   *Abstract of Statistics on Agriculture, Forestry, and Fisheries,*
          1975–83. Ministry of Agriculture, Forestry, and Fishery, Tokyo.
1984   *Japan's Exports and Imports, Commodity by Country.* Japan
          Tariff Association, Tokyo.

H. Government of the United Kingdom

1985   H. M. Customs and Excise Tariff.

I. Government of the United States

Annual   *Current Fishery Statistics — Fisheries of the United States,*
          1973–1984. Washington.
1971   Basic Economic Indicators, Washington, D.C.

J. International Organizations

1. Food and Agriculture Organization (FAO).
    Annual   *Yearbook of Fisheries Statistics.*
    Annual   *International Trade* (1955, 1960–61).
    Annual   *Fishery Commodities* (1955–1976).
    Annual   *Catches and Landings* (1970–1984).
2. International North Pacific Fisheries Commission.
    Annual   Reports. Vancouver.
    Annual   Bulletin, vol. 39. Vancouver.
3. Organization of Economic Cooperation and Development
    Annual Reports. *Review of Fisheries in OECD Member Countries,* Paris.
4. United Nations
    Conference Reports on Law of the Sea (UNCLOS)
    [dates as noted in text].

PART 3

Newspapers, Magazines, Radio

Canadian Broadcasting Corporation (Radio) Interviews with:

Hewison, George	November 1982
Newman, Ed	November 26, 1982
Pearse, Peter	November 26, 1982
Rivard, Jean	November 30, 1982

Current. Newsletter of the Pacific Trollers Association.
Fish Farming International. West Byfleet, England.
Fisherman, The. UFAWU newspaper [dates as cited in text].
Fisherman, The
   n.d. "Private for Profit." United Fishermen and Allied Workers Union
      Newspaper, Special Report [circa 1982].
Globe and Mail, The [dates as noted in text].
Labour Gazette, 1947.
Native Voice, The [dates as noted in text].
Pacific Fisheries Review, The Fishermen's News, vol. 40, no. 5. Seattle, 1984.
Pacific Fisherman Yearbook. Seattle: Washington.
Pacific Fishing
   1982 "Salmon Ranching." July & August.
   1983 "Norwegian Perspectives." January.
Province [dates as noted in text].
Seafood Business Report [dates as noted in text].
Vancouver Sun [dates as noted in text].
Western Fisheries
   1984 "The Future." April.
Western Weekly Reports [dates as noted in text].

PART 4

Companies and Fishers' Organizations: Reports, Newsletters, Correspondence

B.C. Packers
   Annual Reports.
Fisheries Association of British Columbia
   A Status Report on the Commercial Fishing Industry. Vancouver, 1983.
Pacific Gillnetters Association
   "A Fair Share of the Catch." Results of the Salmon Gillnetters Forum,
      November 24 and 25, 1983. Mimeographed.
Prince Rupert Fishermen's Cooperative Association
   Co-Pilot. Newsletter [various years].
   Memorandum of Association and Rules of the PRFCA as amended
      December 12 and 13, 1980.
   Information Bulletin regarding Association Membership [n.d.]
      Annual Reports, 1978–1983.
United Fishermen and Allied Workers Union
   An Open Letter to Fishing Industry Workers from Officers and General
      Executive Board, Nov. 1, 1982.

Convention Report, February 1982.
West Coast Fishermen's Survival Conference, December 10–11, 1983.
Discussion Papers and Background Notes.
The Future of the B.C. Fishing Industry, vol. II. Vancouver, 1984.
Native Brotherhood of British Columbia
Strategy for the Future Agenda. "The Negotiation Process." Working
group discussion, summary, and resolutions from Plenary Session
of Convention, 1982.
53rd Annual Convention Resolutions.
Gitksan-Carrier Tribal Council
Tribal Council response to the Final Report of the Commission on
Pacific Fisheries Policy, submitted to the Honourable Pierre de Bane,
Minister of Fisheries and Oceans, Hazelton, November 26, 1982.
Unpublished internal census, 1979.
Unpublished speech by N. Sterrit to convention, 1982.
Nishga Tribal Council
An Open Letter to Minister of Fisheries and Oceans, January 3, 1983.
Unpublished.
Fishery Management Proposal, 1980. Unpublished.
Southern Southeast Regional Aquaculture Association, Ketchikan, Alaska.
Harvest Management Plan. Unpublished, 1985.
Western Fishermen's Federation
Turning the Tide: The Consultative Process. Final Resolution, Feb.
4–6, 1983.

PART 5

Personal Communication

Bevan, David. Chief Inspector, Pacific Region, Department of Fisheries
and Oceans.
Homma, Elaine. Japan External Trade Organization, Vancouver.
Griffioen, Ward. Former consultant to Southern Southeast Regional Aqua-
culture Association and B.C. salmon-farm operator.
Heggelund, P. Seattle fish broker.
Hilton, Susanne. Ph.D. candidate, Anthropology and Sociology, Univer-
sity of British Columbia.
Lando, Peter. B.C. commercial fisher and graduate student, University of
British Columbia.
Lundby, O. Vice-Consul, Norwegian Consulate, Vancouver.
Mandel, Louise. Legal counsel for Union of B.C. Indian Chiefs.

Index

accidents, 194–195, 222n
accountants, 181-184
accumulation process, 11–14, 17, 18, 28–30
Aemilius Jarvis and Company, 51
agriculture, 14
Agriculture and Rural Development Agency (ARDA), 136, 307–309
Ahousaht, 293, 295, 297, 313–317, 319–322
Alaska, 21, 23, 48, 54, 76, 90n, 92, 93, 102
  aquaculture, 320–322
  Bristol Bay, 80, 88n, 158
  halibut, 66, 140
  processing, 52, 77, 88n, 100, 138
Alaska Boundary Tribunal, 166
Alaska Packers Association, 51, 56
Alert Bay, 256
Alexander Ewan, 49, 51
Allen, G. C., 96, 101
Allied Tribes of British Columbia, 237, 238
American Can Company, 63
Anderson, Alex C., 251, 252
Anderson, E. N., Jr., 257
Anglo British Columbia Co. (ABC), 7, 49–51, 122, 137
aquaculture (see fish farms)
Arctic Waters Pollution Prevention Act, 162
Asiatic fisheries, exclusion of, 111–112
Atlantic region (see also Maritimes), 92, 115, 155, 162, 194–195, 280
  fisheries, 21, 24, 39, 159, 163
  processing, 22, 117, 120, 121, 175, 205, 217
  and unions, 205, 282
Atlantic salmon, 92, 98, 99
Australia, 68, 93

Babine, 252
Bain, Joe S., 88n
banks (also by name), 29, 48, 57, 79, 244
  loans to fishers, 126–135

Bank of Montreal, 48, 51, 52
Barrett, G., 21, 115
Barnes, K., 2, 21, 23, 50, 56, 57
Barricade Agreements, 252
Beagle, Mickey, 284
Beale, Calvin, 346
Beaujot, Roderick, 346
Beaver Cannery, 53
Belgium, 68, 93
Bell, David, 326
Bell, F. H., 39–41, 43, 45n, 91n, 324n
Bell-Irving, Henry, 49, 108
Bella Bella, 238, 261, 285, 295
Bella Coola, 295
Berger, T., 358
Berkes, Fikret, 253
Bevan, David, 105n
bilateral negotiations: U.S./Canada, 154–156, 165–170
Bishop, R., 4
Boldt Decision (U.S.), 165–166, 357, 358
bonus payments (see charter payments)
Booth Fisheries (Seattle), 310
Bossin, Bob, 301
Bristol Bay (see Alaska)
Britain (see United Kingdom)
British Columbia (see also B.C., government of)
  economic importance of fisheries, 14–16
  fish production and market data, 102–105
British Columbia Canning Company, 49
British Columbia Fishing and Packing Company, 272
B.C. Fisheries Commission Report (1922), 111, 112
B.C. Fishermen's Protective Union, 229
B.C. Fishermen's Union, 228
B.C. Gillnet Association, 241
British Columbia, Government of
  Human Rights Code, 288
  Labour Code, 26
  Labour Relations Board, 283
  and Meares Island, 297, 358

and off-shore oil, 162
British Columbia Packers' Association of
New Jersey, 52, 53, 57
B.C. Packers, 3, 20, 116, 123, 152n, 219,
274, 280–281, 287, 307, 309, 355
as dominant firm, 47, 55, 56, 68, 80, 87
117–118, 122, 137, 142–144, 189,
296, 333, 354, 356
history of, 51–53, 55–58, 114,
117–118, 122, 302, 333
and fishing fleet, 40–41, 81, 138,
184–185
plant closures, 53, 57, 118, 137–138,
141, 143
sale of rental fleet, 81, 138, 264–265,
357
B.C. Seiners Association, 229
British Columbia Seafood Exporters
Association, 71–72, 89n
Burns, J., 305
Butedale, 137, 329
buy-backs, 125, 126, 246, 258, 348, 355
Buzan, Barry, 161, 162

Campbell, Kenneth, 326, 327
Canako Sea Products, 296
Canada (*see also* Canada, government of
and separate listings for government
departments)
Supreme Court, 26, 162, 268n, 280
Canada, Government of
(see below for Departments and
Ministries; *see also* Canada, and
State)
and B.C. Packers, 138–139
and conservation of stocks, 145–149
International negotiations and treaties
(*see also* separate treaty listings),
153–170
regulation of fisheries (*see also*
licences), 107–152
subsidies to fisheries, 127–135
subsidies to processors, 136–138
subsidies to shipbuilders, 135, 150n
and United States governments,
154–156, 165–170
Canada, Government, Department of
Environment, 9, 148
Canada, Government, Department of
Fisheries and Oceans (*also*
Department of Fisheries), 3, 9, 10,
36, 42, 88n, 92, 111, 113, 115, 117,
118, 144, 147–149, 150n, 169, 184,

195, 196, 198n, 241, 249, 251, 252,
253, 254, 255, 266, 293, 300, 320
employment data, 16–18
Minister's Advisory Council, 246–247
Salmoned Enhancement Program (*see*
separate listings)
Canada, Government, Department of
Indians and Northern Affairs, 19,
132, 240, 258
purchase of gillnetters from BCP, 81,
138, 240, 264–265, 358
Canada, Government, Department of
Munitions, Supply, 113
Canada, Government, Department of
Regional Economic Expansion, 167
Canadian Bank of Commerce, 4, 51, 130,
132
Canadian Fish Company, 40–41, 43, 55,
68, 75, 80, 84, 87, 117, 122, 137,
138, 142, 151n, 198n, 274, 287,
302, 307, 309, 355
Canadian Trades and Labour Congress,
230
Canfisco (*see* Canadian fish)
canneries (*see* processors)
Cannon, G. Harry, 314
capital investment (*see* investment)
Carmichael, Alfred, 62, 65n
cash buyers, 80, 189, 198n
Cassiar, 65n, 80, 139, 141
catch allocation, 186
Ceepeecee, 54, 302
Central Native Fishermen's Cooperative
(Bella Bella), 139, 142, 238, 246,
263-264
charter payments, 75–78, 192–194
Chinese Cannery Exployees Union, 60
Chinese contract system (and contractors),
59–60, 270–271, 274, 275, 288–289
Chinese Contractors Union, 61
Chirot, Daniel, 327
Ciriacy-Wantrup, S. V., 4
Clark, Scott, 254
Clarkson, Stephen, 160, 161, 168, 169
Clayoquot Sound (and Indians), 294,
295, 297, 299, 300, 303, 310-311,
320
Clement, W., 21, 326
"Clover Leaf," 51, 68, 117, 123
Coast Salish bands, 238
coastal communities
decline of, 141, 328–338
"coastal states group," 159

cold war, 170, 231
collective agreements (*see* UFAWV and Processors)
Columbia River, 47, 48, 169, 333
Columbia River Packers Association, 51, 56
combination gear vessels, 36, 127, 185, 232
Combines Investigation Act, 26, 27, 274
Commercial fisheries (*see* fisheries)
commission merchants (agents), 47, 48, 49, 50
common property (*see also* Property rights), 4–5, 28–29, 187, 243, 353, 359
communal property, 5, 11, 250
communities, 293–349
Communist Party, 230, 271
Conley, J., 108, 110
Connelly, P., 217
conservation, 17, 110, 119, 124, 144, 145 and U.S./Canada treaty, 153–154, 164–166
consultative process, 244, 245, 359
Continental Shelf, 161, 163
contract system (fishers and processors), 78–80, 218, 272–273
co-operatives (*see also* PRFCA)
Co-operative Fishermen's Guild, 190, 246
Copes, Parzival, 170n
corporate concentration, 122–123, 137–144
countervailing duties (U.S.), 167–168
Couture, 27
Cove, John, 250, 268n
crew, 190–194, 223-227, 230-231, 235, 316
share system, 76, 77, 191–194
Cruickshank, Don, 181, 182, 186
Crutchfield, J. A., 69, 86, 88n, 91n
Cuneo, Carl, J., 327
currency values, 101

Davis Plan, 125–126, 137, 263, 314, 316, 317, 339–340, 346, 347–348, 353, 355
Deep Bay Fishing and Packing Company, 272
Deep Sea Fishermen's Federation Union, 229
Deep Sea Fishermen's Union, 41, 58, 229, 230, 234, 235, 281

debt, debt load (*see* fishers)
Department of Environment (*see* Canada, Government of)
Department of Fisheries (*see* Canada, Government of)
Department of Indian and Northern Affairs (*see* Canada, Government of)
Department of Munitions and Supply (*see* Canada, Government of)
Defence Purchasing Board, 113
dependency theory, 327–328
Deschenes, B. H., 195
division of labour
ethnic, 218–220
sexual, 214–218
social, 213–220
technical, 61–64, 208–213
Dixon Entrance, 163, 166
dog fish liver oil, 114
Dolphin Seafoods (Cleveland), 236
Doyle, Henry, 51, 52
Doyle Fishing Supply Company of San Francisco, 51
dragger (dragging), 43–44
Drucker, Philip, 262, 268n, 300
Duff, Wilson, 329

Edmunds, and Walter, 117, 274, 302
education, 176–180, 215–216
Empire Cannery, 54
English syndicate, 50
Esquimalt, 54
"equal pay for equal work," 283–288
Evans, Saford, 124
exports (*see* markets-retail)

families, 173–176, 190–191, 215–216, 220–221, 316
farmers, 177, 179–180
farms (*see* agriculture, fish farms)
Federal Business Development Loan, 136
Federal Provincial Committee on Wage and Price Disputes, 21, 26, 230–232, 237, 238, 241
Findlay, Durham and Brodie, 49
Finlayson Channel, 54
fish farms, 28, 98–100, 245
in B.C., 322
other world locations, 321–322, 356
fish meal, 57, 114
Fish Processors Bargaining Association of B.C., 89n
Fisher, Robin, 251

Fisherman, The, 117, 142, 240, 241, 271,
    272-273, 274, 275, 277, 289n
fishers (commercial) (*see also* by gear
    type), 173-198
  age at entry, 176
  as co-op owners, 227-228
  crew relations to skipper, 190-194,
    226
  crisis of 1980s, 139-141
  division between, 223-225
  difference by gear type, 180-186, 226
  debt load, 127-135, 139-143,
    340-342
  education of, 176-180
  ethnic divisions, 228-229
  families, 173-176, 190-191, 221
  futures, 196-197
  health and safety, 194-195
  income, 149n, 178-184, 198n,
    311-313, 315, 340-343
  as independent (simple) commodity
    producers, 20-25, 28, 110, 111,
    188-190, 225-227, 244, 335
  labour conditions, 225-226
  numbers of, 116, 123
  off-season employment, 177-179
  organization pre-1945, 228-230
  rental, 240
Fisheries Act, Canada, 9, 10, 155, 252,
    253
Fisheries Association, 71, 72, 75, 76, 89n,
    151n, 232, 239, 246, 276, 284, 287,
    353
fishery commission re: traps (1902), 54
Fisheries and Marine Services, 132
Fisheries Council of B.C., 89n
Fisheries Council of Canada, 163
Fisheries Improvement Loans Act (FILA),
    128-135, 150n
Fishery Prices Support Board, 118-119
Fisheries Research Board, 119, 125
Fishermen's Benevolent Society, 228
Fishermen's and Cannery Workers
    Industrial Union, 229, 272
Fishermen's Protective Association, 111
Fishermen's Survival Coalition, 186
Fishermen's Improvement Loans (FIL),
    135
fishing "as a way of life," 190-194, 317,
    319
fishing, economic importance of, 14-16
Fishing Vessel Assistance Program, 167
Fishing Vessel Owners' Association, 229,
    246

Fleet Rationalization Committee, 181-186
Ford, C. S., 105n
Fraser, G. A., 120, 125, 126, 127, 150n
Fraser River, 23, 36, 37, 47, 48, 49, 50,
    52-55, 59, 62, 64n, 93, 107, 108,
    110, 167, 169
  canneries, 329-338
  and Canada-United States relations,
    155-156, 167
Fraser River Canneries' Assocation, 228
Fraser River Fishermens' Protective
    Union, 228
Fredin, R. A., 93
freezing, 64, 70, 210-211
freezer system (vessels), 38
freezer system (processing), 57, 306-307
freezer technology, 73-74
freezer-trawlers, 159
Friedlaender, Michael, 258
Fry, M. G., 160
Fuguitt, Glenn V., 346
fur trade, 22, 25

Garrod, Steve, 199
Gateman, Bill, 272, 273
gear (*see also* separate listings by gear)
  divisions by, 35-39, 180-186,
    223-226, 240-243, 342-343
  gear-types, earnings of, 181-184
  gender distinctions (*see* women)
General Agreement on Tariffs and Trade
    (GATT) (*see also* tariffs), 101
geoduc, 309
Georges Bank, 16, 167
Gillis, D. J., 255
gillnet (gillnetters), 4n, 35-36, 110
  associations, 241-242
  incomes, 181-184
  sale of fleet by BCP, 81, 138, 357,
    264-265
  share of catch, 184-185
  technology of, 35-36, 333
Gitksan-Wet'suwet'en Tribal Council, 10,
    254, 266
Gladstone, P., 61, 109, 228, 231, 237, 262
Goldseal, 68
Gorden, Alex, 277
Gordon, Clarkson, 15, 356
Gordon, H. Scott, 4, 119, 124
Granatstein, J. L., 113
Green, Stephen, 166, 167, 168
Gregory, H., 13, 21, 22, 50, 56, 57
Griffioen, Ward, 322
groundfish, 37, 42-44, 140

Groundfish Temporary Assistance
  Program, 167
Gulf of Georgia, 10, 15, 38, 169, 248, 333
Gulf Trollers Association, 9
Guppy, Neil, 338

habitat, 5–8, 29, 144, 145, 153, 161–162,
  164, 245, 331
Hako Fishing, 139
Hale, B., 194
halibut, 37, 38, 44n, 47, 55, 66, 86–87,
  140
  crews, 40, 233
  exchange board, 41, 87, 229
  high seas capture, 156
  history of fishery, 39–42, 230–233
  international treaties, 156
  lay area agreements, 233
  and PRFCA, 230, 232
  techniques of capture, 232
  and UFAWU, 232
Halpern, J., 324n
Hardin, G., 4
Harper, Janet, 253
Hawthorn, H. B., 22
Hayward, Brian, 24, 43, 91n, 107, 109,
  110, 126, 127, 194
Head Tax, 59
Healey, Julie, 258
Hecate Strait and Hecate Channel, 54
Heggelund, P., 322
Heizer, R., 251, 267n
Hudson's Bay Company, 47, 249, 251

ideology, 144, 186
  of independent producers, 187
  of common property, 29, 353, 359
  of fishers, individualism, 187
  of unionism, 234
immigration policies re: Chinese, 64
Imperial Plant, 55, 117
INCO, 161
Income Tax Act, 135
independent commodity producers,
  fishers as, 20–25, 28, 110, 111,
  188–190, 225–227
Indian, Land claims (*see* property rights)
Indian (Native) (*see also* native
  Brotherhood), 4, 11, 22, 23, 29, 47,
  59, 111, 249–269
  Associations, 236–240
  Aboriginal rights/land claims, 238–239,
  250-252, 268n, 357–359

Communities, 60
fishers, 138, 253–269
fishing rights, 252–255
food fishery, 253–255
licences, 236, 258, 265
religions, 237
shoreworkers, 215, 218–219, 259–261
  and UFAWU, 234, 237–240
Indian Act, 10
Indian Fishermen's Assistance Program,
  132
Indian Fishermen's Economic
  Development Program, 246
interception, 335, 337
International Commission for the
  Northwest Atlantic Fisheries, 159
International North Pacific Sockeye
  Commission, 93
International North Pacific Fisheries
  Commission/Convention, 105n,
  106n, 165
International Pacific Halibut Commission
  (ind. Internat'l Fisheries
  Commission), 40, 44n, 87, 154, 155
International Pacific Salmon Fisheries
  Commisson (also Convention), 154,
  156, 167, 169, 303
Inverness, 53
Investment, 120
iron butchers (also iron chinks), 63, 209
Island Cash Buyers, 139

J. H. Todd and Sons Ltd., 49, 51, 53–54,
  122, 137
Jackson, Ted, 322
J. S. McMillan Company, 139, 310
James, H., 113, 150n
James, Michele, 255, 258, 325n
Jamieson, Stuart, 228, 231
Japan (Japanese) (*see also* Japanese-
  Canadian), 45n, 67, 80, 83, 90n, 92,
  102, 105n, 157, 301
  consumer revolt, 355
  exports, 95–96, 116, 301
  fishers, 111
  high seas fishery, 156, 158, 165
  imports, 88n, 354
  investment in B.C., 18, 84, 91n, 139,
  142, 165, 234, 353–354, 355
  Peace Treaty, 157
  salmon production, 93–94
  workers from, 285
  yen value, 96, 101
Japanese-Canadians, 23, 228

Jarvis, Aemilius (*see* Aemilius Jarvis and Company), 51, 52
Johnson, Barbara, 157, 159, 160, 162, 163
Johnston, Douglas, 156, 157
Jorgensen, Susan, 288
Juan de Fuca Strait, 237

Kanter, R. M., 218
Karjala, D., 194
Keating, Thomas, 166, 167, 168
Ketchikan, 155
Kildonan, 273, 302
Klemtu, 53, 137, 284
Kyuquot, 54, 302, 303
Kyoquot Trollers Association, 302

Labour (*see also* shoreworkers, independent commodity producers, UFAWU)
   definition of, 19–25
   fishers as (*see also* independent commodity producers), 20–22
   shoreworkers as, 19–20, 199–222, 329–330
Labour market, 19–25, 59–61
   shoreworkers, history of, 59–64
   shoreworkers, contemporary, 199–222
Ladner, Edna G., 59, 62, 65n
Ladner, T. Ellis, 48, 65n
landed values (diverse species), 66–67, 81, 340
Lando, Peter, 258
Lane, Barbara, 251, 252
Lane, Robert, 251
Langdon, Frank, 157, 165, 166, 170n, 354
Langdon, Steve, 76, 77, 322
Larkin, Peter, 169, 181
Laurence, Joseph C., 256
Law of the Sea (*see* United Nations)
LeBlanc, Romeo, 165
Ledbetter, Max, 81
Lewis, Michael, 294, 315, 316, 317
Licences (licensing) (*see also* Davis Plan, Indian), 5, 8, 23, 31n, 50, 109, 111, 124–126, 335
   "A" and "B" vessels, 125–126, 150n, 313
   Area, 246
   bidding, 244, 245
   Davis Plan, 125–127, 263, 314, 316
   halibut, 156

herring, 311–313
   held by Japanese fishers (1901), 228
   Limitation Program, 50, 79, 126–127, 180, 331, 339–340, 346
   Pearse recommendations, 243
   salmon, 311–313
Lichter, Daniel, 346
Locke, John, 4
longliner, 39–42
Logan, R. M., 156
Loblaws, 123, 142
logging (*see* forestry)
Lundby, O., 99, 105n
Lyons, C., 52, 53, 55, 299, 330, 333

manganese nodules, 157, 161, 170n
(*S.S.*) Manhattan, 162
Mandel, Louise, 253
Manpower retraining programs, 178
Marchak, Patricia, 182, 187
Maril, R. L., 175, 190
mariculture (*see* fish farms)
marine scientists, 161
"Maritime group," 158
Maritimes (*see* Atlantic)
markets (raw fish), 66–91, 188–190
markets (retail, wholesale, export)
   and B.C. processors, 67–69
   by date: 1900–1914, 55; World War I, 55–56; World War II, 57–58; since 1945, 93–106
   by region: Japan, 234, 287; U.K., 57–58; U.S., 57–58; world, 92–101
   by species: halibut, 86–87; herring, 83–85; salmon, 69–83, 92–106, 116–117, 151n
Marketosis Reserve, 313
Martin, B., 88n
Marubini, 139, 142, 238, 354
Masset, 295
Masset Canners Ltd., 54
Matthews, Ralph, 326
Maud, Rolf, 250
McCann, L. D., 327
McDonald, Jim, 22
MacDonald, M., 217
McDonald, Robert, 48, 49, 52
McGonigle, M., 163
McGovern, T. B., 51
McKay, Will, 140, 255, 258
McMullan, John, 79, 127, 341
Macpherson, C. B., 4, 5, 6, 9, 30n
McQuillan, Kevin, 346

McRae, Donald, 163, 164
Meares Island, 297, 358
Middlemiss, Danford, 161, 162
Milerd Fisheries, 139
Millbanke Industries, 238
Miller, Phillip, 36, 78, 90n
mineral (*see* mining)
mining, 8, 14, 169, 329
Mitsubishi, 354
Molson's Bank, 51
moratorium, 26–27, 80, 84, 124, 245
Morrell, Mike, 254
Munitions and Supply Act, 113
Muszynski, Alicja, 23, 112, 118, 170n,
    278, 289n

Namu, 137–138, 285
Nass River, 64n, 251, 329, 331, 335
    and canneries, 284
National Energy Program, 168
National Sea Products, 118
native (*see* Indian)
Native Brotherhood of British Columbia
    (NBBC), 3, 26, 27, 224, 256,
    262–264
    history of, 236–240
    membership, 237–238, 262–264
    negotiations, 26, 27, 239–240, 274,
    277–278
    reactions to Pearse, 246
    UFAWU, 27, 230, 234, 239–240, 263,
    271, 277
NATO, 163
Nelson Brothers Fisheries Limited, 53,
    54–55, 78, 122, 274, 302
New England Fish Company, 39–41, 55,
    138
Newfoundland, 14
Newman, Ed, 239–240
Newman, Peter, 113
Nichol, Jack, 27, 241, 271, 281
Nicholson, George, 302
Nishga Tribal Council, 238, 246, 255
Nippon Suison, 139
Nootka Bamfield Company, 274, 302
Noranda Corp., 161
Norpac Fisheries, 302, 355
Norr, J. L., 28
Norr, K., 28
North Island Trollers Cooperative
    Association, 58
Northern Native Fishing Corporation
    (NNFC), 81, 138, 258, 264–267, 319

Northern Trollers Association (NTA), 246
North Pacific Cannery, 138
North Pacific Fisheries Convention
    (1953), 157–158
Northwest Atlantic Fisheries Commission
    (NAFAC), 157
Northwest Passage, 162–163
Northwestern Fisheries, 139
Norway, 170
    fish farms, 99–100, 322
    exports of frozen salmon, 99–101,
    105n, 322
Nuu-chah-nulth, 238, 309, 314, 320

Oakland Fisheries, 136, 139, 141
Ocean Dock, 54
Ocean Fisheries, 80, 81, 82
Oceanside (Prince Rupert Plant), 55, 138
O'Connor, James, 13
O'Shaughnessy, Helen, 284
Offe, Claus, 12
oil
    companies, 161, 168
    federal-provincial disputes over, 162
    and Law and Sea debates, 157, 162
    disputes, 162, 168
    tankers, 163
Oregon, coastal fisheries, 102
Oriental licences, 112
Ouellette, Ken, 258
Ostrom, V., 4
overcapacity (processors), 108, 110, 124,
    127, 135, 144, 148
overcapitalization of vessels, 135
overseas relief aid, 58
Overwaitea, 142

Pacific Coast Salmon Seiners Association
    (PCSSA), 223, 242
Pacific Gillnetters Association (PGA and
    UFAWU), 241–242, 246
Pacific Halibut Fishermen's Union, 229
Pacific North Coast Native Co-operative
    (Port Simpson), 136, 354
Pacific Region Fisheries Council, 244
Pacific Rim Mariculture, 139
Pacific Salmon Commission, 169, 320
Pacific Trollers Association (PTA), 166,
    223, 241, 242, 246
Packers Steamship Company Limited,
    The, 53
Paramount Plant, 55
paternalism, 187–188

Pattison, Jim, 68, 142
Pearse, Peter H. (*see also* Royal
    Commission on Pacific Fisheries),
    150n, 192, 193
Philips, Richard H., 36
Phillips, Leslie, 227
Phillips, Paul, 326
pilchard, 36, 37, 42–44, 45n, 302
Pibus, Christopher, 253
Pineo, Peter, 174
Pinkerton, Evelyn, 306, 322
Plourde, C., 4
Pocock, D., 253
pollution (*see* habitat, oil, forestry,
    mining)
Pontecorvo, Guilio, 69, 88n
Portes, Alejandra, 327
Port Edward, 54, 55, 138
Port Hardy, 306
Port Simpson, 239
Port Simpson Co-op, 81
power drum, 36
prices (*see* markets-retail; markets-
    rawfish; labour; strikes; UFAWU)
Prince Rupert (*see also* PRFCA), 43, 54, 55,
    58, 87, 279–282, 342–346
Prince Rupert Amalgamated Shoreworkers
    and Clerks Union, 222n, 235, 282
Prince Rupert Fishermens Cooperative
    Association (PRFCA), 3, 41, 48, 58, 68,
    84, 87, 218, 224, 227, 246, 280, 302,
    309
  and halibut fishery, 233
  history, 229, 230, 235–236
  ideology of, 23, 235
  finances, 236
  labour relations, 235–236, 282–283
  membership, 235–236
  and UFAWU, 281–282
Prince Rupert Fishermen's Guild, 235
Prince Rupert Plant, 55
Prince Rupert Vessel Owners Association,
    246, 280
Pritchard, John, 260
privilege, fishing as, 136
processors (processing companies;
    canneries) (*see also* shoreworkers),
    46–65
  bankruptcies and takeovers since
    1980, 139
  canneries, 47–65, 107–112
  canning and freezing ratios, 69–75
  cost of production, 206–208

modernization period (1940–1967),
    112–123
  plant closures, 284–285
  and retail markets, 67–69
  and raw fish markets, 69–91
  regulation period (1968–current),
    123–144
  relations with fishers, 75–82, 187–190
  state subsidies to, 136–138
  technology, 208–213, 279
property, vessels as, 243
property relations, 18–28
property rights (*see also* State, licences,
    and Indian), 3–11, 28–29, 136
  and Indian land claims, 10, 240,
    357–259
Prosperity Marine, 80
Proverbs, Trevor, 91n, 139, 354, 355
Provincial Cannery Co., 54
Provincial Cooperatives Act, 230
Puget Sound, 37, 52, 54, 88n, 333–335
purse seine (*see* seiners)
putters, 314
pyramiding, 126

Quality Fish, 80, 139
Queen Charlottes, 38, 159, 163, 237
quotas, 169
  and halibut, 156
  and foreign vessels, 159
  and Pearse recommendations, 245

racism, 65n, 228, 229
  and Chinese, 59
Ralston, Keith, 23, 49, 50, 108
Raunet, Daniel, 268n
Reagan, U.S. President, government of,
    168–170
Reciprocity Treaty, 154
reduction plants, 42, 117, 232, 273, 314
refrigeration (*see also* freezing), 64
region, 293–349
  canneries, 328–338
  and centralization, 328–338
  regional inequalities among fishers,
    338–342
  distribution of gears, 342
  trends with respect to fishers, 342–346
regional division of fleet, 338–347
Regional Indian Fishermen's Assistance
    Board, 132
Reid, D. J., 48, 50–53, 108, 135
Restigouche Accords (1982), 253

Riches, David, 267n
Richmond Cannery, 53
Rivers Inlet, 54, 64n, 118
RivTow Straits Ltd., 82
Roberts, Leslie, 113
Roberts, Margaret, 299, 300, 303
roe herring (*see* herring)
Rogers, G. W., 251, 267n
Rohner, R., 90n
Rousseau, J. J., 1
Ryal Bank, 130, 132, 139
Royal Fisheries, 80, 139, 141, 281–282
Royal Commission on Canada's
    Economic Prospects, 119
Royal Commission on Chinese and
    Japanese Immigration, 228
Royal Commission on the Commercial
    Fisheries of Canada, 115, 120, 121,
    124, 125
Royal Commission on Pacific Fisheries
    (Pearse Commission), 5, 28, 29–30,
    144–148, 223, 236, 238, 357, 359
  briefs to, 233, 242–243, 254
  citations to, 36, 43, 119, 126, 135, 140,
    156, 180, 184, 187–188, 189, 198n,
    253, 261, 329, 347
  reactions to, 243–248, 263, 317
  recommendations on fleet
    rationalization, 9, 227, 243–248,
    263
Royal Commission on the Status of
    Women, 284
royalties, 243, 245, 246
rugged individualism, 175, 187, 190

Sacramento River, 47, 333
St. Mungo Cannery, 55, 138
salmon, 23, 35–39, 66–83
  (*see also*: Fraser river, Skeena river,
    Japan, markets, gillnet, seine, troll,
    Canada, government of: treaties,
    processors)
  canning, 70–73, 81–82, 95–99
  exports (markets), 92–106, 116
  fresh and frozen, 73–75, 210–211
  pack size, 334–338
  state regulation of fishery, 107–152
Salmon Canners' Operating Committee,
    273
Salmon Purse Seiners Union, 272
Salmonid Enhancement Program, 17, 79,
    102, 144–146, 168, 180, 348
Sanderson, John, 244

San Francisco, 59
San Juan Islands, 37
sawmills (*see* forestry)
Schmidhauser, John R., 166
Schwindt, Richard, 68, 78, 87, 88n, 89n,
    91n, 184, 187, 189, 192, 198n, 309
Scott, Anthony, 4
Scott, R. Bruce, 301
sea-bed (*see* United Nations)
Seafood Productions, 306
Seal Cove, 138
Seattle
  and Canadian fishers, 155
  and halibut fishery, 87
  raw fish markets, 302
seiners (purse seiners), 36–37, 42–43, 58,
    80
  crew, 37, 77–78
  income, 181–184
  share of catch, 184–185
  relations with processors, 75–82
  technology, 36–37
  and UFAWU, 232
Shaffer, Marvin, 72, 89n, 136, 142, 145
share system (crew shares), 37, 111,
    191–194
Shearer, R., 327
Shearwater plant (Bella Bella), 238
Shinners, Wayne, 169
shoreworkers (*see also* Indians, labour
    market, women)
  early history, 59–64
  contract system, 59–60, 218, 270–271,
    272–273, 274, 288–289
  income, 204–206, 216, 276–279,
    282–288
  personal data, 215–216
  seasonal work, 199–206
  size, 199–203, 222n, 329–330
  and UFAWU, 224, 270–290
shrimp (and shrimp trawling, shrimp
    grounds), 166
simple commodity producers (*see*
    independent commodity producers)
Sinclair, J., 194
Sinclair, Sol, 116, 124, 125, 127
Sinclair, William F., 305, 347
Skeena River, 53, 61, 64n, 252, 266, 274,
    331, 335
skill fishing, 174–175, 178–179
Sloan, Gordon: Report, 124
Small Business Development Bond
    Program, 135

Smith Butchering Machine (see Iron
    Butcher)
sockeye, troll caught (see also salmon),
    72, 81
Sointula, 90n
Sooke traps, 54
Sooke Harbour Fishing & Packing Co., 54
Sorokin, Pitirum A., 175
Southern Vancouver Island Tribal
    Federation, 238
Sparrow, Leona Marie, 257, 260, 314
sports fishery, 15, 169, 245, 247
Spradley, James, 268n
Stacey, Duncan A., 48, 50, 52, 110, 333
state, the (see also Canada, government
    of; B.C., government of), 3–19,
    25–30
    contradictions of, 126–136, 356–357
    and crises, 144–149
    licensing, 124–126
    property rights, 9–11, 124, 127
    and salmon fishery, 107–152
    and monopsony relations, 137–144
Steinberg, Charles, 25, 26
Sterrit, N., 91n
Steveston, 53, 54, 117
Stevens, H., 27
Strait of Juan de Fuca, 156
strikes
    after 1945, 26, 231, 233, 239, 241, 263,
        277, 278–279
    before 1945, 228, 271–272
    1967, Prince Rupert, 233–234,
        279–282
    1973, 283–288
Subsidy Program for Construction of
    fishing Vessels in British Columbia,
    150n
surplus value, 19, 142
Suttles, Wayne, 250, 267n
Swezey, A., 251, 267n
Swiftsure bank, 166–167
Swiftsure Strait, 16

tariffs (also import duties) (see also GATT
    in Japan, in U.S., in U.K.), 68, 95,
    101, 116, 154
Tenant, Paul, 237
Tepperman, Lorne, 326
Thomas, R. J., 214
Thompson, P., 175
Thornton Creek Enhancement Society,
    320

Toba Inlet, 295
Todd, J. H. (see also J. H. Todd & Sons
    Ltd.), 51, 53–54, 302
Tofino, 138, 286–287, 293–325
Tofino Cooperative Fishing Association,
    301
Tofino Fisheries, 136, 309, 310
Toronto Dominion Bank, 130, 132
Torrey Canyon, 162
tourist industry, 15, 247, 327
trade relations (see Canada, Govt. of/U.S.
    relations)
Tradewind Seafoods, 139
"tragedy of the commons," 5, 11, 119,
    244
transfer system, 300–301
Transpacific Fish, 305
traps (trap nets), 53–54, 107, 109–110,
    155
trawlers (trawling), 43–44
trollers, 303–304, 315–316
    associations, 214
    income, 181–184
    share of catch, 184–185
    technology of, 37–39
Trudeau, P. E. (and Government of), 153,
    162
Truman proclamation on coastal
    fisheries, 157
Tulloch, Andrew, 308–309
twelve-mile limits, 154, 159–160, 163,
    165
two-hundred-mile limit, 79, 90n, 105n,
    140, 154, 157, 162–165, 168, 180

Ucluelet, 238, 293, 302, 305
Underwood Tariff treaty, 40
unemployment insurance, 178–180,
    198n, 219–220, 278, 298
Unemployment Insurance Act, 27
Union Steamship, 328–329
United Fish Cannery and Reduction Plant
    Workers Federal Union, Local 89,
    272–273
United Fishermen's Federal Union, 229,
    230, 272, 273
United Fishermen and Allied Workers
    Union (UFAWA) (see also shore-
    workers), 3, 71, 76, 123, 138, 139,
    142, 144, 157, 166, 190, 241, 244,
    308
    pre-history, 229, 271–274
    history, 25, 230–240, 271–290

Canadian Trade Union Congress,
    Canadian Labour Congress, 231, 234
and communism, cold war, 231
ideology of, 234
and independent seiners, 232
and Indians/Native Brotherhood, 27,
    230, 234, 239–240, 261, 263, 271,
    277
membership, 25–27, 222, 230, 231,
    232
and Prince Rupert Fishermen's
    Cooperative Association, 234, 281,
    282
reactions to Pearse, 245, 263
and right to bargain, 26–28, 231
and shoreworkers, 141, 224, 261,
    270–290
strikes, 26, 231, 233–234, 263,
    271–290
and women, 275, 283–288
United Kingdom 47, 57, 67, 68, 92, 101,
    114
United Nations
    Conferences on Law of the Sea
        (UNCLOS), 153, 156–165, 169–170
    Continental Shelf Convention, 154,
        159
    seabed resources, 154, 158–165
    and mining interests, 160
United States (and Governments of), 16,
    23, 51, 52, 67, 92–97
    anti-combines (trust) law, 26, 52
    army, 54, 161, 168
    bilateral negotiations with Canada,
        153–170
    Boldt decision, 165–166
    Conservationist policies, 155–156
    and Fraser river, 165, 167, 333–336
    freight rates, 49
    imports (fish), 67, 116
    and international agreements, 160–170
    mining interests, 161–167
    and Swiftsure Banks, 166–167
    tariffs, 155
    and 200-mile zone, 164
    Washington, State of, 165–167
U.S.S.R./Russia, 92, 93, 102, 156, 162

Vancouver, 37, 48, 49, 87, 342–347
    and Union Steamship, 328–329
Vancouver Island (and West Coast of), 38,
    54, 159, 237, 274, 293–325, 338
Vancouver Sun, 9, 139, 166, 239

Veltmeyer, Henry, 326
vertical integration, 142
vessels, 115–116
    average value of, 343–345
    foreign, 165
    market value of, 340
    numbers of, 123, 344
    seizures and impoundments, 141
vessel-owners
    associations of, 240–342
    debts of, 340–342
    Indians as, 257–259
    and Pearse commission, 242–243
    and UFAWU, 240–242
Victoria, 48, 50, 52, 59
Victoria Canning Company
    incorporated 1891, 49–51
vitamin A oil, 41, 43, 57–58, 114

Wallerstein, Immanuel, 327
War Labour Board, 273
War Measures Act, 112–113
War Supply Board, 113
Wartime Fisheries Advisory Board, 113
Ward, Peter, 111
Warren, Roland, 317
Warriner, Keith, 338
Washington, state of and coastal fisheries,
    21, 102, 165–167
Washington Treaty (1871), 155
Watkins, Mel (and Watkins Report), 355
Webster, Peter, 313
Wellington, 62
Western Fisheries Federation, 186
Western Fishermen's Federation,
    244–245, 246
Western Weekly Reports, 27
Weston's, 47, 68, 122–123, 138, 142,
    151n, 152n, 280
whaling, 301
Wickham, E., 233
Wilen, J. E., 150n, 192, 193
Wilson, George, 51
Wilson, W. A., 177, 178
Wingen, Robert, 308, 309, 319
Windsor Cannery, 61, 62
women (*see also* shoreworkers), 175,
    197n–198n, 199–222
    immigrants, 64, 218–220
    in canneries, 214–219
    and comparisons to men, 214–218,
        287–288
    native Indian, 60–62, 240, 285

reserve army, 199–200, 217–218
wages, 274, 275, 276, 283–288
in UFAWU, 275, 283–288
Workers' Compensation, 27, 278
Workers' Unity League, 230
World War II, federal government and
war bonds, 112–114

Zacher, Mark, 163
Zellner, A., 86, 91n